Comparative Design of Structures

Shaopei Lin · Zhen Huang

Comparative Design of Structures

Concepts and Methodologies

上海交通大學出版社
SHANGHAI JIAO TONG UNIVERSITY PRESS

Springer

Shaopei Lin
Department of Civil Engineering
Shanghai Jiao Tong University
Shanghai
China

Zhen Huang
Department of Civil Engineering
Shanghai Jiao Tong University
Shanghai
China

ISBN 978-3-662-48043-4 ISBN 978-3-662-48044-1 (eBook)
DOI 10.1007/978-3-662-48044-1

Jointly published with Shanghai Jiao Tong University Press

Library of Congress Control Number: 2015951346

Springer Heidelberg New York Dordrecht London

Printed on acid-free paper

Springer-Verlag GmbH Berlin Heidelberg is part of Springer Science+Business Media
(www.springer.com)

Preface

This book "Comparative Design of Structures" is devoted to study the essential of design philosophy, in particular, structural design. The purpose of this book is to explore the essentials of design activity in conceptual and philosophical point of view incorporated by structural design as the carrier. On the other hand, in our information era, it should not be underestimated the importance of using current computer technology to represent designer's intelligent ideas during design processes, by which the past knowledge and experience of human being in design can be stored in computer.

As a matter of fact, design activities are the highest intelligence of human being and can be regarded as a series of decision making under uncertainties for approaching design goal(s). We also understand that design is a process of comparison between past samples stored in the mind of the designers who are doing the comparison, even though unconsciously, during design processes.

To simulate this procedure of the designer by means of computer is actually an artificial intelligence (AI) problem, and to store design knowledge in a knowledge base and to infer the design objective through AI are the tendency of design development nowadays. Therefore, using fuzzy theory for treating uncertain design environment and knowledge engineering for cumulating design knowledge into computer is the right way for accommodating the development tendency of design. In other word, we are entering in a new era, which is characterized by "Softening of the theories and hardening of the knowledge," and among them, design development is including in this range.

How to use computer AI technology to develop future design procedure? How to inherit design experience for future usage? One have to develop a theoretical framework of how design is performed by human being and of how the design philosophies that a designer is adhered. This is why one has to emphasize the principle of "comparative design."

For computerized "comparative design," there are two crucial problems to be studied: First, the implicit comparative design activities are essentially an intelligent process, it must be represented by certain AI formulation for its implementation in computer, and this is what Chaps. 1 and 8 have presented; Second is the understanding of the modeling of design object, i.e., to clarify the procedure of an entity system through its components by composition and decomposition. We fully use the advantages of the complexity of structure for realizing the modeling of composition and decomposition, and this is what the remaining chapters have presented in detail. Through structural design, the authors have modeled a structure into its fundamental elements and then assembled its composition through 1D to 2D and then 3D structure step by step for illustrating the essential processes of comparative choice. The authors have spent a big volume of this book in describing the comparative design choice in different level of structural composition, that enable the readers to make sense of that comparative design choice is the insight essential of the design. On the other hand, the book can be a reference book for design engineer or a reference text book for related undergraduate and graduate students.

Structural design is aimed to provide a good structure which satisfies the entire requirements in architectural aspect, economic aspect, etc. The design process is the integrity of human intelligence. Incorporating artificial intelligent technologies into CAD system, the intelligent CAD system (or ICAD) is then emerged. This is an important tendency for modern development of CAD technique, which represents the state of the art in this area.

This book "Comparative Design of Structures" is devoted to study the essential of design philosophy, in particular, structural design. An outstanding structural design is generated by creative originality of designer based on comparing different similar existing samples following to the law of structural behavior. Since the global layout and overall mechanical behavior are the core in structural conceptual design, it is important to point out that the number and the quality of accessible similar samples to compare, or the past experience of the designer, are the key in design, and doubtlessly, the knowledge engineering application will help in this regard.

The latter part of this book introduces the method of intelligent design based on knowledge engineering with fuzzy reasoning. As an example, the principles of comparative conceptual design of high-rise building are presented; these principles can be used in different fields as one of the generalized methodologies of intelligent design.

The idea of comparative design is based on human empirical cognition and fuzzy identification capability that reflect the creativity of the designer in a superior design alternative; however, it can never realize without the comparison with other similar existing samples in the past, on which the experiences of the designer were built. In an outstanding conceptual structural design, this innovative idea of designer should be compliant to mechanical behavior of structure, where the architectural aesthetics and the rational composition of structural elements are performed. For achieving optimum performance of structural behavior, a process of integration of intuitive experiences is needed rather than the details of computer structural analysis. This is why we also provide in this book a set of "Best Practice" examples of structure and structural components with its merits and demerits for creating a comparative environment.

We address the principles of "understanding the correctness through examining the wrong" and "promoting creativity through comparison of cases." Such a case-based reasoning method in artificial intelligence is proved to be effective by a series of successful examples in intelligent design from practice; certainly, it does work in the field of structural conceptual design.

In global sense, the digitization of comparative design can be concluded as a part of the discipline of "Quantitative Management." Design is the bridge between a subjective function to an objective reality, which is essentially a series of decision making in the scenario context for approaching optimum function performance of reality. The design processes under the support of uncertain experience (or uncertain characteristics of cases) can be regarded as a series of decision making under uncertainties, while fuzzy reasoning capability of human being is most effective. One can utilize the artificial intelligence (AI) technology for digitizing the processes of design or quantized decision making as a problem of quantitative management based on fuzzy logic and AI, while the "Fuzzy-AI Model" will work correspondingly [6–8].

The "Fuzzy-AI Model" is based on simulating the capability of fuzzy reasoning of human being to make decision from corresponding information. For most of the human, intelligence is come from their intensive capability of processing massive fuzzy information, then to simulate such ability of human being could be exactly regarded as the art of artificial intelligence, that is why the model is named. "Fuzzy-AI Model" also reflects the idea of comparison of corresponding information, while the model used for design purpose must be naturally accessed.

Another objective of this book is to unify the concepts of comparative design with the quantitative management by using "Fuzzy-AI Model." An important statement is that essentially, the quantitative management means to use a series of digitized expression of decision making, which is the combination of decision science and AI technology for exploring the insight relation of these two disciplines. On the other hand, design processes are exactly a series of design decision making, and the principles of digital quantitative management make it possible to unify both comparative design and management through computer environment.

This essential theoretical development in methodology would potentially follow with a wide area of different applications, including the comparative conceptual design of structures presented in this book. We look forward to the blossom of perspective applications of digitized theory, both in management and in engineering design; it is the aims of the authors to explore the insights of the problem using structural design as a window.

The authors would deeply appreciate the chief editor of SJTU Press Mr. Guangliang Li for his help and insistent encouragement during the process of writing this book. The authors appreciate sincerely Mr. Tengfei Fu and Ms. Muochi Lu for their efforts on English editing. The authors appreciate the graduate students Mr. Liang Sun, Mr. Guiqiu Huang, Mr. Chang Zhang, and Mr. Pu Wang for their help in drawing of this book.

This book is supported by National Natural Science Foundation of China (51178265). The authors appreciate for its supports.

Shanghai Jiao Tong University
September 2015

Shaopei Lin
Zhen Huang

Contents

Chapter 1
Introduction of Design Philosophy

1.1 The Essentials of Engineering Design

Design is the bridge between a subjective function and an objective reality as shown in Fig. 1.1, which is essentially a series of decision making for approaching optimum function performance of reality.

In engineering design, the "function" used to be a requirement of performance that people expect the future objective entity can be realized. For instance, people want an object for residing or for working shields, and then, a building is assigned as the object of engineering. For each function requirement, the designer should "engineer" corresponding measures, approaches, or efforts for its implementation. The implementation of all function requirements engineered by the designer attached on an objective entity, which is what people expected for.

How the design performs as a bridge between function and objective reality? The designer starts to work firstly on his (her) understanding of an objective entity in which the function requirements can be fully attached and realized. This objective entity can be a real structure or any either form of hardware or even software. The objective entity of design possesses all the function requirements as its attributes; it represents by a "design point" in the n-dimensional design space with n-vectors reflecting the individual independent functional requirements.

In the real world, the exact "design point" actually does not exist, for all the functional requirements cannot be expressed exactly, but can be satisfied fuzzily. Through heuristic intuition based on past experience and memory, which is a set of massive fuzzy information in different attributive aspects to be processed by the designer during design, the fuzzy identification of a design alternative in this fuzzy zone is performed, which means the cognition of the design entity or the concept of the design object is being revealed. This process is much sophisticated that it represents all the intelligence of the designer. Moreover, it is much complicated that it is even impossible to describe the detailed procedures. However, in reality, it is consummated just simply by a heuristic fuzzy decision making by the designer.

S. Lin and Z. Huang, *Comparative Design of Structures*,
DOI 10.1007/978-3-662-48044-1_1

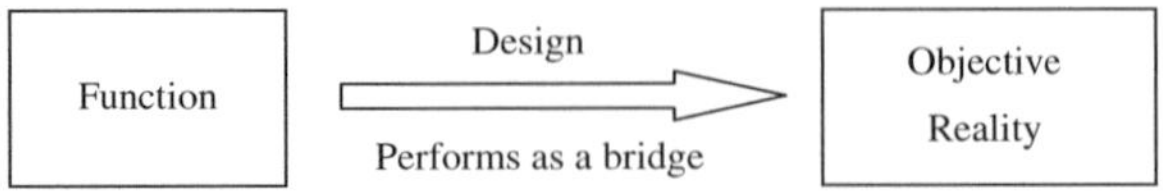

Fig. 1.1 The essentials of design

Why a designer can do that as if it is out of his (her) imagination? Why a designer could recall all his (her) memories in a second and be able to draw out the sketch of expected design entity for satisfying the functional requirements? Let us keep these problems in mind for solution later.

1.2 The Concept of Comparative Design

For satisfying functional requirements, the definition of comparative design can be described as the procedures of drawing out a solution by means of comparison with existing similar samples in a design space S_d. Under the computerized digital environment, the comparative design procedure should be based on the theory of design space, in which the design comparison can be performed through digital approach.

1.2.1 Design Space Theory

Any design alternative can be expressed as a "point" in an n-dimensional space with n-orthogonal axes (vector), representing its individual design functional requirements. Each similar design sample is represented as a "point" in the same design space with the similarity relation between the samples and the design alternative being determined.

Actually, since the design attribute is vague in its evaluation, it exists in a "fuzzy interval" of judgment; therefore, the expression of design alternative will be no longer a "point," but a fuzzy "zone" in the n-dimensional design space. Any point located in this fuzzy zone can be regarded as an acceptable or satisfactory design solution.

Comparative design concept is to determine the design solution compared with a group of similar samples. If the most of the functional attributes of the sample are similar to the expected design requirements, this sample can be chosen as one in the group of comparative samples. The attribute of comparative sample is located in vicinity beside each attribute of design solution, and the difference between them can be measured by "fuzzy distance." Through weighted average of the sample attributes counter proportional to the "fuzzy distance," the individual attribute of final design solution is thus identified. The above-mentioned statement can be expressed analytically as below.

Denote S_d the design space with n-attributes represent n-functional requirements of the design.

S_d is composed of n-independent vectors as the orthogonal axes to form a n-dimensional space, where a defined design solution represents a n-dimensional fuzzy zone occupied in this design space S_d. According to the orthogonality principle of space theory, the axis vector in design space S_d is perpendicular with each others, and the projections of design point P_d in each axis is P_{di} ($i = 1, 2, 3, \ldots, n$), respectively, represent the quantitative valuation of each functional requirement of the design.

A design solution is defined by its particular design attribute (functional requirement) F_{ri} quantitatively, and it is represented by the projection P_{di} ($i = 1, 2, 3, \ldots, n$) of the design point P_d to n-attribute vector X_i ($i = 1, 2, 3, \ldots, n$) as shown in Fig. 1.2. As mentioned before, actually P_d is no longer a point, but a fuzzy zone in design space S_d, correspondingly, and its projection $P_{d,i}$ ($i = 1, 2, 3, \ldots, n$) in each axis X_i occupies a fuzzy interval in vicinity of it.

For fuzzy expressing the differences of two samples $\underset{\sim}{A}$ and $\underset{\sim}{B}$, the "fuzzy distance" between them $D(\underset{\sim}{A}, \underset{\sim}{B})$ is introduced by Eq. 1.1:

$$D\left(\underset{\sim}{A}, \underset{\sim}{B}\right) = \sum_{i=1}^{n} w_i * d_i\left(\underset{\sim}{A}, \underset{\sim}{B}\right) \tag{1.1}$$

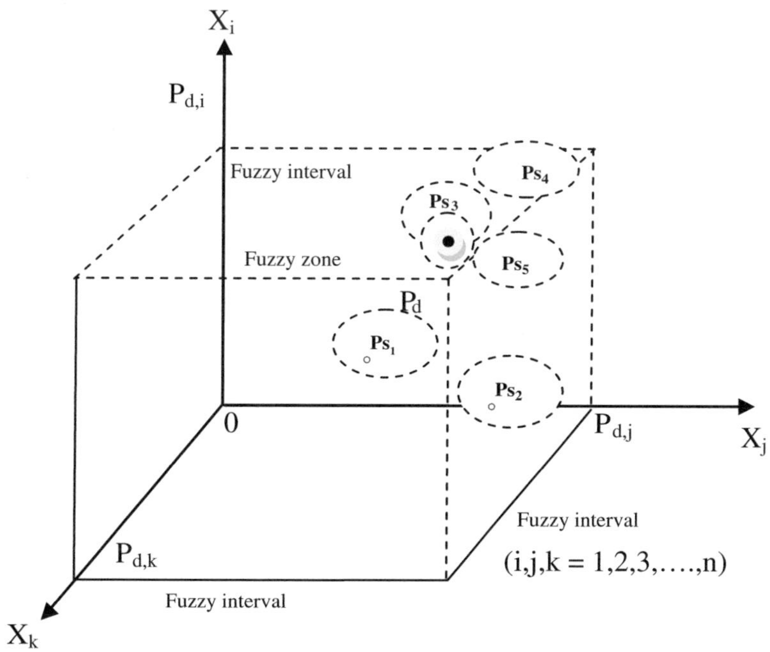

Fig. 1.2 The fuzzy comparative design philosophy

where n is the dimension number of the space of state $\underset{\sim}{A}$ and $\underset{\sim}{B}$; w_i is the weight of ith attribute.

$$d_i(\underset{\sim}{A}, \underset{\sim}{B}) = \left|\mu_{\underset{\sim}{A}}(t_i) - \mu_{\underset{\sim}{B}}(t_i)\right| \quad \left(\text{if } d_{-i} < d_i < \bar{d}_i\right) \tag{1.2}$$

The weighted average sum of the attributes between two comparative samples is defined as the final solution of individual attribute, where the weights are counterproportional to fuzzy distance d_i of each ith attribute, where $\mu_{\underset{\sim}{A}}(t_i)$ and $\mu_{\underset{\sim}{B}}(t_i)$ represent the membership function values of state $\underset{\sim}{A}$ and $\underset{\sim}{B}$ in membership function curve $\mu_i(t)$ of ith attribute.

The approach of determining the design solution P_d through fuzzy comparative design philosophy can be shown in Fig. 1.2: Where there are chosen p samples noted as P_{sj} (j = 1, 2, …, p), we need to find the design solution P_d through comparison from these p samples (associated with n-dimensional fuzzy zone).

1.2.2 *The Fuzzy Expression of* $\mathbf{P_d}$

The P_d is determined by its n-projections P_{di} (i = 1, 2, 3, …, n) in each attributive axis, which is determined by the average of weighted sum of fuzzy distance between the ith attribute of design solution P_{di} (i = 1, 2, 3, …, n) and the ith attribute of jth sample $(P_{sj})_i$ (i = 1, 2, 3, …, n; j = 1, 2, 3, …, p) in ith axis.

To determine fuzzy distance between the projections P_{di} and $(P_{sj})_i$ in ith axis, one can offer first approximation by defining a preliminary design solution projection in ith axis P_{oi} which is the mean of the projection of P samples:

$$P_{oi} = (1/P)\{\} \tag{1.3}$$

The second approximation is performed by the weighted projection of jth sample in ith axis, and then, the projection of design solution in ith axis P_{di} can be expressed by weighted average sum of the projections of all jth samples P_{sj} (j = 1, 2, …, p):

$$P_{di} = \{\} \quad (i = 1, 2, \ldots, n) \tag{1.4}$$

where w_{ij} represents the weight of jth sample in ith axis;

$$w_{ij} = \left\{\left[\mu_i(P_{oi}) - \mu_i(P_{sj})\right]_{i^2}\right\}k \tag{1.5}$$

μ_i is the fuzzy membership function of ith design attribute, so $[\mu_i(P_{oi}) - \mu_i(P_{sj})]$ represents the fuzzy distance between P_{oi} and P_{sj} in ith axis

$$k,\ \text{an unification factor}\ k = 1/\{[\mu_i(P_{oi}) - \mu_i(P_{sj})]_{i^2}\} \tag{1.6}$$

Substituting Eqs. 1.5, 1.6 to Eq. 1.4, we obtain the projection of design solution along ith axis P_{di}. Thus, a design solution matrix D_{ds} can be formed as follows:

$$D_{\text{ds}} = \{p_{di}\}\ (i = 1, 2, 3, \ldots, n). \tag{1.7}$$

1.3 From Empirical to Intelligent Design

Let us take structural design as an example; in preliminary design stage of an intelligent design, both numerical computation and symbolic inference are involved. While a case-based reasoning design decision is being adopted for the structural type selection, a quantitative design decision is being made for approximate sizing of structural members. Since the preliminary design is aimed to solve global arrangement of structural component layout and its interrelationships of sizing, this is the most important stage in whole design processes, and the rationality of the preliminary design will relate, even dominate to the quality of the design. It is proved by practice that the fuzzy model is most appropriate for such reasoning operation.

The design processes are characterized of a series of design decisions making. In empirical design, the decision making will mainly depend on the heuristic judgment of the designer based on his (her) experiences. It is a process of subjective mental manipulation of designer's personal knowledge, experience, and expertise to form the kernel of his mental activities in design for solving problems. Nevertheless, these personal experiences are difficult to share between each other since human being is a poor knowledge transferor, due to the fact that they can act far more than they can express; it means that to acquire knowledge from domain expert is really a tedious and troublesome task, which implies the main obstacle and bottleneck in building a design expert system for sharing experience. Thus, to incorporate AI technology for accommodating such difficult situation seems appropriate. Since AI can be regarded as "to simulate human intelligence by computer," in the fields of science and technology of information era, the tendency of "Softening of the Theories and Hardening of the Knowledge" [1] is revealed evidently; simulation of the mental activities of experienced designers through AI approach can greatly facilitate and improve the design capabilities by sharing accessible experiences from all others; in design processes, this tendency of transition is called as "From Empirical to Intelligent Design."

The most popular approach for intelligent design is case-based reasoning method, the advantages of which can be concluded as follows:

- It automates the knowledge transferring process in any design attributes and can make the knowledge acquisition process faster and less error prone.
- It provides more qualified and comprehensive knowledge than individually extracted from domain experts manually.
- It provides a "group transfer" of knowledge through "best-practice" samples (cases), and the knowledge can be well packaged and effectively used for drawing a "best" new design.
- It accumulates knowledge as "memory" and serves for future applications, which is in such an amount that is impossible to be carried by any individual human expert through his (her) memory.

We recognize the world is full of uncertainty and the engineering design does so. The designer must study the uncertainty characteristics of design environment in order to select the proper mathematical tool accommodating to the characteristics of design problem. Nowadays, there are two main branches of uncertainty studies in engineering design: the probability-based theory and the fuzzy mathematics-based theory. Though there are commonalities between these two, however, their difference is essential. Each of them deals with events with uncertainties in different natures.

The probability theory is suitable to treat those uncertain events, while it is in uncertain occurrence; however, the fuzzy mathematics theory is suitable to treat those events, while it is uncertain in degree. It is recognized that the design alternative is defined by its attributes which vary and possess the nature of uncertain in degree rather than uncertain in occurrence. For compatible reason, it is improper to use a mathematical tool which violate to the nature of event; thus, it concluded that the intelligent design is appropriate to apply fuzzy reasoning approach rather than to use any kinds of probability theories.

During preliminary stage of structural design, the uncertain information includes the selection of reasonable structural type, the evaluation of seismic zone, and the reasonable structural safety degree. Obviously, the designers have not yet mastered the details of the interrelationship between those uncertain design attributes, so it cannot be treated as a random event, but an event of uncertain in degree. Taking the advantages of fuzzy approach into consideration, one could conclude that it seems reasonable to adopt fuzzy reasoning approach in preliminary design of structures.

According to above analyses, a conclusion can be made that Fuzzy-AI Model is suitable for intelligent design. Starting from system approach and software engineering, the basic intelligent design philosophy can be achieved through case-based reasoning approach, which essentially is a comparative design method based on comparison of similar samples and approach to final design alternative through various intelligent reasoning, such as above-mentioned fuzzy reasoning, neural network reasoning, and genetic reasoning. Certainly, this methodological approach can be used in broad areas, including civil engineering, mechanical engineering, management science, as well as events of sociology and economics.

Introducing AI technology into design has explored a new era of computer-aided design; one cannot underestimate its importance and far-reaching inference to the development of engineering science. Intelligent design is an interdisciplinary technology which involves AI, network technology, parallel engineering, and intelligent information processing technologies. As one of the core techniques of computer-aided engineering, intelligent design theory is accommodating to recent growing tendency of systemization and computerization of huge engineering systems.

On the other hand, there are a lot of nonlinear and non-structured problems exist in the complex engineering system, intelligent design idea could cumulate and share human intelligence in the past for current usage, and it escapes those sophisticated and troublesome relations of nonlinearity and other mathematical difficulties and approaches to final problem solving straightforwardly by reasoning approach.

The most popular approach in intelligent design is "case-based reasoning (CBR)" method [2, 3], and the essentials of CBR are "solving current problem by referring to past experience (or samples)." Due to the fact that inference logic is the basic thinking philosophy of human being, CBR has nowadays widely used in different areas for practical problem solving.

Preliminary design of an engineering object can be realized by a series of good cases (best-practice samples) in the past in a *n*-dimensional design space, using "fuzzy distance" to evaluate the similarity of any sample with the designed one, that is, exactly an experienced engineer performs during their design procedures. From a user's point of view, the flowchart scheme of intelligent design by case-based reasoning (CBR) method can be shown in Fig. 1.3, it starts from input initial data or

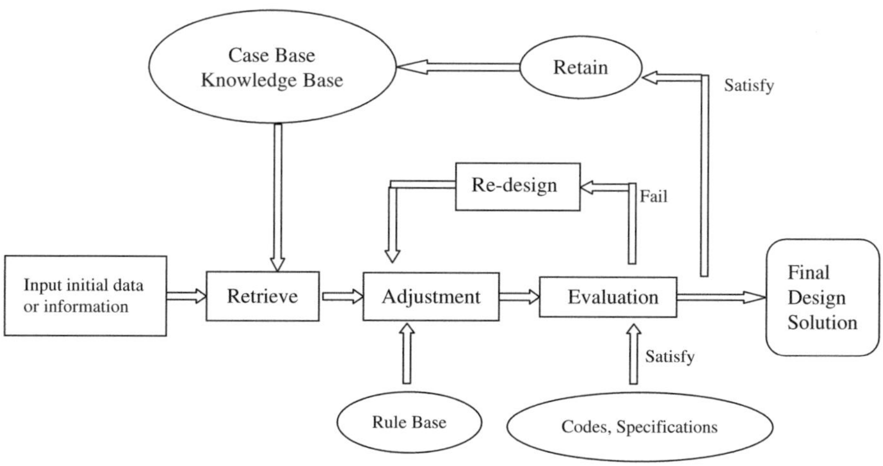

Fig. 1.3 Flowchart scheme of CBR method

information about the design object, the designer tries to retrieve similar cases or samples from his memory or from case base to obtain an initiative design alternative, he(she) follows with adjustment through different rules in the rule base, then this design alternative can subject to evaluation by codes and specifications. The final design solution can be obtained if it is satisfactory; otherwise, it goes to redesign for next run of adjustment. If the evaluation is successful, the design solution will retain in the case base (or knowledge base) for future design application.

1.4 Functional Aspect of Design

We may call the "engineering design" the "design." For "engineer" is not only a person who involves to create something in use as it is a noun; however, "engineer" means to scheduling something as it is a verb. The design object could be any kinds of real-world entities, including a variety of engineering entities such as bridge, building, automobile, airplane, ship, and other non-engineering consuming or service products.

The design is defined as a process of "from function to structure" which starts from an initial concept of satisfying people's demands with specific items of function; any design is to realize these functions into reality and serve for people. Obviously, those demands could be tangible and intangible, for product is not necessarily to be a real entity as a tangible object as we discussed before, it could be a virtual product that satisfying people's needs in different services; such intangible objects can be various financial products of stock, mutual fund, and hedge fund. Moreover, human's spiritual enjoyment is another kind of intangible products, such as the design of tremendous products in huge media industry for our daily life.

Why function is so important to design? It is obvious that the philosophy of comparative design is to satisfy the function of designed "product" (which is tangible or intangible) through the comparison with numbers of similar "products." In other words, the designed "product" is just an entity of isotopic carrier attached with different functions. For distinguishing two different "products," one should just examine the differences of its different attached functions. Therefore, the function difference is the only criterion for distinguishing one "product" to the others.

While a sample base is built with detailed classification of different function attributes of sample "products," it will play crucial role during design process by which the function of the designed "product" can be generated one after another through the sample base along its classified function attributes.

How to satisfy the function requirements of the "product" is essential, and it is of first place of importance in design. The processes of implementation for each function requirement is that the designer searching from the sample base for matching the available function attributes of similar sample attached on an objective entity, which is what people expected to.

1.5 On Sustainable Development

The concept of sustainable development should become the standards of human life in twenty-first century. This earth is facing serious challenges of environmental pollution, resource exhaustion, ecological degradation, population explosion, and economic degradation. Therefore, how to value the harmonious development of human and nature in different engineering design and promote the sustainable development of human society will be an important problem which directly affects the survival of future generations. "Engineer is at the center of the society, who can promote the sustainable development of society through knowledge, skills, and expertise." The progress of society partly depends on how to use the knowledge and skills of engineers to improve the human living environment and gives the prospect of sustainable development.

The concept of "sustainable development" has different definitions in different perspectives. British G.H. Bruntland suggested in 1987: "What sustainable development has described is a kind of process, it satisfies current needs but not damage the future generations to meet their needs." It be understood as that whenever you consider the needs to meet your current need, it should not affect the needs of next generation in their living and survival. Therefore, we should follow the principle of "scientific outlook on development" in our life, and we should implement the rules of energy conservation, emission reduction and make full use of renewable energy, terminate chaotic exploitation of natural resources, reuse of existing material resources, implement the idea of green culture, insist of ecological and environmental protection, etc. These should become the social responsibility of every designer nowadays.

For instance, as a designer in civil engineering, one needs to keep sustainable development idea in mind, as far as possible to reduce the negative effects and impact to environment. We should firmly take corresponding measures in design and construction, embodied sustainable development idea in "four savings" for environmental protection:

(1) Energy saving: increase the efficiency of energy consumption, adopt reproducible energy (such as solar and wind energy), use heat preservation and heat insulation materials, use intelligent technology, etc.;
(2) Material saving: use less or do not use materials that consume more resource (such as steel and clay brick), use light structure to save material, use recycling materials, etc.;
(3) Water saving: establish rainwater collection system and water recycling system and reduce the consumption of natural resources;
(4) Land saving: save the land occupation in architectural planning and reduce land occupation through optimizing planning traffic and high-rise residence;
(5) Environmental protection: eliminate the use of harmful materials and make full use of natural ventilation and natural lighting, recycling use of construction garbage, etc.

If applying the above steps in every project, then we can reduce construction cost, maintenance cost, and the total life cycle cost including structure demolishing and improve structural net assets value, the quality of infrastructure, and effective serving life.

Designer is at the center of a new world which is created by human through transforming nature by applying scientific rules; Each designer has the responsibility to do contribution to the sustainable development of the earth in their daily work, let the earth become more beautiful tomorrow.

1.6 On Aesthetics in Design

1.6.1 Principles in Consideration of Aesthetics

Designing a new product or evaluating an existing object must be judged by whether it satisfies three basic elements of design (function, structure, and aesthetics). A successful design must be the perfect work which is created by both designers and engineers; it is also the perfect combination of art and technology.

Besides the concept of sustainable development discussed in Sect. 1.5, function satisfaction and mechanical design with aesthetic consideration form the main line throughout the whole processes of design. For instance in architectural design, while not only the mechanical design of structure should be reasonable, but also aesthetics design is more critical, while upon the satisfaction of function requirements the latter one plays an important role in deciding whether the architectural scheme can win the bid. It is why the structure engineers should possess some aesthetics knowledge for strengthening personal aesthetic accomplishment and make sense in architecture. Pay attention to the application of aesthetic rules in architectural design, such as harmonious and unified, concise and fluent, symmetrical balance, and color performance, and incorporation to the conceptions of ratio, rhythm, and sequence, an excellent architectural design can then be achieved.

In design, one have to keep in mind the principles that "Structure with appropriate mechanical property is beautiful," "Nature will not do idle work." In reality, a big tree is beautiful; only because the ratio between its trunk and branches meets the natural law of mechanics. The height of branches and the size of crown must be related to trunk's endurance of self-weight and resisting to wind load.

A building should firstly meet with its functions, ensure the safety of structure, it needs to resist the external load and deadweight in decades or hundreds of years during its service life. So the building with smooth and harmonic mechanical performance is appropriate and seems beautiful. At this point, high-rise building has certain similarity with the tree.

Many ancient buildings appear symmetrical balance, having harmonious proportional relation and rhythm sense, the cohesions of components are skillful and rigorous, it makes the force flow concise and lively, there is no extra design and redundant parts, and they become historical and cultural constructions spread cross thousands years.

The structure which is harmonious with surrounding environment is beautiful. Structure design has technical means (application of mechanical concept) and also has the creative space of art (aesthetic application). Aesthetic requirement is that structure is harmonious with the surrounding environment. Building should be able to blend in nature landscape and not undermine the coordination of surrounding environment, such as the requirement of skyline design in architecture.

Swiss Salginatobel Bridge was named the most beautiful bridge in the twentieth century [4]. It was designed by Swiss engineer R. Maillart in 1930 compatible with surrounding environment. All parts of the bridge are properly perfect. It is a sickle concrete arch bridge, set in the valleys of Alps; simple white bridge appears unique with the background of blue sky and green hills and gives people a feeling of "the outstanding one." Walking on Salginatobel Bridge is a real spiritual enjoyment; you are so close to mountains, white clouds, and blue sky. It constitutes a wonderful landscape of Alps. World experts praise the bridge "wonderful article excelling nature," "the best combination of force and beauty," "like a picture of scene in Alps." The bridge has become an attracting place of Switzerland, which is visited by streams of people.

1.6.2 National Characteristics in Aesthetics

The design with national characteristics is beautiful. The national one is exactly also the international one. Many ancient buildings in the world have distinct national characteristics and became the landmark building in various countries.

The architectural design of Chinese Imperial Palace has distinct Chinese characteristics. It is grand, magnificent, dignified harmony, representing the essentials architectural art of ancient China, reflecting long traditional Chinese culture and the outstanding workmanship of the artisans in building five hundred years ago.

The Imperial Palace is strictly built by capital building principle. In building layout, the Imperial Palace uses the technique of shape change, ups and downs, to combine into a whole. In function, it fits the hierarchy of feudal society and at the same time, reaches the artistic effect of balanced and change of shape.

The roof form of Chinese architecture is rich and colorful, and in the Imperial Palace building, there are more than 10 different forms of roof. If three main halls are taken as an example, the roofs are not identical. The roofs of Imperial Palace building are paved with all kinds of glazed tile pieces; the main temple is given priority to yellow; green is used for the residential buildings of prince, and other blue, purple, black, emerald green and peacock green, sapphire, etc., and colorful azure stones are more used in garden or azure stone wall. Each end of the main ridge in the roof of supreme harmony hall has glass kiss animal facilities, sedate effectively, and hold big ridge. The model of kiss animal facilities is beautiful; they are components and decorations. Part of the tile model three-dimensional animal images, such as dragon and phoenix, lion, and hippocampus, which symbol propitious and majesty, play a role in adornment in building. The design and buildings of Imperial Palace are unique masterpiece with most Chinese national features.

To sum up, while a building meets the using requirement, it provides people with a beautiful feeling inspirit. The architectural beauty is a problem of which different people have different views, but it also has certain rules. This section contains some aesthetic principles of structural design, which are well applied in many famous buildings. Architectural design is a comprehensive create activity, which needs to consider all aspects of environmental culture, if copy some aesthetic design principle, designed buildings may give a kind of inflexible feeling to person, and is quite different from the expected aesthetic feeling. Therefore, how to apply aesthetic rule flexibly, and combine with the local history humanity social feel, design excellent architecture which not only can meet the using requirement of people but also can satisfy people pursuit of beauty, this is the problem to think in architectural design.

1.7 Conclusions

This chapter is devoted to discussing basic law of design philosophy than exploring how the basic procedures of a "product" (whether tangible or intangible) are designed. We emphasis the design is the process of "from function to structure" means we catch the essentials of design is just a process of how a designer through comparison attach the function requirements onto an isotopic carrier to form an entity (real or virtual) of "product."

In the past, without the help of computer, the process of comparison can be completed just by the heuristic consideration of the designer, or just by the experience of the designer to fulfill the design requirements through the memory of past samples in his (her) mind. At present, by the help of computer and artificial intelligence technologies, it is possible to use knowledge engineering technique to store number of past samples through digitization in the sample base; then, the design process can be implemented through "case-based reasoning" to the sample base to get appropriate function attributes for the designed "product."

Most of the chapters in this book are devoted to conceptual design of products, for the purpose of achieving practical effectiveness, and we will mainly use the "product" as a structural design to describe concretely the processes of how a designer treats the conceptual design of structure by a comparison of past experience.

Chapter 2
Introduction to Structural Design

2.1 The Features of Modern Structures

The purpose of modern structural design is to provide the structure with up-to-standard operational functions. Physical integrity serves as a prerequisite for a functional structure to withstand external interference. The physical integrity is provided by load-bearing components, which comprised of numerous specific materials. Thus, modern structural design will need to follow the principle of "take it as s a whole while start from the subsystem." It means when we design for the subsystem we still have to take into account the mechanics, aesthetics, and the need of sustainable development of the entire structure. The modern structural design should focus on three main parts: mechanical concepts, aesthetic concepts, and sustainable development, which are independent while interrelated at the same time.

One feature of modern structure is to meet growing functional demands. These demands are not confined to the increment of height (the record of the highest building has been constantly rewritten), the prolongation of bridge span (the record of the longest span has also been constantly rewritten), and the addition of new structural systems (new structural systems and new materials have sprung up and been applied to actual construction). Newer demands are embodied more in the aspect of "green" sustainable development.

The development of modern structural engineering has to coordinate the social, humane, and natural development. Modern high-rise structures, bridges, spatial structures are not in blind pursuit of being impressively high, wide, or giant but have put the emphasis upon "functional requirement," "energy conservation," and "being environmentally friendly." Therefore, engineers need to contemplate upon aesthetics and sustainable development on the basis of the overall conception; meanwhile, they are supposed to employ mechanics and structural systems to compose and decompose structure, turning a complex structure into simplified three-dimensional framework, two-dimensional plane substructure, or even one-dimensional linear or nonlinear components and composing these components

S. Lin and Z. Huang, *Comparative Design of Structures*,
DOI 10.1007/978-3-662-48044-1_2

back to a modern structure which could meet diversified functional requirements of being ultra-high, long-span, spatial, etc.

1. Being ultra-high: Due to the limitation of urban land, ultra-high or high-rise structures have become one of the trends nowadays. Following Shanghai Jin Mao Tower (88 stories, 420 m high) and Shanghai World Financial Center (101 stories, 492 m high), an even higher skyscraper known as Shanghai Center is under construction and will be completed in 2014. For ultra-high structures, proper design needs to focus on seismic performance, as well as wind load, wind induced lateral deformation, and overturning moment at the foundation.
2. Long span: Long-span bridges have naturally built over wandering rivers. Typical types include cable-stayed bridge, suspension bridge, and varied arch bridge. Take the recently completed Jiangyin Yangtze River Bridge for instance, its main span stretches over 1385 m. Both Shanghai Lupu Bridge and Chongqing Chaotianmen Yangtze River Bridge are steel arch bridges with spans of 540 m and 552 m, respectively, and are currently on top of the world's largest span arch bridge list. Along with the increase of the span, issues such as wind vibration, earthquakes, structural dynamics, and stability will become prominently significant. Those concerns will pose great difficulty and challenge upon the structural design and calamity prevention.
3. Novelty: With growing constructions in twenty-first century, more and more novel structures and new technologies have been emerging. The features of novel structures lie in the peculiarity of the presence and the complexity of its stress state, which for the moment is devoid of available code to abide by, in terms of effective approaches of analysis and design. As a result, performance of such structures might not be predictable and controllable as it is supposed be.

Last century had witnessed an enormous amount of ultra-high-rise structures, ultra-long-span bridges, and a collection of novel structures with distinctness. To analyze their performance under severe conditions is extremely complicated and challenging. However, as long as the fundamental mechanical principles are followed, and reasonable models are applied, it is possible to effectively evaluate different preliminary designs then find the optimum one.

When it comes to ultra-high-rise structures, they could be simplified as cantilever beams in the preliminary design. In that case, stress state could be analyzed; structural systems and configurations compared; approximate dimensions of major components determined; and construction costs of different structural systems estimated. Then questions can be raised in selecting different design schemes, such as reinforced concrete or steel, reinforced concrete core wall system or steel frame, a simple rectangle layout or a complex H layout, and a simple cuboid with identical section or a trapezoid with variable sections. All these questions could be answered through the application of conceptual design and analytical methods illustrated in this book. Figure 2.1 shows the Shanghai Jin Mao Tower and its simplified model of a cantilever beam.

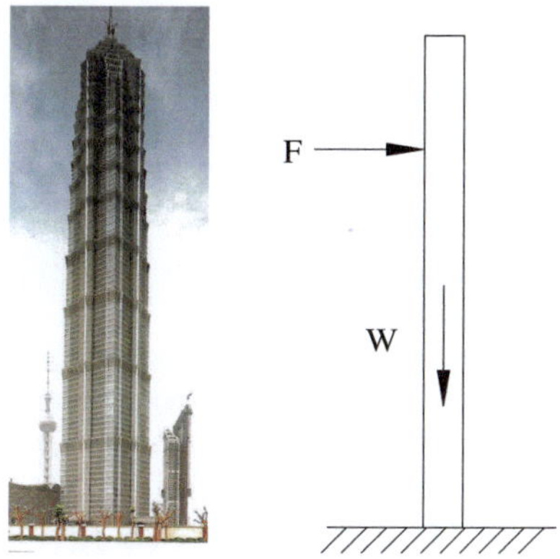

Fig. 2.1 Shanghai Jin Mao Tower and its simplified model of cantilever beam

As for long-span structures, especially bridges, they are simplified as simply supported or continuous beam in the preliminary design. In that way, stress states could be analyzed; approximate dimensions of major components determined; and construction costs of different structural systems estimated. For instance, the adoption of simply supported beam or continuous beam, and the selection of reinforced concrete structure or steel structure should result in significant difference regarding costs and configuration. Figure 2.2 shows the Donghai Bridge of Shanghai and the simplified model of its cable-stayed bridge (over the main channel).

When approaching novel spatial structures, they could be simplified as arch structure or suspension structure or a combination of several structural systems in the preliminary design. Similarly, stress states could be analyzed; approximate dimensions of major components determined; and construction cost estimated. Figure 2.3 illustrates the steel structure of "Bird's Nest" stadium in Beijing and the simplified model of its portal-rigid frame. From Fig. 2.3b, we know that the "Bird's Nest" actually is achieved through laying 48 portal-rigid frames around the oval opening in the roof; Fig. 2.3c shows the simplified model of one portal-rigid frame.

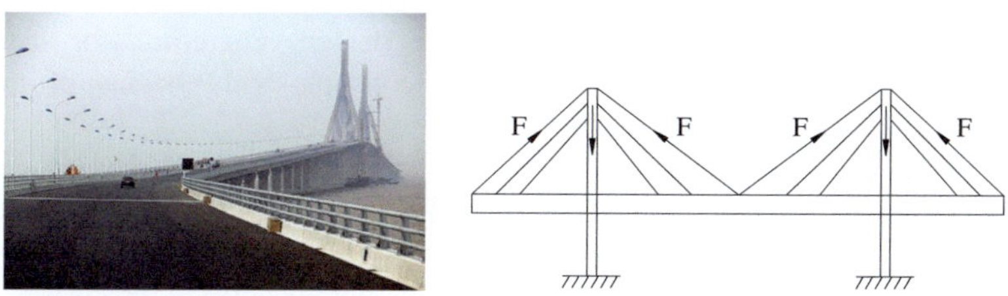

Fig. 2.2 Shanghai Donghai Bridge and its simplified model of cable-stayed bridge

Fig. 2.3 Beijing "Nest" stadium and its simplified model of portal frame. **a** Beijing "Nest" stadium, **b** the structures, **c** portal frame

2.2 The Processes of Structural Design

As Wilson adapted from an unknown author in his book [5]:

> Structural engineering is the art of using materials that have properties which can only be estimated, to build real structures that can only be approximately analyzed, to withstand forces that are not accurately known, so that our responsibility with respect to public safety is satisfied.

Therefore, structural engineering is the combination of technology (mechanics) and art (aesthetics). It requires our structural engineers and architects to ensure the reliability of designed structure through technical means. Meanwhile, it also requires realizing aesthetics and operational functions through artistic means. In other words, with the guarantee of structure integrity, art is also given a creative role in design. And the need of sustainable development should be also taken into account.

All the above tells us that there is no such thing as the only correct solution in structural design but satisfactory design. Besides, structural design requires estimation of properties of material, approximate analysis of structural model, and estimation of potential external loads, in order to ensure the safety and satisfactory operational functions of the structure. Moreover, architectural design is in essence a "conceptual design" from schematic to preliminary and finally to detailed design. Conceptual design does not only play an important role in schematic design but also provide guidance for preliminary and detailed design. Therefore, for structural engineers and architects, learning about the concepts and systems of modern structural design is of considerable significance.

Traditional civil engineering education puts too much emphasis on detailed design of structure and precise analysis of mechanics while ignoring the application of a global perspective of mechanics and systems. Modern structural design expects designers to propose satisfactory structural system and concepts in the early stage of design. Outstanding conception and schematic plan could not only guarantee the feasibility of the overall design but also optimize the plan and economize the costs of construction, in which way physical integrity and aesthetic concern are delivered and requirement of sustainable development is met.

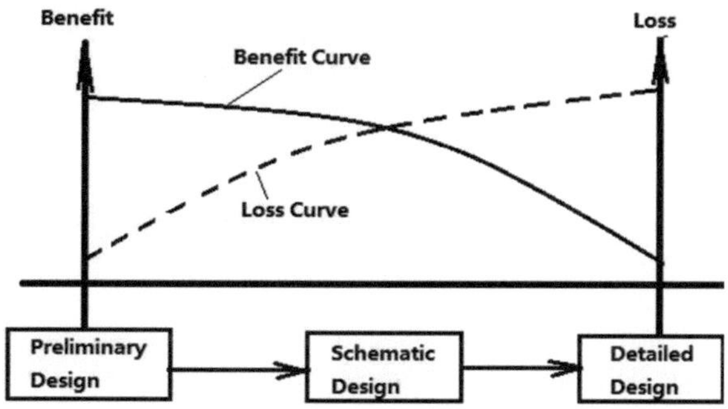

Fig. 2.4 Conceptual design in all stages of structural design

Architectural design consists of three consecutive stages: schematic design, preliminary design, and final detailed design. And conceptual design could be applied through the entire process of structural design. Figure 2.4 illustrates the significance of conceptual design in all stages. If a better design is selected in the earlier state of design based on conceptual design, it would maximum the benefits in terms of structural rationality, artistic configuration, and economized costs, all of which would results in a favorable winning bid. If the conceptual design is well applied in the stage of preliminary or final detailed design, especially on key issues, certain benefits are still expected. However, compared to the early stage, the benefits gained in the latter stage would be relatively smaller. Similarly, tiny errors in the earlier design stage might results in significant loss at the later stage. Numerous lessons have been learned.

This book aims to provide the readers with concepts and deeper knowledge of mechanics as well as its application in all structural systems so that readers would be able to analyze the key issues of structure from a global perspective and well apply the concepts and systems of the structure in schematic design as well as in preliminary design. In this way, benefits are maximized and loss in later design and construction is avoided.

2.3 The Process of Architectural Design

2.3.1 The Elements of Architectural Design

Before the introduction of architectural design, some nomenclatures concerning architectural and structural design have to be clarified [6].

1. Civil engineering: the summation of all permanent artificial structures that are made of physical materials in order to improve the living conditions of human beings.

2. Structural engineering: the summation of all structural systems that are solid-made, capable of force bearing, and manifest in diversified configurations and in the purpose of improving the living condition of human beings.
3. Architecture: a course about interior and exterior arrangements of structure so as to achieve operational functions. The arrangement does not only have to meet the requirements of spatial combination but also have to provide an artistic configuration as well as a sense of harmony with the surrounding environment.
4. Structural engineer: Structural engineer is responsible for plan, design, construction (quality, cost and control of construction period included), maintenance, and rehabilitation of the structure.
5. Architect: Architect is responsible for schematization (skyline of the group of structures included), spatial composition of the structure, aesthetic design and management of the internal and external configuration, and decoration and coordination with surrounding environment.

To design a new structure or to evaluate an existing one, three essentials need to be given proper concerns:

1. Functions. Structures should possess certain operational functions to meet the need of human activities;
2. Structural integrity. Structures should be constructive, structurally reliable, and economically feasible.
3. Aesthetic experience. Structures should provide a symbol of social and aesthetic values for the users, and harmony with the surrounding environments.

From the above three essentials of architectural design, we could see that architects are more concerned with the functional and aesthetic performance. Structural engineers are more concerned with the structural integrity. A successful architectural design has to be a comprehensive work created together by architects and structural engineers, which all the essentials are fulfilled in a collectively optimum manner.

2.3.2 The Relation Between Structure and Architecture

Unlike other industrial or agricultural product, architecture is a unique product displaying spatial arrangement as well as the surrounding environment. Each one of them is unique in a way of its district nature and how it interacts with its surroundings. There are no two identical leaves in the world, and so it is true for architectural products. Structures are supposed to be a creative product completed by architects and structural engineers together. The capability of architectural design of architects and that of structural design of structural engineers should be tightly integrated and harmoniously complemented.

Figure 2.5 shows the "Arabian Tower" project [8] which was awarded by international structural engineer association in 2000. The intention is to display a yacht sailing on Arab sea, providing a special natural feeling for direct senses and

Fig. 2.5 Tower of Arabs [7]

indirect imagination. The entire configuration of the structure is like a canvas, connected to the giant curved pillar frames on the front and back. The mutual support provides a significant horizontal stiffness to the main structure. The extension of giant curved pillar frames takes the form of the mast of a canvas sail; the building is built upon the artificial island at the bay, on the basis of which the podium looks like the hull of. The whole work embodies a perfect combination of architectural and structure design and synthetically reflects the achievement in all aspects of aesthetics, society, humanity, architecture, and construction technology. Among these factors, architectural design plays a leading role in meeting requirements of aesthetic expression and operational functions while structural design guarantees the feasibility of construction and reliability of the structure.

For infrastructure design, most projects are structural engineering-oriented. Figure 2.6 shows the two-story tunnel of Shanghai Fuxing Eastern Road, which is a two-way, "double-barreled" tunnel of eight lanes. When projects of such nature are concerned, so as to meet functional requirements of the project, the emphasis of construction will be inevitably put on matters of civil engineering including the selection and layout of the route, structural design of underwater and underground structures, excavation techniques, and ventilation. Architect has a relatively limited role in infrastructure construction projects.

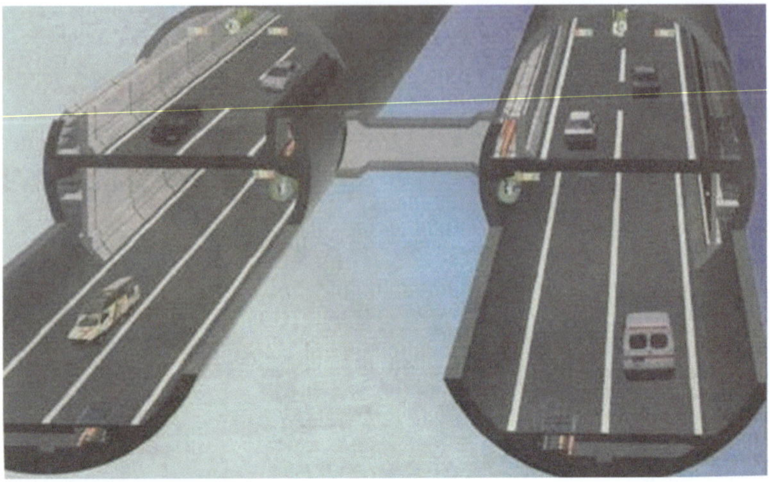

Fig. 2.6 Double-layer tunnel in East Shanghai Fuxi road [8]

Therefore, architects and structural engineers relay on each other in a complementary way. As for different projects, primary role shifts between them. However, due to the different backgrounds, architects and structural engineers are sometimes found to have difficulty in cooperation. This is due to that the architecture is a complex of tangibility and intangibility and that of abstract and concrete notions. Architects put more emphasis on general and global aspects, such as spatial forms and atmosphere created. On the other hand, structural engineers focus more on tangible and concrete structures and detail aspects. Different ways of thinking sometimes result in difficulty in communication.

Architects deal with issues in a global and top-to-bottom way. On the basis of experience, they take primal conditions, restraints, and available resources into consideration so as to conceive a spatial arrangement with the emphasis upon the entire structural system. This will provide not only the operational functions but also an expression of aesthetics. Howbeit, traditional education instructs that structural engineers take a local and bottom-to-top approach and start from learning the treatment of details regarding specific beams and columns with certain neglect of an overall performance of structures and requirements of sustainable development. In this way, a chasm is formed between architects and structural engineers. As a result, communication and creative cooperation, which are of great significance in schematic design and preliminary design, are restrained.

In the stage of schematic design, issues regarding the entire structure rather than details of local components are the major concerns to be dealt with for both architects and structural engineers. Only a profound understanding of relations between the different spatial forms of the architecture could better enhance the comprehension of the demand for local components. Thus, the architecture is first regarded as an integral spatial form, in which case structural analysis of the entire structure comes as the priority and derives substructural systems and key components; then concepts and knowledge of fundamental mechanics could be applied to attain a perfect structure possessing all three essentials of architectural design.

2.3.3 The Process of Architectural Design

The process of architectural design takes three steps as follows:

1. **Schematic design**

The first stage of architectural design is schematic design. In this crucial stage of design, an architectural design of the entire structure would be visualized in accordance with requirements from the owner. And the interaction between major substructures, the construction feasibility, and costs are to be considered. After owner is satisfied, approval from local bureau of planning and administration is also needed. A schematic design is generally required to provide a three-dimensional model, overall schematic plan, simple plane and elevation layout, estimation of total costs, and construction plan. Schematic design sets the tone of overall quality of the architectural design, and thus probability to win the bid.

2. **Preliminary design**

The second stage is preliminary design. In this stage, improvement shall be made to the plan of choice. Preliminary design is expected to determine the dimension of substructural systems, that of major components and the relation between major components together with their geometric and spatial dimensions. At the completion of preliminary design, a detailed version of the overall plan, plane and elevation layout, dimensions of major components, the estimation of total cost, and plan of construction shall be expected. Preliminary design, after being submitted, could be employed as a reference in construction bid.

3. **Detailed design**

The third stage goes to detailed design. In this stage, preliminary design selected by owner and design-related professionals would update the plan to final detailed plan for construction. Priority of this stage is given to detailed design of substructures and all structural components, including the selection of specific materials and determination of construction technology, and detailed design for construction. Detailed design offers guidance for construction, in which case any modification shall need the approval from designers and owner.

What we emphasize is that architectural design is an iteration process with levels and steps, in which architects and structural engineers provide and depend upon feedbacks from each other. From the very first stage of schematic design, repeated modification shall be anticipated. An optimized edition of design would be eventually accomplished through interactions of feedbacks and modifications.

T.Y. Lin once said "We must learn to distinguish principal contradiction from numerous contradictions." To design from a perspective of overall performance is of utmost importance for both architects and engineers. It does not mean that detail design is not important, and inappropriate construction of details actually led to many structural failures. Generally, architects and structural engineers shall first establish a comprehensive understanding of the overall spatial form and relevant operational functions. And then they focus on the detailed design of specific overall system, substructures, and components to achieve an optimized design.

2.3.4 The Application of Structural Concepts in Architectural Design

Concepts of mechanics shall be brought into architectural design as early as possible in the design process. A reasonable consideration of structural concepts could lay the necessary foundation for the following preliminary and detailed design. In other words, architects are supposed to cooperate with structural engineers in the early stage of architectural design.

It is possible and necessary to introduce concepts of mechanics in the early stage of design. Structures are required to provide all sorts of services which entail vertical transportation, horizontal transportation, and systems indispensable in daily life such as water, electricity, gas, heating, ventilation, air conditioning, acoustics, lighting and waste disposal. Space for vertical and horizontal transportation is embodied in structures as elevator shafts and corridor of all sorts. Since all these facilities mentioned above have to be allocated within the specified geometry from the perspective of operational functions, coordinated arrangement is required. From the perspective of mechanics, vertical transportation such as elevator shaft often constitutes vertical components with fortified integrity which undertakes self-weight and horizontal forces. This is particularly notable in high-rise structures, for that vertical transportation of elevator shafts often constitutes the core tube commonly seen in high-rise structures and comprises "tube-in-tube" structure with outer frame-tube (Fig. 2.7).

Figure 2.7 shows a typical "tube-in-tube" high-rise structure; the outer tube is an external wall framework being comprised of close columns and horizontal beams; as for the inner tube, it is the concrete shear wall elevator shaft. The inner tube is connected to the outer tube through the slab on each floor, in which case an integral "tube-in-tube" structure is realized with high strength and stiffness to resist horizontal loads.

As for other indispensable systems for daily life such as heating, ventilation, air conditioning, water supply, power supply, and gas supply, they could be integrated in horizontal and vertical pipes of variable sections and installed. And their spatial arrangement has to be integrated with mechanical system of the structure. For

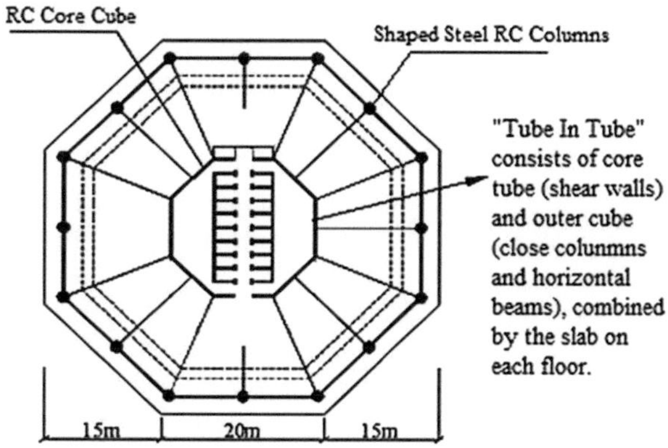

Fig. 2.7 "Tube-in-Tube" structural system [9]

instance, horizontal pipes for equipment could be arranged within the space between secondary beams or on the transfer floor.

Moreover, despite that the influence of acoustics and lighting is categorized as architectural physics, it still has everything to do with structural configuration. For instance, acoustics has a decisive call upon the spatial configuration of roof. Specifically, an arched roof could produce converging effects, which is quite fit for concert halls and opera houses; as for dish-shaped roof, it has diverging acoustics, which is adoptable for exhibition halls and libraries.

In sum, in early schematic stage of architectural design, architects and structural engineers are supposed to work collaboratively and bring in concepts and systems of mechanics as early as possible. A successful architectural design shall be the fruit of wholehearted cooperation between architects and engineers as well as a quotable case of successful application of mechanics and structural systems.

2.4 The Structural Behavior and Design Principles

2.4.1 The Structural Behavior and Energy Principle

1. **The energy principle**

To ensure that the structure provide operational functions as expected, a physical integrity has to be kept under external interference, which specifically indicates that structure would not fail under the combination of all external loads.

The law of motion of all matters is correlated with certain change of energy in its nature. Take downward-flowing water as an example. It is the function of gravity potential energy. The service state of structures reflects the mode of motion of material systems; it relates to certain change of energy within the structure; the actual state of structure more often than not indicates the extreme state of certain energy [9]. In other words, investigation of the energy changes of the structure could lead to the determination of stress state of the structure. And this would be of great help when it comes to dealing with complex issues such as strength, dynamics, and stability of the structure.

2. **Static equilibrium and the principle of minimum total potential energy**

Under the action of external load, the interior of structure would produce stress and strain response to exterior displacement. Strain energy is generated inside the structure and external work is done by external work upon displacement. In light with principle of least action in mechanics, nature produces nothing that is not necessity. As for structures, the least action is the total strain potential energy within the structure and this strain potential energy generated within is the minimum potential energy produced by the work done by the internal stress upon the internal strain. If supports of the structure are not displaced, reaction force of structure does not do work; in accordance with the principle of energy conservation, total potential energy of the structure is the algebraic summation of internal strain energy and the work done by external loads. Assuming that the structural displacement concerning total potential energy is virtual displacement and the internal strain of structure is virtual strain, the extreme value of total potential energy could be calculated (which is to conduct variation) and the equilibrium equation of the structure could be attained. Thereby, the essence of force method applied to static indeterminate structure in mechanics is consistent with that of the principle of virtual work expressed by virtual displacement and the principle of minimum total potential energy. Similarly, the essence of displacement method applied to static indeterminate structure in mechanics is consistent with that of the principle of virtual work expressed by virtual force and the principle of minimum total complementary energy.

3. **Dynamic equilibrium of structure and the principle of minimum total potential energy**

According to the principle proposed by D'Alembert, dynamic equilibrium could be regarded as equivalent static equilibrium posterior to the introduction of inertial force. As a matter of fact, the load bore by the structure and the internal force constantly change with time. The result of this simplification turns dynamic issues into static issues of keeping dynamic equilibrium at any moment. In the sense of an instant, the principle of minimum total potential energy holds true to dynamic equilibrium of an instant when the structure as a whole is concerned.

Therefore, in the sense of an instant, Rayleigh approach, equivalent to the principle of minimum total complementary energy, could also be applied to dynamic equilibrium of structure in an instant. Rayleigh approach establishes the equation between maximum strain energy and the maximum kinetic energy. When the structure is in its natural vibration state, Rayleigh equation could provide the natural vibration frequency of the structure.

Natural vibration frequency is an intrinsic nature of structure, which varies with the distribution of stiffness and mass but not to the magnitude and type of external loads. Natural vibration frequency serves as the foundation of research upon dynamic response of structure in the early stage. However, in that stage of structural design, components are not decided yet the deadweight and distribution of stiffness. As a result, natural vibration frequency could only be estimated through simplified calculation.

An accurate calculation of natural vibration frequency together with dynamic analysis might be excessively detailed, involving theories of structural analysis of depth, and would not be elaborated in this book. Readers with an interest in those approaches could refer in Refs. [5, 9].

4. **Stability and the principle of minimum total potential energy**

Some structures remain in equilibrium of stability under relatively small load. As the load gradually increases to certain extent, although the ultimate strength or even the elastic limit of material has not been reached, deformation has largely increased to the extent that structure fails to bear any more loads. In this case, overall instability or local buckling takes place.

Overall instability in nature originates from fundamental change in equilibrium state, being transited from stable equilibrium to unstable equilibrium. Unstable equilibrium temporarily meets the requirements of equilibrium, or from the energy point of view temporarily suits the principle of minimum total potential energy. Nonetheless, a small "tip" from the outside would break the equilibrium and results in overall instability.

During the transition from stable equilibrium to unstable equilibrium, a critical state exists and that is the so-called neutral equilibrium state. Load applied in such a state is designated as critical load of instability. When it comes to structural design, load designed for the structure to bear shall be smaller than critical load of instability so that instability is prevented. Since issues regarding instability could be summed as that of transition between equilibrium states, extreme value of total potential energy within a structure at its critical state could be calculated so as to determine an equation of critical equilibrium and critical load of instability.

Local buckling is the instability of components and is similar in nature to overall instability.

2.4.2 Overall Stability of Structure and Stability in Its Geometric Composition

When schematic design is in process, overall stability and that in geometric composition have to be checked; otherwise, conception and analysis of strength and stiffness would deem meaningless.

1. **Overall stability of structure**

All permanent structures are supported by the foundation. The structure as a whole needs to have effective support from the foundation.

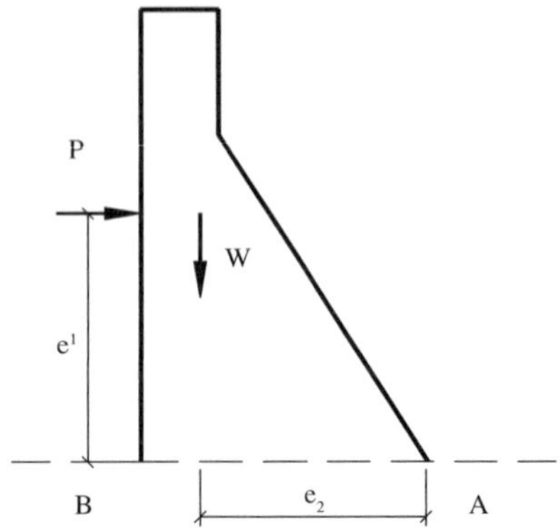

Fig. 2.8 Overall stability of the dam

As for the dam shown in Fig. 2.8, two fundamental requirements for overall stability are given as follows:

(i) Stability regarding horizontal load: $P < W \cdot f$
P is the lateral water pressure upon the dam; W is the deadweight of the dam; f is the friction coefficient between the dam and the foundation.

(ii) Stability regarding overturning moment (relative to point A at the bottom): $P \cdot e_1 < W \cdot e_2$
e_1 and e_2 represent the length of arm of force, as shown in Fig. 2.8.

2. **Stability in geometric composition of the structure**

Structure is comprised of many components and only when these components compose in a rule-abiding geometric way could the structure stay stable. Otherwise, when the structure itself is geometrically unstable, tiny little interference from external load could lead to considerable (infinite in theory) displacement. Hence, geometrically unstable structure is not qualified to take load and has to be excluded before schematic design starts. This is especially significant when the design is for spatial truss and latticed truss comprised of numerous bars.

Fundamental principle of geometric composition for a stable truss structure could be described as: Truss structure is formed by many a "triangle discoid" which is a geometrically stable unit composed by three hinged components. Truss structure itself could be regarded as geometrical constant and is supported by the foundation through three or more non-convergent connecting bars.

Figure 2.9 shows a stable truss structure; it consists of eight "triangle discoid," number 1, 2, and 3 of which connects with the foundation. As for the truss in Fig. 2.9b, although it has the same number of bars as the structure in Fig. 2.9a, it is a geometrically variable structure due to the difference in geometric composition.

Figure 2.10 shows a geometrically unstable truss structure getting support from the foundation through three bars converging at point O. Since these three bars could not prevent the structure from rotating around point O, this truss is considered as geometrically unstable.

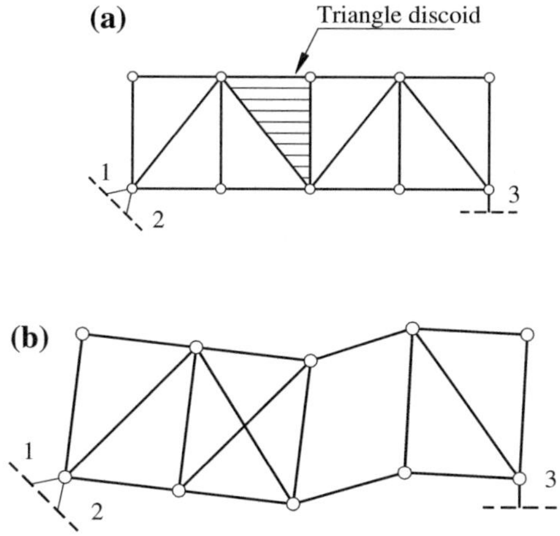

Fig. 2.9 Comparison of stable truss and geometrically variable structure, **a** the stable truss, **b** the geometrically variable structure

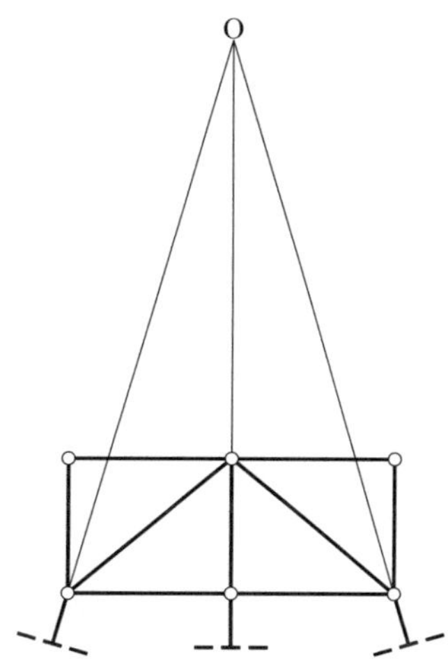

Fig. 2.10 Geometrically unstable truss

From the perspective of mechanics, geometrically unstable structures have no resistance in at least one direction which enables the structure to move without restraint in that specific direction. If looking at the matrix of overall stiffness matrix of a geometrically unstable structure, at least one zero elements would exist among all diagonal elements.

2.4.3 The Upper- and Lower-Bound Theorem in Limit Analysis of Structure

An important step of structural design is to conduct limit analysis, which is to find the ultimate bearing capacity and the failure mode. Since materials are capable of producing plastic deformation to some extent, the ultimate state of structure generally is its plastic limit state. Limit analysis is to study the ultimate bearing capacity of structure in its plastic limit state. When the load working on the structure augments to a certain limit, the material of structure transforms into a plastic state, in which case deformation infinitely multiplies and the structure fails to bear loads. And this state is called plastic limit state of structure. The goals of plastic limit analysis of structure are: (1) to determine the ultimate bearing capacity of structure; (2) to determine the stress distribution that meets the requirements of boundary conditions in the limit state of structure; and (3) to find out the most probable failure mode.

To determine the ultimate bearing capacity of structure, one has to know external forces, geometric boundary constraints, and the material properties. As mentioned before, the properties of material, external loads, and geometric boundary constraints are not accurately known, therefore reasonable estimations are needed. Conditions needed in the meantime are: (1) yield strength and constitutive model of materials; (2) equilibrium condition; and (3) compatibility condition. Yield strength and constitutive model of materials are based on experiments, while equilibrium and compatibility have to be satisfied in limit state.

Solution that meets all three conditions mentioned above is considered as a complete solution, which in other words is the complete solution to the ideal bearing capacity of structure which simultaneously meets sufficient and necessary condition for structure failure. Due to the complexity of load bearing and the uncertainty concerning properties of material, external loads and geometric boundary constraints, complete solution is not easy to attain which makes simultaneous meeting of three conditions impossible for limit analysis. Nevertheless, two theorems are developed from the theory of limit analysis: the lower- and upper-bound theorem. The upper-bound theorem proposes sufficient condition for failure which indicates that the structure will surely fail once such load is imposed upon the structure. Since the bearing capacity proposed by upper-bound theorem is generally larger in magnitude than the potential load that structure might have to bear, it is unsafe to design with reference to assumed failure of the structure derived from upper-bound theorem. Lower-bound theorem proposes necessary condition for the prevention of structure failure, meeting the condition of which means the structure is safe under the action of such load. Therefore, it is conservative applying the lower-bound theorem in design.

Current code indicates that limit design for normal section of reinforced concrete beam employs assumed stress distribution on the normal section and a flexure failure mode. It assumes that tensile reinforcements begin to yield and compressed concrete is up to its compressive strength. The equilibrium between the internal and the external force on the normal section could be employed to determine ultimate flexural bearing capacity of the beam. This is a typical solution derived from the lower-bound theorem, which meets only the equilibrium condition and material capacity with the exception of deformation compatibility.

In sum, when designers conduct ultimate bearing capacity (strength) design for structures, it should be noted that limit analysis does not get the complete solution which meets all three necessary and sufficient conditions of failure; instead, it usually gets a lower-bound or upper-bound solution which meets two out of three conditions. On one hand, a good limit analysis of structure shall have a most likely mode of failure anticipated and in the meantime have a relatively lower (under given mode of failure) upper-bound load designed. On the other hand, a good limit analysis of structure shall also find out the most probable mode of equilibrium between the internal and the external force and as a result of having a relatively higher (under given mode of failure) lower-bound load designed, which enables a more accurate estimation of the bearing capacity of the structure.

2.4.4 Instability and Structural Dynamics

Modern structural design features structural dynamics and structural stability as factors that have to be considered. Estimation of dynamic features such as fundamental frequencies is usually needed in early schematic design.

Frequent earthquakes and other dynamic loads directly threat the safe operation of structures. For instance, resistance to strong seismic movement holds the key to a safe design for ultra-high-rise structures; in the meantime, vertical seismic effect is threatening the safe operation of ultra-long-span bridge. Dynamic response of spatial structure under the action of complex and variable dynamic loads also poses difficulty to structural design.

Under complex combination of loads, light-weight and thin-walled structures (especially spatial structures) are especially prone to local buckling and overall instability issues, such as local buckling and overall stability of thin-walled component in compressed areas. As for long-span structures, overall buckling is more likely to happen. Since the slender components are prone to buckling failure, sometimes reinforcement is needed to provide sufficient stiffness, which keeps the slenderness ration within a reasonable range and avoid local buckling.

To illustrate consideration of buckling stability in structural design, Fig. 2.11 shows an example of a reinforcement design process for U-shaped steel beam. Figure 2.11a shows a U-shaped steel beam under the point load *P*; the bending

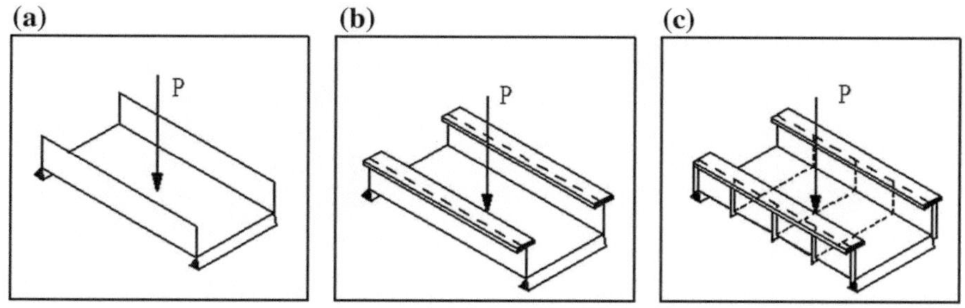

Fig. 2.11 Process of reinforcement design for a U-shaped steel beam [9]. **a** Reinforced at end of beam, **b** reinforced at beam flange, **c** reinforced at the web

moment produces longitudinal tensile stress on the bottom surface and compressive stress on the upper surface, the latter of which might induce local buckling on the wing. Figure 2.11b illustrates the approach to locally reinforce the transverse stiffness of the wing so that local compressive buckling of the wing is prevented, and thus the overall bearing capacity of the beam is enhanced. As load increases, buckling failure could happen to the web or bottom plate due to lack of restrains. Figure 2.11c shows that horizontal reinforcement is installed to prevent local buckling at the web or bottom plate, and the overall bearing capacity of beam is further enhanced.

Although structural dynamics and structural stability are not of same nature, they are apparently interactive with each other. A typical example is the famous Tacoma Narrows Bridge, which collapsed due to overall instability under strong wind in 1940, several months after opened to traffic.

The deck of Tacoma suspension bridge consists of a variety of light-weight and thin-walled structures. Figure 2.12a shows the bridge prior to collapse [6]. Under a

Fig. 2.12 Overall instability failure of Tacoma suspension bridge [6]. **a** Before failure, **b** after failure

series of complex aeroelastic flutter, the slender deck system is prone to lateral buckling. However, the suspension cables could not effectively prevent lateral deflection (actually designed to allow the bridge to sway in wind), which eventually caused the collapse. Since this accident, wind tunnel testing became a necessary experiment to examine the aerodynamic performance for new long-span bridges. Figure 2.12b shows the moment that the bridge was collapsed.

Modern structural design takes into consideration the factors of dynamics and those of stability without exception [10]. Unfortunately, structural engineers still have to rely on complicated mathematical formulae and modeling methods (or other numerical discretization methods), in the help of computers. Due to the scale of complexity and time consumption, this approach is difficult to be implemented in the earlier stage of design. Therefore, it is necessary to find some simplified models and approaches to estimate and evaluate overall structural stability and dynamic performance.

2.4.5 *Fundamental Principles of Structural Design*

Structural engineering is based on practical experience as well as theoretical mechanics. Three fundamental principles shall be paid attention to in regard of conceptual design and systematic composition of structure:

The first fundamental principle is that the ductility of structures (capacity to withstand plastic deformation) should provide considerable tolerance for excessive loads, which forms the first "perimeter" to prevent any possible errors from inevitable simplification, or even mistakes from engineers as human beings. The second fundamental principle is that details like joints should be carefully designed so that premature failure is avoided. The third fundamental principle is that building more redundancy through static indetermination is also effective to prevent premature structural failure.

1. **The utilization of ductility of structures**

Ductility describes the ability to withstand plastic deformation under extreme loads. It induces ductile failure and functions as a storage capacity to resist extreme external loads.

One of the three fundamental principles for structural engineering is that "structural ductility could provide considerable tolerance for minor errors of engineers." That is to say, small errors from simplified calculation or in design will be mitigated by redistribution of stress within materials. Then the structure would continue to bear load without catastrophic failure even when it is overloaded to certain extent. Granted that it is possible and effective to use simplifications in structural design, however, when dealing with brittle materials, engineers should carefully take the brittleness into consideration to avoid sudden brittle failure.

Many designers favor the "lower-bound theorem" to control structural safety. It refers to a stress path (also known as force flow) design. If force flow is in balance with external loads and the stress of material along the path does not exceed the material strength, the structure is deemed safe. In this way, the service load is always lower than strength so the safety is achieved.

Recently, endeavors have been made to include equilibrium conditions and material constitutive conditions into structural design. American Concrete Institute (ACI) has already applied the lower-bound theorem in its design for joints, torsion, and shear design in its design code. However, special attention should be given to steel components, especially steel connections between column and beam. Despite its ductile nature, steel might turn to a brittle material under triaxial loading conditions. In this case, in order to achieve desired safety level, a proper design would come from the lower-bound theorem viewing steel as a brittle material.

For main load-bearing components, the force flow should be designed to be as clear as possible. The force flow substructures should also be clear so that the corresponding contribution could be easily determined. The "force flow" within the structure should take the shortest path to the support. Moreover, the main "force flow" of structure shall be as smooth as possible and take the form of a natural streamlined curve. A disturbed line of force flow inevitably leads to local stress concentration which would not only produce additional deformation but also raise the cost due to higher material volume to achieve desired safety level.

In addition, the ductility could enable much more energy absorption so that more deformation before failure is tolerated and the failure process is prolonged. This is of special significance in terms of seismic performance. And steel structure is quite outstanding in this respect.

When it comes to structural design, there is no so-called the only correct solution. Designers, facing a lot uncertainty, should be able to identify the relatively satisfactory solution or ideally the optimal solution.

2. **Connection and joint design**

The second fundamental principle of structural engineering is that "The second fundamental principle is that details like joints should be carefully designed so that premature failure is avoided." The significance of joints should be emphasized because they are at the "key positions" and crucial to the overall structure. Numerous structural failures were caused by joints failure. Apart from steel structures, a great many reinforced concrete structures have been improving joint ductility to improve seismic performance. More often than not, the joints of reinforced concrete structures would be densely reinforced in order to achieve better seismic performance due to improve joint ductility. Much more energy will be absorbed during a seismic event which enables the structure to be: (1) intact under minor earthquakes; (2) functional and repairable (non-yielding) under moderate earthquakes; and (3) non-collapsing under major earthquakes.

To ensure normal functions of the structure, joints are supposed to be reliable; however, large amounts of joints failure are induced by small defects in material itself, welding sequence and quality, weather conditions, and supporting conditions. It should be given careful consideration to joint design so that the joint would live up to expectation.

As a matter of fact, the cost of joints accounts for considerable percentage of the overall cost, especially when novel spatial structures are concerned, in which case costly cast steel joints are needed. And joint design has to meet comprehensive requirements of strength, stiffness, ductility, and budget, so that joints are stronger than connecting components. When the structure is overloaded, plastic hinges are forced to develop in components so better structure stability is achieved and premature collapse is prevented.

3. **The utilization of static indetermination**

The third fundamental principle of structural design is that building more redundancy through static indetermination is also effective to prevent premature structural failure. A robust system is required to take into consideration that some components might fail due to reasons which are not designed for (such as terrorist attacks). It is also important to correctly evaluate the impact of accidental failure upon the overall bearing capacity of the system. Therefore, redundancy is provided through static indetermination to deal with contingencies such as the damage of local components inflicted by fire, earthquake, wind vibration, and terrorist attacks.

Static indetermination (redundancy) is an effective way to improve the robustness of the structure, which means the risk of premature overall structural failure due to accidental failure of certain components is mitigated. Redundancy allows more room for stress redistribution and serves as another "perimeter" for the safety of structures. From this perspective, "vulnerability" of structure could be diminished.

As for a frame structure, an increment of static indeterminations usually will not impact the cost and construction techniques significantly. Although static indeterminate structures generate additional stress under temperature gradient and uneven subsidence, frame structures more often than not are still designed and constructed with a great many static indeterminations so that the integrity and reliability of the structure are enhanced.

Attention should also be paid to the distribution of static indeterminations within the structure. Excessive concentration of static indeterminations as well as insufficient redundancy should both be avoided in design. Figure 2.9b shows an example: The second-to-the-left joint of the truss is one of static indeterminate and the third-to-the-left joint is geometrically unstable due to the lack of one diagonal bar member. The example also explains that local insufficient static indeterminations might result in inadequate safety and local premature failure.

2.5 The Loadings Analyses

The accuracy of load estimation has a direct impact upon the safety and economic efficiency of structural design. Underestimation puts the structure in danger while overestimation leads to an economically inefficient and oversized cost of construction. We will discuss how to reasonably select a probable load of design in this section.

All structures undertake corresponding loads in their service condition, and therefore, the selection and determination of load become the very first factor to be considered in structural design. Since an actual load could hardly be precisely known, load estimation becomes a significant issue in the early stage of design. Therefore, a reasonably estimated design load is needed so that major errors in early age calculation could be avoided.

All external interference that induces stress and deformation could be collectively designated as "load." Loads upon the structure could be divided into two types: direct and indirect loads. And the former is then divided into dead loads and live loads. Dead load refers to the load that does not change over time, such as structure self-weight and permanent facilities. As for the live load, it refers to the load that varies with time, such as service live load, wind load, seismic load, snow, vibrations due to mechanical facilities and other high-frequency vibrations, and impact load. Uneven foundation settlement and temperature gradient, classified as indirect loads, might generate stress and deformation as well. This type of loads is difficult to estimate and usually ignored in the early stage of design; nonetheless, some construction measures would take place to compensate the effect, such as the installation of settlement joints or expansion joints.

In the early stage of design, major loads that call for consideration include vertical loads (deadweight and vertical live loads), wind load, and seismic load. Content followed will introduce simplified estimation of loads; as for precise load selection, please refer to the "Load code for the design of building structures (GB50009-2012)" [11].

2.5.1 *Vertical Load*

1. **Vertical dead load**

Vertical loads are mostly comprised of vertical dead load and live load. Dead load mainly refers to the self-weight of the structure. In the early stage of design, a convenient way to estimate the self-weight is by distributing the overall deadweight of structure to the unit of floor area and calculating the equivalent uniform load. Dead load of the structure usually consists of self-weight of floor, roof, ceiling, wall, partition, column, and affiliated facilities such as corridor, transfer floor, escalator, and elevator shaft. The estimation of dead load could be conducted according to known density of the material and potential dimensions of components; as for concentrated weight of walls and columns, they could be turned into equivalent uniform load. For instance, wall area of residential buildings is estimated as 2–3 times the internal floor area, while that of commercial buildings is estimated as 1–2 times the internal floor area. In this way, as long as density of the material and potential dimensions of components are known, deadweight could be quickly estimated. For instance, the density of steel is 7850 kg/m^3, that of concrete about 2400–3000 kg/m^3 and that of brick masonry about 1000–2000 kg/m^3.

In the early stage of design, in light of gained experience and material of choice, equivalent uniform load of deadweight of structure could be approximately taken as:

Next, dead load shall be adjusted in accordance with dimensions of components from preliminary design. When it comes to final detailed design, dead load shall be further adjusted with reference to detailed dimensions of components. In this way, rough estimation of deadweight in beginning could finally be turned into reasonable determination of dead load.

Since structural stability usually favors structural deadweight, overall estimation of dead load could help designers to better assess the integrated impact generated by wind and seismic load, and thereby a better estimation of effectiveness in resisting lateral load could be achieved.

2. **Vertical live load**

Vertical live loads refer to service live loads of all kinds. Since some live loads are distributed on a small surface and others in a large area and might vary with time, it is difficult to precisely calculate. In the early stage of design, live loads could be approximately taken as equivalent uniform static load; as for the magnitude, reference could be made to corresponding entries in "Load code for the design of building structures (GB50009-2012)" [11]. Table 2.1 lists the standard value of some common live loads.

3. **Vertical snow load**

Snow load is a vertical load acts upon roofs. In cold regions, snow load sometimes becomes the control load in structural design. When that is the case, careful calculation has to be conducted with reference to relevant design code [11].

Table 2.1 Standard value of some common service live loads [11]

Application	Classification	Standard value (kN/m^2)
Floor	Residences, hotels, office buildings, classrooms, etc.	2
	Auditoriums, theaters, cinemas, etc.	3
	Malls, exhibition halls, station halls, etc.	3.5
	Bookstores, archives, storerooms, etc.	5
Roof	Non-accessible roof	0.5
	Accessible roof	2
	Roof garden	3

In the early stage of structural design, uniform snow load is usually calculated with reference to the standard value of local snow load and to different types of roof; the result refers to the uniform snow load on horizontal projection plane. For instance, the standard snow load of Shanghai urban area is 0.2 kN/m^2; when the angle of roof is less than 25°, the snow load of that roof on horizontal projection plane is 0.2 kN/m^2.

When it comes to the calculation of vertical live loads on the roof, the larger of snow load and service live load shall be selected; there is no need to do summation, because the probability of simultaneous occurrence of these two types of loads is quite small.

4. **Example of vertical load estimation**

A 10-story commercial office building takes the form of a frame-shear wall-reinforced concrete structure. The plane layout of its standard floor is as shown in Fig. 2.13. Live load on the roof is assumed to be identical with floor live loads. Estimate the vertical uniform load for design.

Solution

1. **The deadweight of floor of a standard floor**

The area of this floor is $20 \times 40 + 20 \times 20 = 1200\ m^2$. Thickness of slab is estimated as 10 cm and that of the ceiling plastering and floor pavement is about 3 cm. If the density of concrete is taken as 3000 kg/m^3, then the deadweight of estimate slab of this floor would be estimated as $(0.1 + 0.03) \times 30 = 3.9\ kN/m^2$, and total deadweight of this floor would be $1200 \times 3.9 = 4680$ kN.

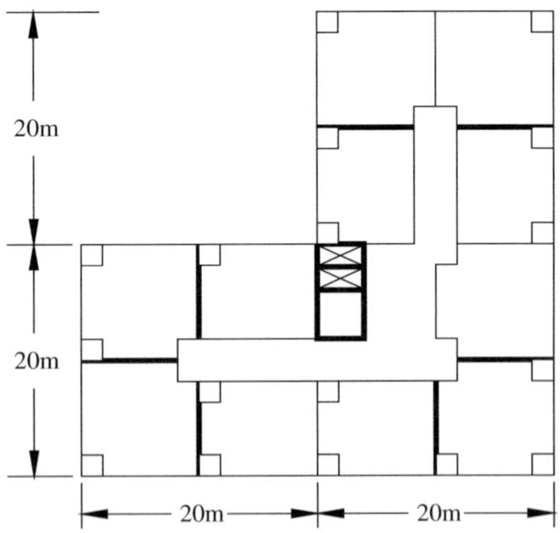

Fig. 2.13 Layout of a commercial building

2. **Deadweight of the wall of a standard floor**

The height of this floor is 3 m; the length is about 400 m; and that of the shear wall is 100 m. Assuming that the shear walls have identical density with common walls, then the surface area of wall on this floor is about $3 \times 400 = 1200$ m^2 (in this commercial building, the ratio between surface area of the wall and that of the floor is (1); the deadweight per unit surface area (columns included) is estimated as 2.0 kN/m^2; and total deadweight of wall on this floor is $2.0 \times 1200 = 2400$ kN.

3. **Total deadweight of the building**

Total deadweight of the building is the summation of the deadweight of slabs and walls of ten standard floors.

$$10 \times (4680 + 2400) = 70{,}800 \text{ kN}$$

4. Average deadweight of each floor is attained by dividing total deadweight of the building by total surface area.

$$(4680 + 2400)/1200 = 5.9 \text{ kN/m}^2 = 590 \text{ kg/m}^2$$

This estimation is within the range of aforementioned common deadweight of reinforced concrete structure, being between 5 and 9 kN/m^2.

5. The value of vertical load of design is attained by multiplying average deadweight of floor by a dead load factor of 1.2 and then be added with standard live load on that floor multiplied by a live load factor of 1.4. Since it is an office building, standard live load 2 kN/m^2 could be acquired in Table 2.1. Assuming that the live load on the roof is the same with that on the floor, so in global design and calculation of structure the structural design calculation vertical use shall be

$$1.2 \times 5.9 + 1.4 \times 2 = 9.88 \text{ kN/m}^2 \approx 10 \text{ kN/m}^2$$

2.5.2 Wind Load

In non-seismic regions, wind load is the main lateral load. Windward side of the building directly bears the wind pressure while other sides might have wind pressure or wind suction to bear.

1. **Determination of wind load**

The magnitude of wind load that structure has to bear correlates with windward area, speed of wind, air density (varies with height and temperature), configuration, and surface roughness of structure, as well as factors concerning surrounding environment of the structure. In the early stage of design, wind load bore by the structure could be simplified as the sum of standard value of wind load upon windward side and that upon leeward side. Standard value of wind load is related to local basic wind pressure.

Basic wind pressure p_0 and basic wind speed v_0 are directly converted by expression $p_0 = v_0^2/1600$. Basic wind speed v_0 is measured by meteorological administration through specifically stipulated field sampling, in which case data of average annual maximum wind speed are collected from the height of 10 m with duration of 10 min and then deducted with a once-in-fifty-year probability.

Under normal circumstances, basic wind pressure p_0 could be found in the diagram of national basic wind pressure listed under "Load code for the design of building structures" [11]. For instance, in Shanghai area, p_0 is 0.55 kN/m^2; Guangzhou is 0.50 kN/m^2; Beijing is 0.45 kN/m^2, and so on. When it comes to design for high-rise structures, wind load is selected with reference to once-in-a-hundred-year probability. In this case, simplified calculation of wind load shall raise basic wind pressure (once-in-a-hundred-year) by multiplying a factor of 1.10.

In the early stage of design, standard value of wind load could be estimated with basic wind pressure. In the next stage of preliminary design, wind load is calculated according to the load code with careful consideration of specific parameters such as the factors of shape, height variation factor of wind pressure, wind vibration factor, terrain roughness factor. As for structures of significance, wind tunnel testing is needed to simulate actual wind environment so as to accurately determine parameters such as distribution of wind pressure and shape factor. An example is given to demonstrate the estimation of wind load.

2. **Example of wind load estimation**

A 10-story commercial office building located in downtown Shanghai is 30 m high with plane layout displayed in Fig. 2.14. Basic wind pressure of this region is $P_0 = 0.55$ kN/m^2. Estimate the overturning moment M_w and shear force Q_w induced by wind load upon the foundation.

Solution

Assuming that maximum intensity of wind pressure of this region goes from right to left, as shown in Fig. 2.14. The shape coefficients of wind load are given as:

that of the windward side $\mu_1 = 0.8$

that of the leeward side $\mu_2 = -0.5$ (minus sign means suction)

Assuming that the building bears uniform wind pressure p_0

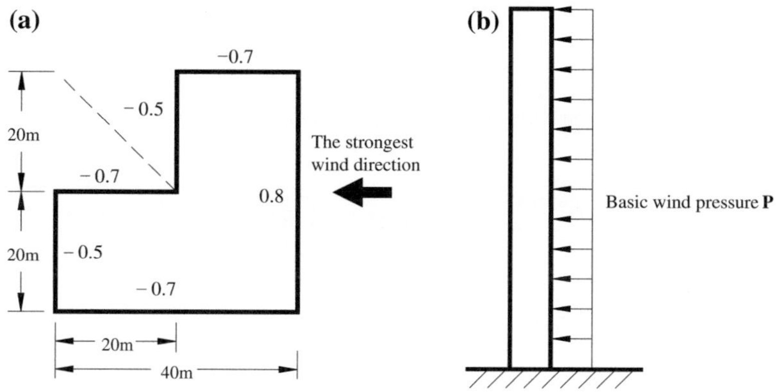

Fig. 2.14 Shape coefficients of wind load and distribution of basic wind pressure, **a** shape coefficients and main direction of wind, **b** distribution of basic wind pressure

Lateral distributed load of wind along the height of structure is

$$p_{\mathrm{wl}} = \mu_{\mathrm{s}} \times p_0 \times 40 = 1.3 \times 0.55 \times 40 = 28.6\ \mathrm{kN/m}$$

Resultant force of wind pressure acts upon vertical centroid of the structure

$$F_{\mathrm{w}} = 30 \times 28.6 = 858\ \mathrm{kN}$$

Shear force on the foundation $Q_{\mathrm{w}} = F_{\mathrm{w}} = 858$ kN
Overturning moment induced by wind load upon the foundation

$$M_{\mathrm{w}} = 858 \times 15 = 12{,}870\ \mathrm{kNm}$$

If the building is higher than 30 m, the influence of height of structure and wind vibration factor upon wind load has to be considered. Vertical distribution of wind load is no more simple uniform no more, but rather like an inverted trapezoid with big end up. As for specific values, please refer to relevant code [11].

3. **Wind tunnel testing**

There has not been a universal calculation equation for shape factor that could be applied to all types of structures and "Load code for the design of building structures" provides only some typical configurations for reference. Therefore, when it comes to particular structures, shape factor of wind load is often determined by wind tunnel testing. Take the Shanghai Pudong International Airport Terminal as an example (Fig. 2.15), the configuration of which embodies a seagull, flying over the sea. Surrounded by giant tilted glass walls, the building sends forth a strong sense of rhythm and stands as a fine example of modern architecture. When it comes to

Fig. 2.15 Shanghai Pudong International Airport

the wing of the structure, it is composed of an arched roof and corresponding truss components. At the turn of each summer, under the action of typhoon from southeastern coastal area, the arched roof generates an enormous amount of airlift force which is possible to lift off the entire roof system. To avoid it, bar system is needed to secure arched roof system. But how would the design of these bars go? And the calculation of wind load? Evidently, available theories could not serve as a solid foundation for the establishment of relevant calculation model for airlift wind load with such complexity. Therefore, specific experiments are needed to determine the state of stress of both the entire structure and the bar system. Specifically, a model of scaled-down seagull-shaped terminal was made and it was tested in a wind tunnel to determine the state of stress of the roof induced by wind load form all directions. Then the test results were used in the design of the entire structure and the bar system.

2.5.3 Seismic Load

1. **General estimation of seismic load**

Seismic load is a dynamic load in nature. Its magnitude correlates with intensity of the seismic event, distance to seismic epicenter, site conditions, quality and natural vibration period of structure, and many other factors; therefore, it is extremely complicated and difficult to precisely anticipate and calculate seismic force, especially when it concerns some potential issues that might arise even more difficulties in the seismic force estimation. Unsynchronized ground motion is one particular concern for large lateral structures such as bridges, which might cause differentiate seismic responses from one end to another. Another issue is epicenter area earthquakes which will induce significant impact loads on structures. All abovementioned issues are still popular research topics.

In the early stage of structural design, since dimensions of components are not decided yet, seismic response could not be accurately calculated and analyzed.

Therefore, simplified calculation is usually applied. In later stages of structural design, when dimensions of components are decided, an accurate computer-program based further analysis of seismic response could be conducted.

Since seismic force inflicted upon structures is proportional to the mass of structures, percentage of seismic force to total mass of the structure G_e is employed in simplified calculation. For instance, in regions with six-degree seismic fortification intensity, seismic force only accounts for 2–5 % of total mass of structure while in regions with seven-degree or higher seismic fortification intensity, seismic force could be up to 10–20 % of total mass of structure. Keep in mind that the seismic load acts upon structures as a lateral load, although a lot less in magnitude than general vertical load, it is still the major reason that causes damage and even failure to structures. In the early stage of schematic design and its relevant calculation, simplified method as followed could be used to estimate seismic force F_{eq}:

In regions with seven-degree seismic fortification intensity

$$F_{\mathrm{eq}} = 0.1G_e \tag{2.1}$$

In regions with eight-degree or higher seismic fortification intensity

$$F_{\mathrm{eq}} = 0.2G_e \tag{2.2}$$

As for the total deadweight of structure G_e, when the structure is simplified as one mass point, it should be taken as 100 % of total deadweight; when the structure is simplified as a series of mass points, G_e should be taken as 85 % of the total deadweight.

As we discussed, structures usually has a lower resistance to later load compare to vertical loads resistance. Therefore, more structure failure is due to seismic loads. In the stage of schematic design for high-rise buildings, vertical movement of the ground could usually be ignored and horizontal seismic force induced by horizontal movement of the ground is the only factor to consider. As for ordinary structures, seismic force is assumed to be vertically distributed as an inverted triangle; mass of each floor is assembled to its centroid and thereby point of application of overall seismic load F_{eq} is determined.

As for long-span or spatial structures, since their vertical stiffness is relatively smaller, vertical vibration often occurs under the action of seismic forces. In this case, vertical resistance to seismic forces needs to be calculated; the method is similar to that of estimation of horizontal seismic forces: 10–20 % of representative value of total gravity load of the structure is taken [as indicated in Eqs. (2.1) and (2.2)] and the only difference is the vertical direction.

2. **Estimation of natural period of vibration**

In the early stage of structural design, to prevent resonance vibration in structures and foundations, dynamic response of structure, in other words fundamental natural period of vibration T_1, needs to be estimated. The approach listed below is simple and convenient to apply but could only be applied to frames or shear walls with rectangular plane layouts, such as reinforced concrete frame-/shear wall structures.

$$T_1 = 0.33 + 0.00069\left(H^2/B^{1/3}\right) \tag{2.3}$$

$$\text{或 or} \quad T_1 = (0.04 \sim 0.09)N \tag{2.4}$$

In the equation, H refers to the total height of structure, B the total width, and N the number of floors.

3. **Seismic load distribution along height**

In the estimation of lateral seismic effects upon high-rise structures, the mass of each floor could simply be assembled to the elevation of each floor; so it is with seismic force. Equivalent assembled mass consists of total vertical constant load of deadweight of the structure and equipment on each floor along with 50 % of vertical live load.

Seismic load originates from inertial force generated by ground acceleration of mass of structures. The magnitude of acceleration is proportional to the vibration amplitude of the structure. As for structures no higher than 40 m, the distribution of seismic force could be approximately taken as a vertical inverted triangle as shown in Fig. 2.16a. Figure 2.16b shows a diagram explaining the seismic force on each floor. When the structure is higher than 40 m, vertical distribution of amplitude

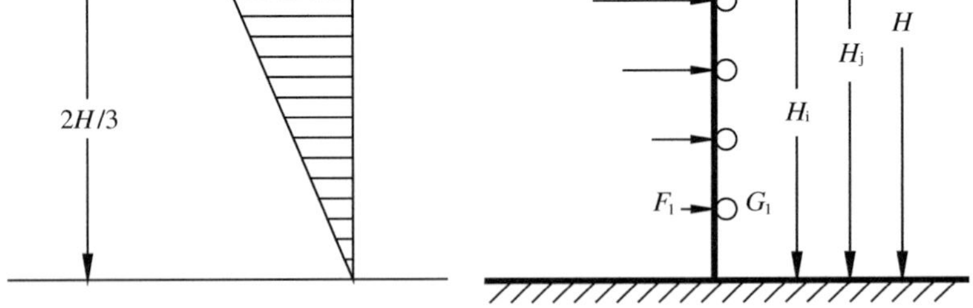

Fig. 2.16 Diagram of distribution of seismic force. **a** Distribution of seismic load, **b** simplified model of calculation

starts to show clear nonlinearity. Nonetheless, in case that height and stiffness between each floor of structure is relatively evenly distributed, distribution of seismic force could also be simplified as a vertical inverted triangle in the early stage of design.

Standard value of seismic force on ith floor F_i:

$$F_i = \frac{Gi \cdot Hi}{\sum_{j=1}^{n} Gj \cdot Hj} \quad F_{\mathrm{eq}}(i, j = 1, 2, \ldots, n) \tag{2.5}$$

F_{eq} in Eq. (2.5) refers to total lateral seismic force upon the structure, which could be calculated by Eqs. (2.1) and (2.2) with reference to seismic intensity fortification criteria. In Eq. (2.5), n refers to the total number of floors, G_i is the sum value of constant load and 50 % of live load on ith floor; H_i is the height of ith floor.

4. **Structural configuration and load effect**

To sum up, configuration of structure is related to lateral load induced by earthquake or wind. Table 2.2 is the comparison between effects of wind and seismic load in accordance with typical configuration of structures [6]. It shows that the overturning moment (2*Ph*/3) induced by wind load of inverted pyramid configuration is as much as two times as it (*Ph*/3) of pyramid configuration. As for overturning moment induced by seismic load, there is 6 times difference between two configurations amounts to six. This is a good example of the significance in configuration selection.

5. **Example of estimation of seismic load**

A 10-story office building, located in Shanghai downtown, is 30 m high and its plane layout is as shown in Fig. 2.13. Seismic fortification intensity in this region is 7°. Please estimate the shear force Q_{eq} and the overturning moment M_{eq} induced by seismic load upon the foundation.

Solution

Height of the building is 30 m; that of each floor is an identical 3 m; and the plane layout is a 40 m × 20 m L-shape. Total estimated weight of the structure is 70,800 kN, in which case 15 % of total weight could be deducted for the structure is a multi-story building. Total horizontal seismic force is calculated by Eq. (2.1):

$$F_{\mathrm{eq}} = 0.1\ G_{\mathrm{eq}} = 0.1 \times 0.85 \times G_e = 0.1 \times 0.85 \times 70{,}800 = 6018\ \mathrm{kN}$$

In accordance with the plane layout of structure, diaphragm is of large stiffness within the plane and therefore could be assumed to be rigid and free of deformation. Thus, mass of the same floor is assembled to a vertical virtual axis that stems from centroid of the structure and the mass of the whole structure is formed along the axis in the form of a string of mass, which is the so-called multi-node model.

Table 2.2 Comparison between effects of wind and seismic load in accordance with typical configuration of structure [6]

		inverted pyramid	inverted prismoid	cuboid	prismoid	pyramid
Effect of wind load	windward surface	inverted triangle	inverted triangle/rectangle	rectangle	rectangle /triangle	triangle
	Distribution of load	concentrated force P arm of force 2h/3	concentrated force P、P arm of force $2h/3$、$h/2$	concentrated force P arm of force $h/2$	concentrated force P、P arm of force $h/2$、$h/3$	concentrated force P arm of force $h/3$
	Bending moment	$M_w=2Ph/3$	$M_w=2Ph/3$、$Ph/2$	$M_w=Ph/2$	$M_w=Ph/2$、$Ph/3$	$M_w=Ph/3$
Effect of seismic load	Distribution of mass	inverted parabolic triangle	inverted triangle	rectangle	triangle	parabolic triangle
	Distribution of load	concentrated force $3P/2$ arm of force $4h/5$	concentrated force $4P/3$ arm of force $3h/4$	concentrated force P arm of force $2h/3$	concentrated force $2P/3$ arm of force $h/2$	concentrated force $P/2$ arm of force $2h/5$
	Bending moment	$M_s=6Ph/5$	$M_s=Ph$	$M_s=2Ph/3$	$M_s=Ph/3$	$M_s=Ph/5$

Evidently, an L-shaped layout inevitably generates eccentricity and corresponding eccentric moment Me between the center of mass and the centroid on the same floor. However, these eccentric seismic force and moment are relatively smaller compare to lateral seismic force and overturning moment and therefore serve as subsidiary factors. In the early stage of design, influence of eccentricity and eccentric moment could be temporarily ignored.

Vertical load on ith floor G_i is calculated as follows:

$$G_i = (5.9 + 0.5 \times 2) \times 1200 = 8280 \text{ kN}$$

Horizontal seismic force F_i, distributed to each floor, could be attained by Eq. (2.5) and is now listed in Table 2.3.

Seismic shear force on the foundation $Q_{eq} = \sum_{i=1}^{10} F_i = F_{eq} = 6018$ kN

Table 2.3 Horizontal seismic force on each floor

No. of floor i	1	2	3	4	5	6	7	8	9	10
Height of ith floor H_i (m)	3	6	9	12	15	18	21	24	27	30
Seismic force F_i (kN)	109	218	327	436	545	654	763	872	981	1090

Seismic overturning moment on the foundation $M_{\mathrm{eq}} = \sum_{i=1}^{10} F_i H_i = 125{,}895$ kNm

Seismic force is assumed to be distributed as an inverted triangle as shown in Fig. 2.16; in that case, the overturning moment upon the foundation is given as:

$$M_{\mathrm{eq}} = F_{\mathrm{ek}}(2H/3) = 6018 \times 2 \times 30/3 = 120{,}360\ \mathrm{kNm}$$

Compare the overturning moment M_{eq} derived from two approaches, the error rate is:

$$(125{,}895 - 120{,}360)/125{,}895 = 4.4\ \%$$

Since the difference derived from these two methods is within 5 %, both of these two methods can be used in simplified calculation of seismic force in the early stage of design.

2.5.4 *Other Loads and Effects*

All factors that would generate indirect load effect and result in internal stress and deformation are designated as indirect loads in engineering. Major indirect loads that structures would encounter include temperature effect, differential subsidence, long-term deformation (creep), and stress loss (stress relaxation).

1. **Temperature effect**

As a fundamental physic law, temperature change could induce expansion and contraction upon materials. If such deformation is not restrained, no stress would generate inside the material. However, actual structural deformation is more often than not restrained or confined; therefore, internal stress consequently generates within the structure. For instance, external change of temperature leads to changes of dimension of structure from the outside, while the internal temperature of structure roughly remains the same; as a result, temperature-induced strain and stress would arise within the structure. Furthermore, when it comes to the casting of mass concrete, the interior of concrete would generate huge amounts of heat through the process of hydration; since the swelling of its volume is constrained by cooler external concrete, massive temperature-induced stress or even cracking of concrete would develop as a result. Thermal expansion and contraction of material is an objective existence, which is quite difficult to calculate to an accurate extent. For the moment, to solve this problem, engineers usually rely on special construction design. For instance, the outer frame is designed to be slightly deformable relatively weak columns are installed in the corners while high-stiffness walls and columns are placed in the center. Through connecting a ductile exterior frame with high-stiffness interior frame, temperature-induced internal stress is reduced. Moreover, expansion joints and construction seams are usually needed over a section along the length of structure.

2. **Differential settlement**

Due to the inhomogeneity of foundation soil and the unevenness of vertical loads, when the foundation takes up a relatively large surface area, differential settlement usually occurs; as a result, additional stress would build up in structures. Differential settlement more often than not leads to unfavorable consequences such as failure of components, cracking especially on walls, and more seriously damage to structures. To prevent such occurrences, detailed investigation shall be carried on the soil condition of construction site and sufficient measures such as improved soil consolidation or pile-intensified foundation shall be added to the foundation design.

Another conventional method is to engineer planar partition in accordance with different heights and sections of structure and to install settlement joints so as to disconnect the foundation and enable free settlement between different parts of the structure.

3. **Change of internal stress and strain induced by long-term loads**

Under continuous action of long-term loads, creep develops over time, which is a phenomenal of increased deformation under constant load in materials. Such an occurrence might result in excessive or unexpected deformation. For instance, excessive defection develops in bending members under long-term loads, which might cause trouble for steel crane girder to remain “flat” during service life. As for prestressed structures, both methods of pretensioning and post-tensioning suffer creep-induced loss of prestress, which is a major concern in prestressed structures design.

Solution usually is to estimate potential deformation or creep. In the early stage of construction, to reserve room for deformation or to apply additional prestress could be used to solve prestress loss or excessive deflection problems.

2.6 The Acting Loads and the Structural Deformation Response

When the structure is at normal working condition, loads will results in deformation. Displacement of certain key point of the structure is usually taken as the measure of structural response to loads. Thus, *P*–*δ* curve as shown in Fig. 2.17 is also known as load–displacement curve [6], which in essence reflects the state of structure under different loads. Generally speaking, the principle of structural design is to make sure that structures keep working in an elastic state under combination of loads (dead load, live load, wind load, and seismic load). The wind load in this combination refers to potential maximum wind load determined by the load code for the region of construction while the seismic load is determined by regional seismic fortification criterion stipulated in the load code. As shown in Fig. 2.17, the structure, under the action of designed load, shall still have a comparatively large reserve of bearing capacity.

Under the action of vertical dead load, vertical displacement could happen to all locations of the structure. For instance, vertical deflection could happen to floor slabs, primary and secondary beams while vertical displacement could happen to walls and columns. In those cases, vertical displacement is often very small, and the lateral displacement is even smaller. After vertical live load is applied to the structure, due to the same applied direction with vertical constant loads, vertical displacement response of structure shall increase somehow but still maintain within an elastic range.

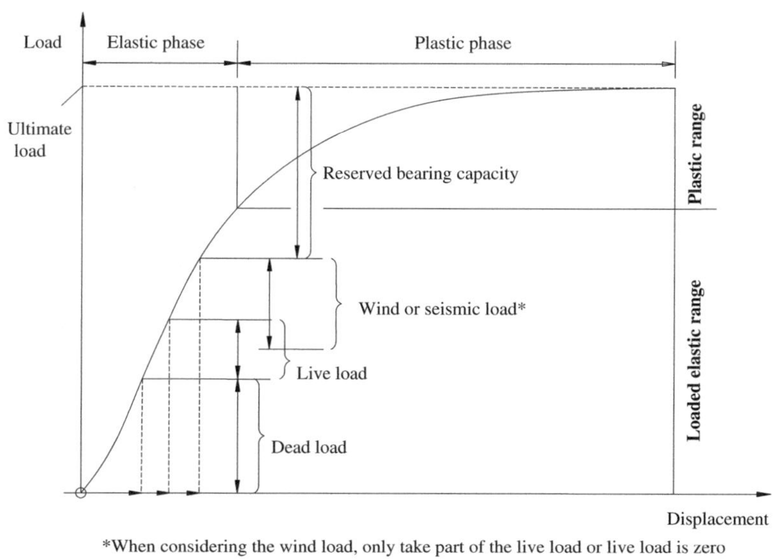

Fig. 2.17 Load–displacement curve of structures [6]

In practice, design value of wind load and that of seismic load shall not be combined, for that the probability of simultaneous occurrence is quite small. Similarly, wind load or seismic load shall not be combined with 100 % vertical live load but with 50 % vertical live load in general. Lateral displacement induced by wind or seismic load is much greater than that generated by dead load and live load. Under the action of wind or seismic load, relatively large lateral displacement could develop and generate significant stresses within the structure, which serves as the major cause of structure failure. Nonetheless, wind or the seismic load is not ever-lasting but short-term dynamic load. Therefore, under the action of load of such kind, allowable stress of material could be raised. For instance, allowable stress of material could be multiplied by a dynamic load coefficient of 1.33 [6].

Structures are required to remain in an elastic state (the upward segment of inclined straight line as shown in Fig. 2.17) under the action of unfavorable load combinations of all kind. Prior to failure, structures still need to go through an elastic–plastic state (the upward segment of the curve as shown in Fig. 2.17). As for most ductile structures, they still need to go through a sheer plastic state, in which case the bearing capacity of structure remains unchanged while displacement displays a rapid development (horizontal segment of the straight line as shown in Fig. 2.17). The area enclosed by the curve and the straight line signifies the energy (work done by external loads) to produce plastic deformation. This energy is stored within the structure in the form of deformation energy and become reserved bearing capacity after the depletion of elastic bearing capacity.

From Fig. 2.17, we could see that with respect to the ultimate displacement of the structure, the displacement induced by dead load, that induced by a combination of dead and live load and that induced by wind or seismic load are relatively small. It indicates that when the structure is made from brittle material and does not have reserved plastic zone after the depletion of elastic performance, the structure will suffer an immediate loss of bearing capacity at the completion of its elastic performance, which is quite dangerous. At moment like this, displacement of structure is relatively small and the structure suffers a sudden failure before it could absorb a large amount of energy, which might lead to a catastrophic loss of personnel and property. Therefore, the energy reserved in plastic deformation is tremendously beneficial and serves as significant safety reserve against sudden collapse.

Chapter 3
Global Design of Structures

3.1 The Classification of Structural Systems

Force-bearing system of structures varies in accordance with geometric and functional requirements. By geometric shape, structures fall into three major categories: multi-story and high-rise structures, single-story large-span structures, and spatial structures. They have different force-bearing systems and mechanical characteristics. Table 3.1 lists the classification and mechanical characteristics of different structural systems.

The figure of vertical high-rise structures usually is vertical slender cuboid or cone, such as various multi-story and high-rise buildings, towers, and chimneys. As for this system, lateral load and deformation serve as the control load and deformation of the structure. Special attention is called for to the design of vertical sub-structural system as well as to the enhancement of lateral force bearing and deformation control in the conception of a high-rise structure, for the structure would be equipped with higher strength and stiffness to resist lateral loads.

As for lateral-long-span structures, they are lateral slender cuboids, such as bridges, portal frame factories, and arch-type large-span stadiums. Vertical load and deformation are the control factors in this system. Special attention is called for to the design of lateral sub-structural system as well as to the enhancement of vertical force bearing and deformation control. This type of structures would require higher strength and stiffness to resist vertical loads.

Most spatial structural systems are lateral-large-span structural systems with three-dimensional force transmission. Key points of design are similar to that of lateral-large-span structural system. Since forces are transferred in three dimensions, this type of structural system has a better force-bearing capacity compared with lateral structural systems.

S. Lin and Z. Huang, *Comparative Design of Structures*,
DOI 10.1007/978-3-662-48044-1_3

Table 3.1 Classification and mechanical characteristics of structural systems

Structural systems	Structural types	Substructural decomposition
Vertical high-rise structural system	Brick and concrete (masonry) structure	Brick wall is vertical substructure; floor is lateral substructure
	Frame structure	Beam-column framework is vertical substructure; beam and floor are lateral substructure
	Shear wall structure	Shear wall is vertical substructure; floor is lateral substructure
	Tube structure	Tube is vertical substructure; floor is lateral substructure
	Mixed structure: Frame-shear wall structure Frame-tube structure Shear wall-tube structure	Frame mainly resists vertical loads; shear wall mainly resists lateral loads Frame mainly resists vertical loads; tube mainly resists lateral loads Shear wall and tube resist vertical and lateral loads together
Long-span structural system	Simple beam or continuous beam structure	Simple beam or continuous beam is lateral substructure; column or support with high stiffness is vertical substructure
	Truss structure	Simple truss or continuous truss is lateral substructure; column with high stiffness or truss is vertical substructure
	The arch or portal-rigid frame structure	The arch or portal-rigid frame is lateral and vertical substructure
	Suspension cable or cable-stayed structure	Suspension cable or cable-stayed structure is lateral substructure; column with high stiffness or truss is vertical substructure
	Mixed structure: Arch and suspension cable structure Arch and cable structure Truss and suspension cable structure	Arch and suspension cable structure may offset side thrust Arch and cable structure may prolong span Truss and suspension cable structure may offset side tension
Spatial structural System	Spatial truss structure	Spatial truss is lateral substructure; column with high stiffness or truss is vertical substructure
	Spatial latticed shell structure	Spatial latticed shell structure is lateral and vertical substructure
	Membrane structure	Membrane structure is lateral and vertical substructure
	Mixed structure: Chord-tension structure Spatial retractable structure	Chord-tension structure is the combination of prestressed cable and truss Retractable structure is the combination of spatial structure and mechanical control system

1. Vertical structural system

Vertical structural systems are divided as brick and concrete (masonry) structural system, frame structural system, shear wall structural system, tube structural system, and varied mixed structural systems (such as frame-shear wall structural system and frame-tube structural system). Vertical structures give priority to the resistance of overturning moment, shear forces, and deformation induced by lateral loads in order to ensure structural safety and functionality.

(1) Brick and concrete (masonry) structural system

For most of the low-rise residential buildings, the adoption of brick and concrete (masonry) structural system with bricks, concrete, and wood as raw material is an economically advisable choice. The most common and economical structural combination should be brick wall and precast or cast-in-place reinforced concrete floor. Floor could also be made from other materials such as wood, and the wall be made with precast concrete wallboard. Under most circumstances, brick and concrete (masonry) structures are comprised of lateral beam–slab substructure and vertical brick substructure. The design of the latter is critical in brick and concrete (masonry) structural system, and it should be able to integrate with lateral substructure to provide the structure with sufficient strength and stiffness to resist lateral loads. Ribbed beam floor of brick concrete structural system, as shown in Fig. 3.1, is comprised of primary beams, secondary beams, and slabs. To support primary beam, columns are set underneath. If columns could at the same time be added to and rigidly connected to the ends of the primary beam, part of vertical loads on the wall would be bored by columns and vertical frame along the direction of primary beam will be constituted, which means the force-bearing system is under the transition to that of a frame structural system.

(2) Frame structural system

When the story of the structure multiplies, the deadweight augments as well and the brick wall in the above-mentioned brick and concrete (masonry) structural system would not be able to bear the vertical loads that increased with the multiplication of

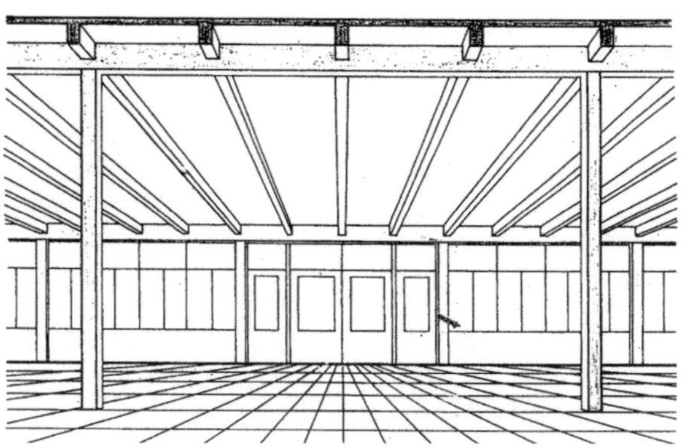

Fig. 3.1 Brick and concrete (masonry) structural system

the stories. Without the selection of another structural system, the thickness of the wall will have to be increased considerably. Otherwise, substructural members such as reinforced concrete columns or shear wall will have to be adopted in order to bear the deadweight of the structure. As shown in Fig. 3.1, if the primary beams are supported by columns at both ends with a rigid connection, a brick and concrete (masonry) structural system is naturally transformed into a frame structure (see Fig. 3.2).

Frame is composed of beams and columns. Components should be rigidly connected with each other so that the bending moment could be transferred. In other words, joint will only produce rigid deformation under loads, while no relative rotation is allowed between members connected at the joint, which enable the frame to remain geometrically intact.

Frame structural system excels in force-bearing capability and functionality fit for division of large space. Compared with brick and concrete (masonry) structural system as well as shear wall structural system, frame structural system has a light deadweight. Since joints connecting beams and columns are rigid, negative moment in beams near connections will offset some positive moment in the mid-span, resulting in a more reasonable and efficient way of exploiting the load-bearing capacities. On the other hand, the construction of frame structural system is well facilitated. Therefore, it is quite fit for assembly construction, with all components being beams and columns, which can be prefabricated in factories with concrete joints cast on site. Or, steel members could be used as beams and columns, in which case the construction of a frame structure becomes even more ready and convenient.

(3) Shear wall structural system

With the overall height of a structure is increased, more lateral strength and stiffness are needed correspondingly. Under lateral loads (wind or seismic), frame structure

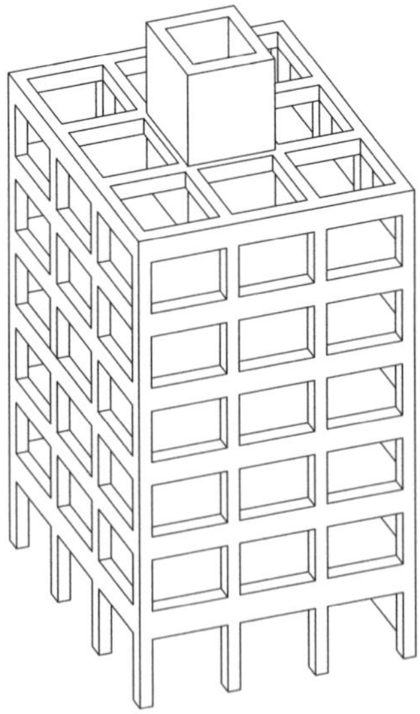

Fig. 3.2 Frame structural system

will produce comparatively large lateral displacement due to lack of lateral stiffness. To solve this problem, considerable increment of dimensions of beam and column members is unwise. Instead, the solution is to install a shear wall in a suitable position, and utilize the in-plane shear stiffness to resist lateral loads.

Figure 3.3 illustrates a frame-shear wall structure. Under wind load, the structure will produce a lateral displacement and give rise to an overturning moment at the foundation. When the shear wall is installed at both ends of the structure, lateral rigidity of the shear wall will surpass that of the frame alone. At the same lateral displacement level, shear walls will attract a majority of lateral loads, resulting in a much lower lateral load in the frame and much lower displacement level for the whole structure.

Shear wall structures are applicable to multi-story or high-rise buildings. In such cases, the outer and inner walls will not only enclose and divide the internal space but also undertake an enormous amount of vertical loads and two-dimensional lateral loads. Simple frame structure is surely fit for the division of large space regarding structural functionality; whereas, when it comes to the mechanical properties, this type of structure more often than not has insufficient lateral stiffness. When part of the frame is reinforced with shear wall, a frame-shear wall structure arises, which substantively enhances the structural capability to resist lateral loads. Moreover, it is more economical to adopt frame-shear wall structure in construction.

(4) Tube structural system

In a shear wall structure, when shear walls are connected as a tube, the structure will not only be able to resist shear forces or lateral displacement but also provide higher resistance to the overturning moment and certain amount of stiffness to resist torsion. In high-rise structures, the inner tube is often integrated with vertical elevator shaft to serve as vertical force-bearing and shear-resisting system. When the inner and the outer tube work together, they are able to resist larger lateral load and deformation. Tube structures commonly seen in high-rise buildings include frame-tube structure and tube-in-tube structure. The former derives from frame

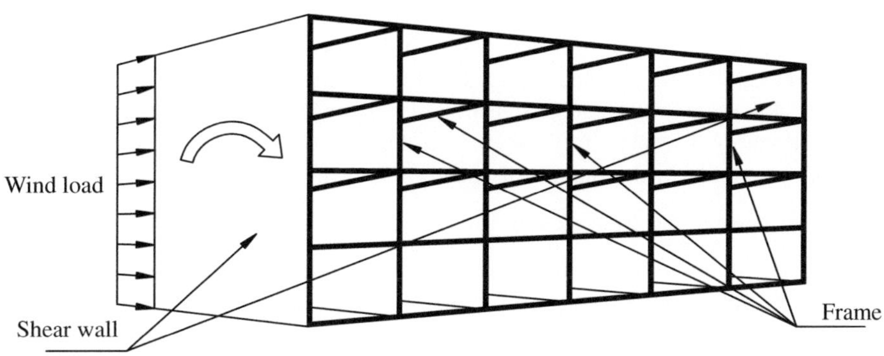

Fig. 3.3 Frame-shear wall structural system

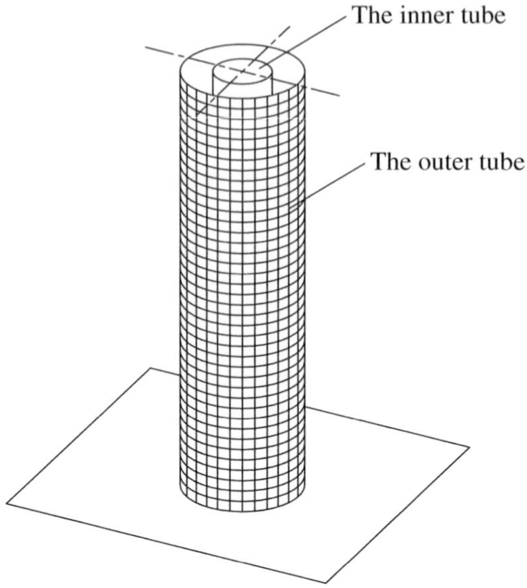

Fig. 3.4 Tube structure

structure. It centralizes the shear wall in the frame-shear wall structural system to the interior of the structure and forms a core tube, which will be in concerted work with the outer frame. That is how a frame-tube structure is composed. As for the tube-in-tube structure, the outer columns with small spacing are usually seen as an outer tube with a certain equivalent wall thickness and work with the inner tube to function as a tube-in-tube structure.

Figure 3.4 shows a high-rise tube-in-tube structure, comprised of an outer and an inner reinforced concrete tube. The inner tube handles a majority of vertical loads and lateral shear forces, while the outer tube mainly provides the resistance to the overturning moment generated by lateral loads.

2. **Lateral structural system**

Lateral structural systems can be divided into simply supported beam and continuous beam system, truss system, the arch and portal-rigid frame system, suspension cable and cable-stayed system, and varied mixed systems, such as the arch and suspension cable structure, the arch and cable-stayed structure, and truss and suspension cable structure. The core question to answer in terms of structural integrity and functionality is how to deal with the bending moment and vertical deformation, which is inevitable in the pursuit of spanning over large space.

(1) Beam system and truss system

The most commonly seen lateral structural systems should be solid-web beams and hollow-web truss systems. Many a bridge could be simplified as solid-web simply supported or continuous beams. To further lessen the deadweight, redundant material is removed and beams become hollow-web truss structures. When solid-web beams are replaced by hollow-web trusses, simply supported or continuous truss system is in shape.

Beams could be molded into thin solid types, such as T-type, I-type, or box-type. As long as the shear capacity is satisfied, web section or hallowed section could be used to reduce the deadweight and save material. One shared merit of steel and reinforced concrete thin-web beam is the convenience to manufacture and install. The greater the height of the beam is, the more advantageous to the resistance of the bending moment and vertical deformation. However, the height of one structural story will accordingly be increased. The deduction of the height of the beam surely lowers the height of the structural story but will require more reinforcements. The application of prestress can balance the bending moment induced by the deadweight of the beam itself and enhance its capability to resist bending moment. Due to the large deadweight of thin-web reinforced concrete beam, the applicable span is comparatively short (about 6–12 m), whereas that of a prestressed thin-web concrete beam could reach 12–18 m and that of a I-type steel beam is up to 30–50 m.

Steel structures naturally are thin-web beam structures. Commonly seen types include H-type, I-type, box-type, and composite type. Reinforcement ribs are generally used to avoid the local buckling of webs and wings. Figure 3.7 is a steel bridge deck supported by a box-type beam. Lateral reinforcement ribs have been installed to the web, while vertical ribs installed to the upper and lower wings. Diagonal supports have also been installed so that an overall lateral stiffness of the beam is ensured. Diagonal supports have furthermore been added to the lateral frame of the web so that cantilever sections on both sides of the deck are secured and the width of the bridge is guaranteed (Figs. 3.5 and 3.6).

As the span increases, greater bending moment will take place and a larger height of the beam is needed to resist the bending moment. As a result, the deadweight of the beam builds up, which makes a solid-web beam an uneconomical choice. When a long span is needed, the height of the beam is raised to resist the bending moment. Meanwhile, a massive amount of web material has to be removed to lower the deadweight until only necessary connecting members are left. In this case, a solid-web beam has been turned into a truss.

As shown in Fig. 3.8, the bending resistance of a beam, as a matter of fact, comes from a concerted work of compressive arch and tensile chain within the beam. With the help of the analysis of force flow (as for the definition and application of force flow, please refer to Sect. 3.3 of this chapter), the upper chord is found to bear

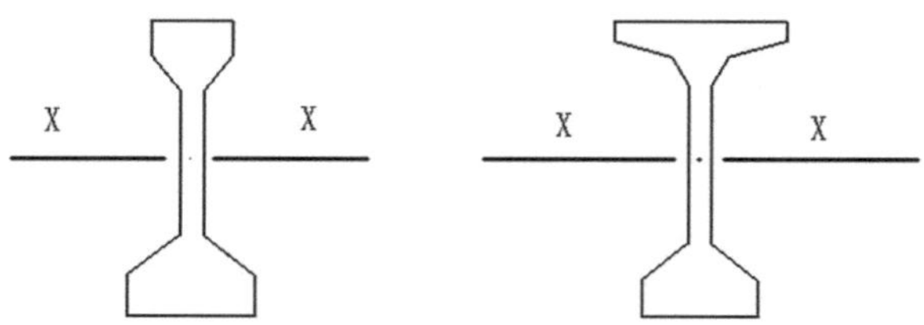

Fig. 3.5 Reinforced concrete beams with different sections

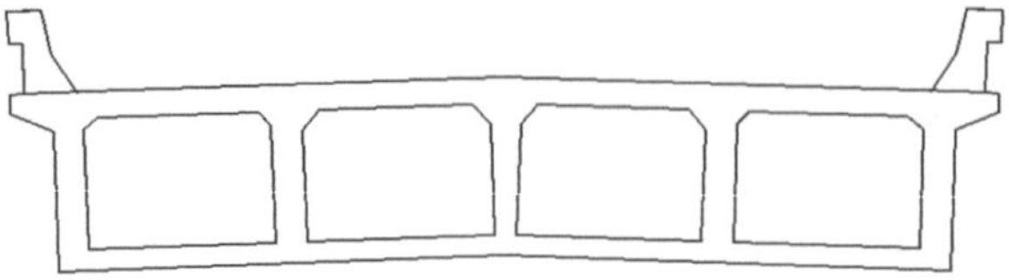

Fig. 3.6 Box section reinforced concrete beam

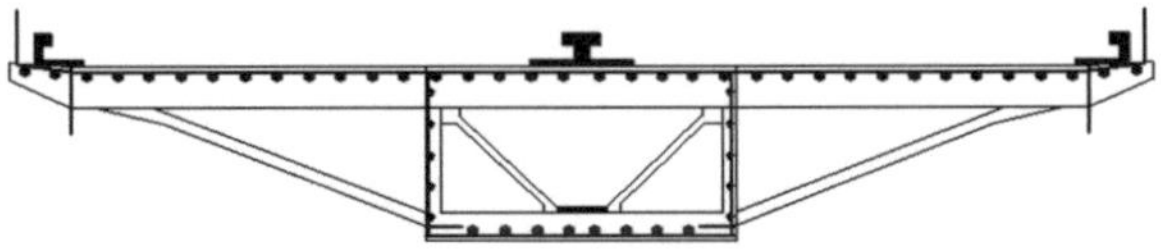

Fig. 3.7 Box-type beam supporting the deck of a steel bridge

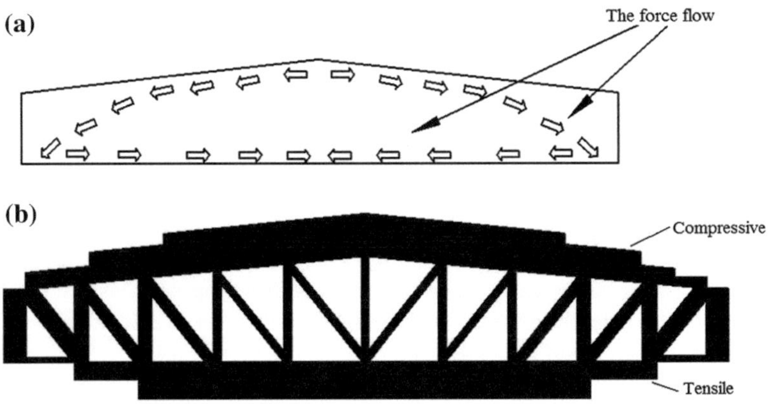

Fig. 3.8 Comparison between beam and truss. **a** The force flow within the beam. **b** Compressive and tensile chords in a truss

compressive force and the lower chord tensile force. The very pair of moment formed by the compression and tension in the upper and lower chords of the beam will counterbalance the bending moment imposed upon the section by the external loads. Thus, truss, in the essence, is a beam only without redundant web material. The upper and lower chords will bear the bending moment, while diagonal web bars take the shear force.

(2) Arch and portal-rigid frame system

From the force analysis of the arch as shown in Fig. 3.9, we know that the supports on both ends of the arch provide not only the vertical reaction V but also the lateral reaction H. Thus, this lateral reaction H provided by end supports can replace the tensile effect generated by the tensile chain within the beam and constitute a

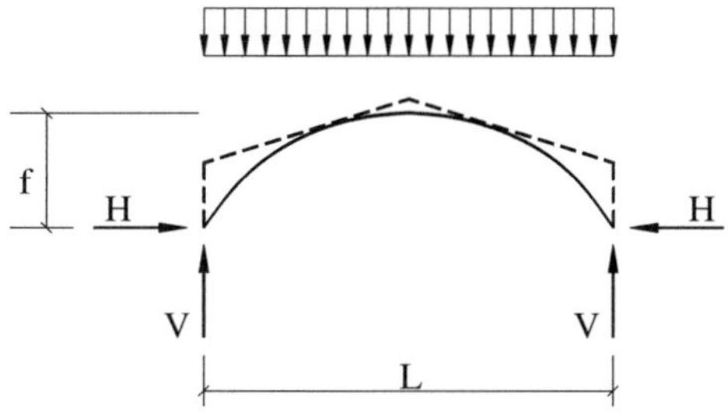

Fig. 3.9 Arch structure and portal-rigid frame structure

counterbalancing bending moment with the value of $H \cdot f$, in which expression f is the rise of the arch. Since the rise of arch is much larger than the height of the beam, the arch has the capability to take on more bending moment and is therefore able to span over a larger space. When the rise-to-span ratio is under certain limit, the arch mainly bears compressive force. Therefore, the advantage of concrete having high compressive strength could be utilized, and indigenous stones could be used in the construction. Compared to beam structures, arch structures have a more optimized force-bearing system and a lower cost. Thanks to the excellence in force-bearing capacity, the grace of the contour and the economy of construction, arch structures have long been one of the favorite choices in practice.

With the beam and columns on both ends rigidly connected, portal-rigid frame is a derivative from the arch. Similarly, lateral reactions from supports on both ends provide portal-rigid frame with the resistance to the bending moment produced by the external loads. Rigid connection between the beam and columns effectively restrains local bending moment within the rigid frame, for which reason this type of structure could effectively serve as lateral large-span structure. Portal-rigid frame consists of few members which enables easy production. Commonly seen reinforced concrete portal-rigid frame structure spans about 20 m, while steel portal-rigid frame spans over a larger space of 50–80 m.

(3) Suspension cable system

Suspension cable system is like an inverted arch structure under loads (Fig. 3.10). However, the axial force in an arch is compressive force, while that in a suspension cable is tension. In addition, suspension cable does not bear any bending moment. High-strength steel is generally applied for tensile cables, in order to take advantage of the lightweight and high strength of the steel. Buckling would not happen to a suspension cable the way it does to an arch. Therefore, the strength of material used in a suspension cable structure will be fully exploited to ensure a longer span over space. This has been fully proved in bridge design for that the longest bridges in the world are all suspension cable structural system.

3. **Spatial structural system**

Lateral structural systems mentioned in the previous section belong to plane system. In other words, loads are transferred within one plane, and the strength and rigidity of such systems within the plane are greater than that outside the plane. If restraints

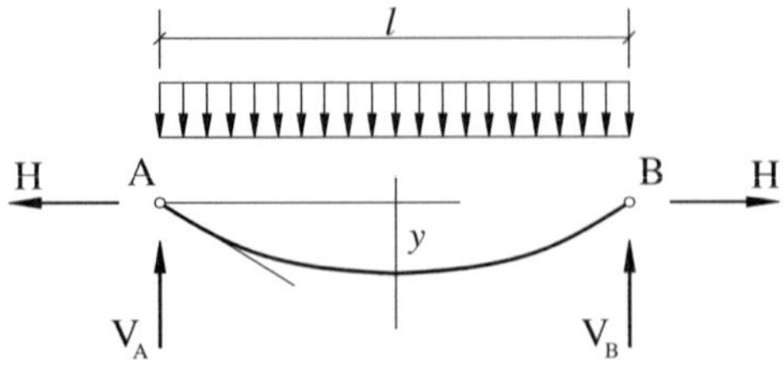

Fig. 3.10 Suspension cable structure

and path of force transmission are added from outside the plane to a lateral structural system, such a system will transform into a spatial structural system. Loads being transferred in three dimensions, spatial components will be able to support and complement each other, and the force-bearing capacity of the structure will be greatly elevated as a result. Spatial structural system could be regarded as a three-dimensional or orthogonal lateral structural system. Similar to a lateral system, major concerns of spatial structural design is the inevitable large bending moment and vertical deformation from spanning over large space.

When components from different planes constitute to the same joint and then form and function as a spatial structure, the efficiency of the material will be better utilized. Such structures include thin shell structure, spatial truss structure, membrane structure, and various other new spatial structural systems. With the increasing applications of steel in architectures, steel structures are often employed to span over large space.

(1) Spatial truss structures

Load-bearing capacity of spatial truss structures comes from out-of-plane restraints (Fig. 3.11a). The composition of plane trusses from different planes allows components to support each other so that the structure would be able to transfer the forces in a three-dimensional space. Due to the high overall rigidity and stability, spatial truss structure has become an ideal choice when a large-span structure is concerned.

The internal force of steel members in spatial truss is either axial tension or axial compression. The members bear the forces in a rational way and the strength of material could be fully utilized. Spatial force transmission further utilizes the strength of the materials (saves nearly 30 % in total material usage), lowers the deadweight of the truss, and spans over a larger space compared to plane truss. The spatial interconnections raise the number of static indeterminations and prevent the structure from crumbling down when individual members are compromised, which helps with the robustness and safety reserve of the structure.

(2) Spatial latticed shell structures

Fundamentally, spatial latticed shell structures are arch structure or suspension cable structure in three dimensions. Shell structures are also bionic in essence and have a reasonable force-bearing system. Such structures consist of curved plates, either single- or double-curved surface. With the restraint of certain rigid frame (or support), diversified spatial structural forms are capable of composing a large space with creative aesthetic senses.

To solve the issue of bending moment inflicted upon the mid-span, spanning structures have evolved from linear within-the-plane beam (in which case the section bears a bending moment) to single-curved arch (in which case the section bears a compressive force), which fully utilized the materials and lowered the cost. Compared to an in-plane arch, shells have a two-way curvature, which enables them to play the role of two-way arches. Thanks to this two-way arch effect, shell structure bears compressive force in each and every direction and therefore has a thin crust, which is more economic by using a lot less materials. For instance, an ideal reinforced concrete shell could span over a distance of 30–50 m with a thickness of only 8–10 cm. However, despite the above-mentioned advantages, several disadvantages such as complicated design and construction process and high cost in construction need to be considered. Meanwhile, due to the small thickness, shell structures have some other limitations such as insufficiency in heat insulation and as heat preservation.

Figure 3.11 illustrates some single-curved shells. In this case, lateral curvature fortifies longitudinal bending rigidity and enables the shell to span over a larger space. Figure 3.11 shows the application of folded-plate shell (Fig. 3.11a) and cylindrical shell (Fig. 3.11b). Reinforced concrete folded-plate shells generally span over 15–25 m, while reinforced concrete cylindrical shell is applicable when the span is between 20 and 35 m. When prestress is applied, the spanned space could be extended to 50 m. Single-curved shells (folded-plate or cylindrical) are commonly used large-span structures. The enhancement of longitudinal bending rigidity highlights the merit in force bearing of shell structures.

In sum, shell structures have been generally applied in construction, for they have a thin curved surface and comparatively rigid supports at the same time. Local buckling at the rim of the thin shell is prevented by rigid supports, as shown in Fig. 3.11. Meeting this stability requirement, the architectural contours of thin shell structures could be highly varied.

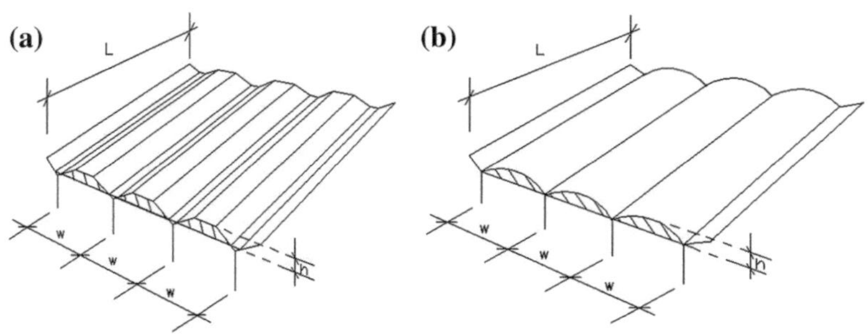

Fig. 3.11 Single-curved folded-plate shell or cylindrical shell

From the perspective of contours, latticed shell structures could be divided into double-curved latticed shells as shown in Fig. 3.12b and single-curved latticed shells as shown in Fig. 3.12c [9]. Truss and latticed shell structures are quite fit for mass production and on-site assembly/lifting/installation. The recent emergence of varied customer-manufactured ball joints has largely expedited the assembly of truss structures and has further lowered the costs and labors of the construction.

Suspension cable structures could also be transformed to spatial latticed shell structures: one-way and two-way suspension cable structures.

Figure 3.13 shows a two-way cross-saddle-shaped suspension cable structure, comprised of two groups of inhaul cables with opposite curvatures. The downward-convex cable bears the forces, while the upward-convex cable stabilizes the structure. These two groups of inhaul cables are fixed to a pair of intersectant inclined arches at the edges, constituting a saddle-like curved surface as the roof, which undertakes the weight of the roof system as well as the external loads. In addition, the supporting system of cables is comprised of two inverted V supporters and the tension rod at the foundation. The tension within the tension rod would offset part of the reaction force at the end of the arches.

Figure 3.14 shows the single-main-cable suspension roof system applied to ChaoYang Gymnasium in Beijing. The main suspension cables in the middle are fixed to and supported by subsidiary frames (shear wall) on both sides. Since the

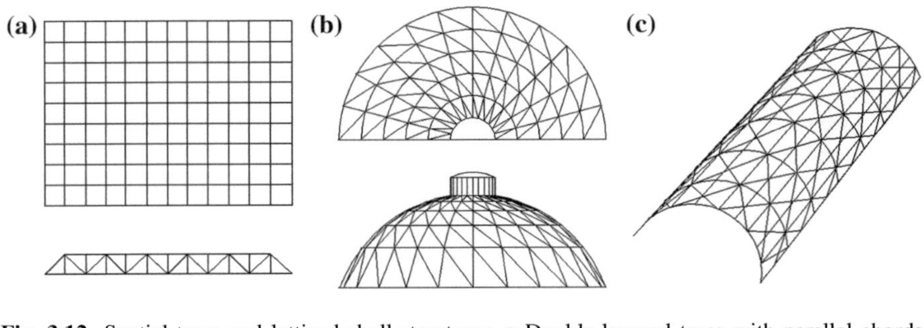

Fig. 3.12 Spatial truss and latticed shell structures. **a** Double-layered truss with parallel chords. **b** Double-curved latticed shell. **c** Single-curved latticed shell

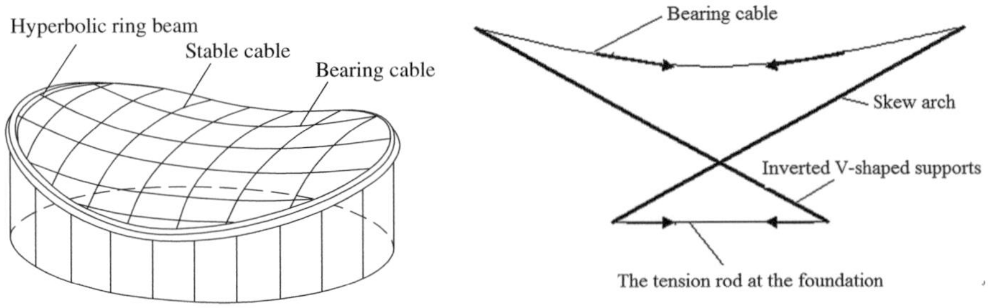

Fig. 3.13 Intersectant saddle-like suspension cable structure

Fig. 3.14 Single-main-cable suspension roof system in ChaoYang Gymnasium in Beijing [9]

main suspension cable provides support for arched trusses, it serves as the central supporting component in the roof system. The other end of the suspension cable is fixed to the inclined arch structure in the outer surface.

(3) Membrane structure

Since membrane structure is constructed by stretching membrane over a plane or frame, it has the least and smoothest surface under given boundary conditions [9]. Similar to steel cables fixed at both ends, membrane structures could bear normal-to-plane loads with axial tensile force (in this case, two-way bearing capacity due to two-way curvature). Although membrane structures could not handle compression and are not as stable, when proper prestress is applied, such structures would be able to bear loads in a stable way (Fig. 3.15). The ultralight weight enables such structures to be widely applied in various kinds of temporary structures. Take the erection of a tent for instance. To ensure the stability of a tent, we only have to support the tarpaulin with a rigid post and fasten them with inhaul ropes. Another form of membrane structures take is inflatable structure, which does not need steel keel for support and is equipped with force-bearing capacity by gas pressure. Figure 3.16 is a tire-like inflatable structure. The supporting ring beams

Fig. 3.15 Membrane structure [9]

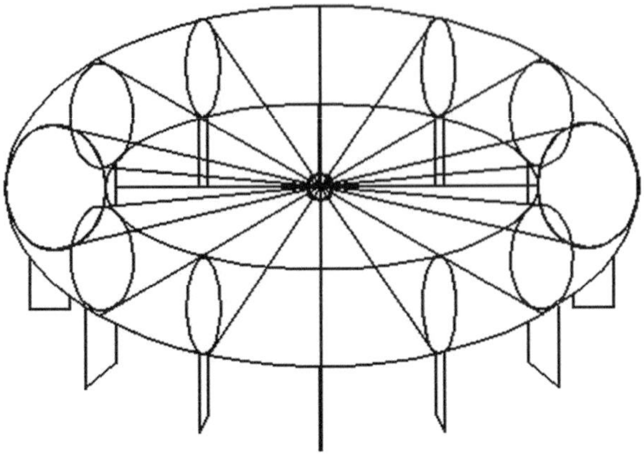

Fig. 3.16 Tire-like inflatable structure

fixate and support the top of the membrane as well as the floor section, which along with the membrane itself could constitute a spatial membrane structure.

(4) Other spatial structural systems

With development of building material and technology, a great number of new spatial structures have recently emerged, such as beam string structures and spatial deployable structures. Beam string structure is a special form of spatial truss structure, which replaces the bottom chord with prestressed steel cable, thus further reducing the deadweight of the structure. Shanghai Pudong International Airport [9] (one-way beam string structure as shown in Figs. 2.3 and 3.17) and Beijing National Stadium (two-way beam string structure) are the two representatives of beam string structures.

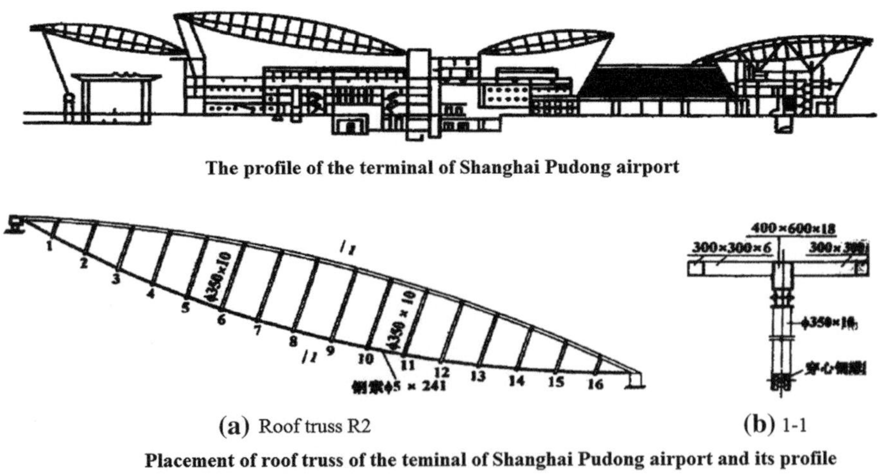

Fig. 3.17 The beam string structure in Shanghai Pudong International Airport [9]

Spatial deployable structural systems could be divided into spatial retractable roof structures and spatial deployable/foldable structures. The former is often applied to large-span roof structures to open or close the roof horizontally in a push–pull manner. As for the latter, spatial deployable/foldable structure, it has been widely applied to architectures, daily items such as umbrellas, tents, and even aircrafts or medical supplies. Deployable/foldable structures, such as umbrellas, which open and fold as required, take up only a small space when folded and are easy to move. The design and construction of such structures will need the cooperation of professions such as architecture, mechanical, and electronic controls, in which case an integrated will be required.

3.2 Methods of Global Analysis of Structures

In order to perform the early-stage design, one must fully consider the overall dimension and support conditions of the structure. One also much ensure that the overall strength, stiffness, and stability (including sliding stability and overturn stability) of structure (including foundation) meet the requirements of design under all load conditions. The overall force analysis of structures can be conducted as follows:

1. Simplifying the calculation model

According to structural system classification discussed in Sect. 3.1, we need to first simplify the actual structures into models, such as simply supported beam system or continuous beam system.

2. Estimate potential loads

With reference to functional requirements of structures, the estimation of vertical and lateral loads should be conducted with the approach introduced in Chap. 2. Vertical loads include the deadweight of the structure (dead load) and vertical live loads. Vertical loads on the structure will be directly transferred to the foundation and serve as the basic loads in foundation design. Lateral load generally refers to wind load or seismic load, which will bring shearing effect to the structure and generate the greatest shear force in the foundation which affects the overall slip stability of the structure. Meanwhile, the overturning moment that lateral loads imposed upon the foundation gives rise to a reaction force from the foundation, which counteracts the aforementioned overturning moment. Thus, the reaction force generated by the foundation to resist the overturning moment becomes another basic load in foundation design.

3. Analyze the overall stress

Vertical and lateral loads, when working synthetically, conjure up various combinations of loads for the structure to bear. And the structure will have to satisfy the equilibrium requirements at all times as follows:

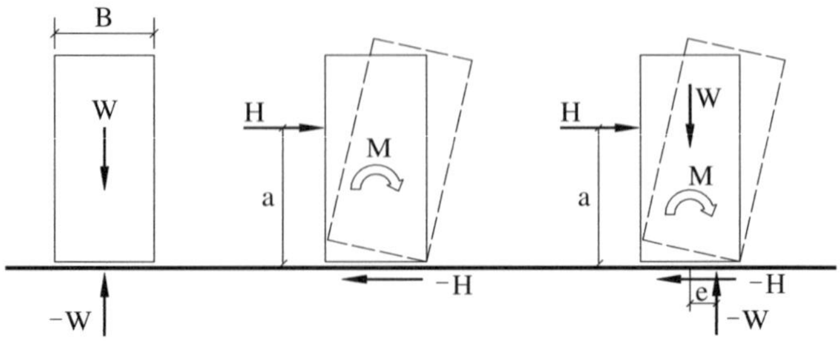

Balance: $W+(-W)=0$ $M=Ha$, $H+(-H)=0$ $M=Ha=We$, $H+(-H)=0$, $W+(-W)=0$

Fig. 3.18 Overall force analysis of the structure

(1) The overall structure and the foundation are stable under the action of static and dynamic loads. In other words, compressive strength of the structure is sufficient;
(2) The overall lateral slip stability of the structure (and the foundation) is ensured, which means the structure has sufficient shear strength;
(3) The overturning stability of the structure is guaranteed, which means the foundation has sufficient bending moment resistance.

Figure 3.18 shows the overall equilibrium of the structure. Vertical load (W) is in balance with the resultant force at the foundation ($-W$). Lateral load (H) is in balance with the shear force at the foundation ($-H$). In order to resist the overturning moment M produced by lateral load H, the resultant force at the foundation ($-W$) has to be eccentric with vertical load (W) with e as the eccentricity. Due to the existence of e, the stress state of the foundation is more complicated and the width of the structure (B) will need to be adjusted so that the stress state of the foundation will meet the strength and stability requirements. Specific discussions will be unfolded in following sections of this chapter.

3.3 Facade Allocation Analysis Through "Force Flow" Method

Force flow is a vivid expression of load transfers and load magnitudes, from loading points to supports. Force flow has two properties: direction and intensity. The direction refers to the geometrical locus of loads being transferred within the structure and the magnitude of the force. Force flow describes the stress state of the structure at service state. Whenever there is force acting upon the structure, there is force being transferred within the structure. The structure will bear the deadweight of itself and corresponding external loads. Force flow will accept, transfer, and release all loads through its force-bearing capacity in way of force flow.

Force flow bears a natural resemblance with "water flow," transferring loads from acting points to foundation via the shortest and smoothest geometric path. This can easily be explained by the principle of least action. Since the force flow produces strain in material and corresponding strain energy as well, the structure, when functioning in a normal, balanced, and stable way, should have a minimum amount of strain energy. Extra length and redundant turns of the path will inevitably increase the strain energy reserved, which render the design not the optimum one. In light of this, in preliminary design, the task engineers who undertake should design or plan the direction of force flow and find out the shortest and smoothest path. This is the basic concept of elevation arrangement in structural design.

Hence, conceptual analysis regarding elevation arrangement could be conducted through the spatial path and the intensity of the force flow. In the early stage of design, designers, with reference to the functions of the structure, should plan for a shortest path that the force flow will take, to ensure that the force will be smoothly transferred to the foundation. At the same time, being aware of the force flow, variable cross sections could be used to increase the load-bearing efficiency. Several structural systems with different force flow patterns are summarized as follows.

1. Vertical load-transferring structure

Typical vertical load-transferring structure is high-rise structure. Under the action of wind or seismic load, vertical deadweight and live loads, high-rise structures vertically transfer these loads downwards to the foundation (Fig. 3.19). The intensity of the force flow increases with the accumulation of stories. Thus, there are different

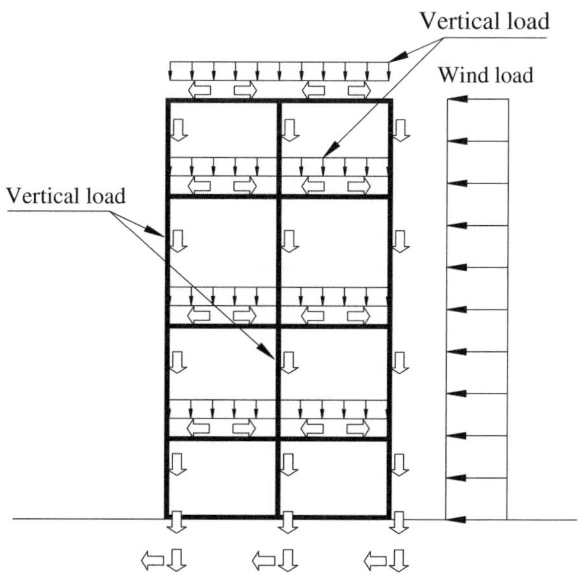

Fig. 3.19 Vertical load-transferring structure

force flow intensities within such vertical structure. For instance, due to that the upper columns bears less force than bottom columns, the upper columns could be designed to be have less reinforcements than bottom columns.

2. Lateral load-transferring structure

Typical lateral load-transferring structure comes to beams and slabs. As shown in Fig. 3.20, a beam could transfer vertical loads via the internal force flow to supports. From the figure, we know that tensile reinforcements should be allocated in major tensile direction at the lower part of the member as marked out with solid lines. While the concrete in major compressive stress as marked out by dotted lines, it aligns with the direction of force flow.

3. Broken-line-typed load-transferring structure

Typical broken-line-typed load-transferring structures go to all truss structures (Fig. 3.21). External loads are transferred through the members of truss structure in the manner of a broken line to the support.

4. Streamlined load-transferring structure

The feature of a streamlined load-transferring structure is that the force flow takes its own streamlined path and directly goes to supports. There are plane and spatial streamlined load-transferring structure. Figure 3.22 shows us the force flow of a deck arch bridge, in which the loads on the deck are transferred to the main arch via the vertical bars and then to supports. If the structure has a curved surface, the loads will be transferred through the curved surface and then to supports (Fig. 3.23).

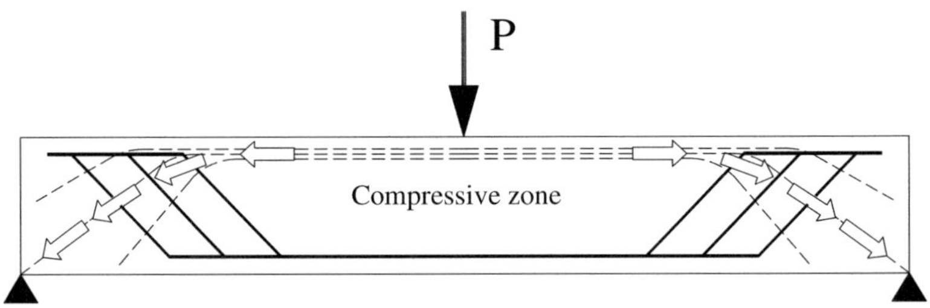

Fig. 3.20 Lateral load-transferring structure

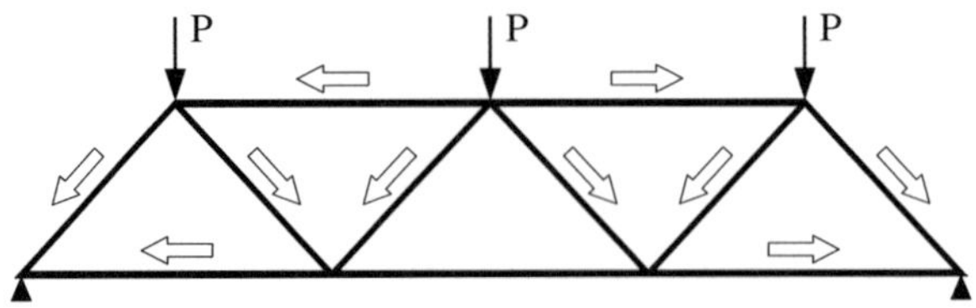

Fig. 3.21 Broken-line-typed load-transferring structure

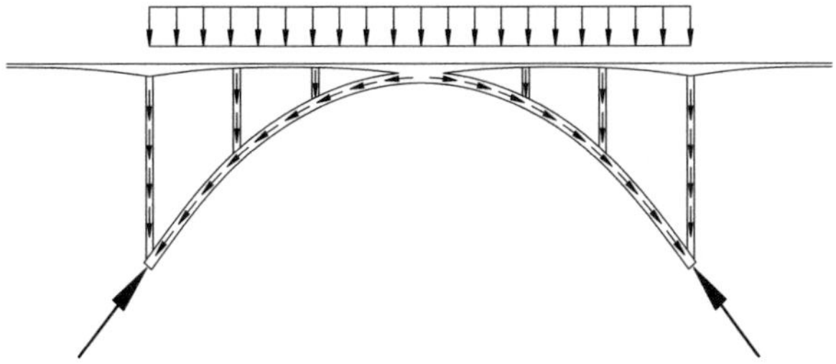

Fig. 3.22 Plane streamlined load-transferring structure

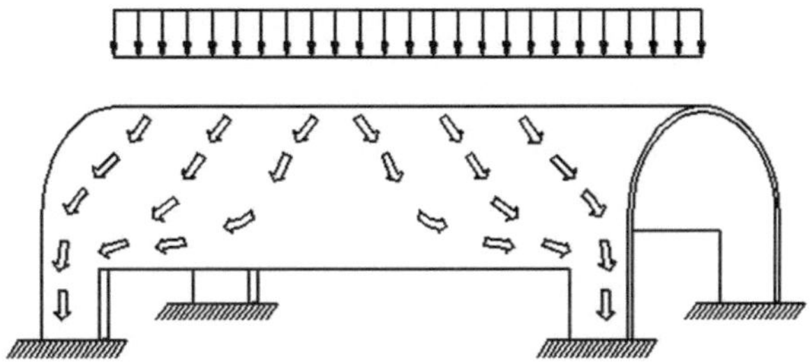

Fig. 3.23 Spatial streamlined load-transferring structure

Streamlined load-transferring structure shows the convergence of the force flow in a clear way and therefore enables the design of a variable section and reduces material in accordance with the change of intensity of a force flow. Take the arch shown in Fig. 3.22 for example. The intensity of the force flow at the top of the arch is low and correspondingly the section of the arch can be comparatively small; the intensity of the force flow at the foot is the highest, and the cross section area should be accordingly the largest. The design of a variable section should agree with the intensity change of the force flow.

In sum, when it comes to the selection of elevation arrangement and sectional area of major members, designers can refer to the least action principle and find a smoothest path for the force flow, as long as the functional requirements are met. Meanwhile, the design of a variable section could be conducted in accordance with the intensity change of the force flow, which reduces the material usage to the optimum.

3.4 Planar Layout Analysis Through "Footprint" Method

"Footprint" is another name for the plane layout of the foundation of a structure. Since the force flow will ultimately reach the foundation, the plane arrangement of the foundation of a structure plays an important role in overall analysis of the structure. Like the way people stand, when there is large space between two feet, greater side thrust will be needed to push the standing person over; otherwise, when there is no space between the two feet or the person is standing on one foot, a comparatively small side thrust will suffice to push the one over. The force acting upon the structure is like the side thrust and the arrangement of the foot/feet is like the plane layout of the foundation. The arrangement of the foundation looks very similar to the footprint of the structure. In this book, it is proposed to refer to the plane layout of the foundation as the "footprint" so that readers could readily understand and apply.

"Footprint" analysis analyzes the relation between the plane layout of the foundation and the overall stress state of the structure. Then it estimates the stress state and reinforcement allocation of major components at the foundation, as well as the cost of the whole structure. "Footprint" analysis optimizes the plane layout of the structure and improves the overall force-bearing capacity of the structure [6].

1. Assumptions in overall analysis

In the early stage of design, designers will need to conduct an overall analysis of the structure so that designers will have a comprehensive understanding about the features of the structure. Then sufficient strength, stiffness, and operational effectiveness could be ensured.

In this stage, structures are simplified as rigid bodies so as to analyze the relation between the stress state of the structure and its surrounding environment in an overall perspective. The assumptions and principle to abide by in the overall analysis in this stage are as follows:

(1) Structure is regarded as a rigid body (the impact of openings is ignored) embedded to the ground through the foundation;
(2) Loads, which are transferred to the ground through the foundation, include vertical deadweight, vertical live loads (equivalent evenly distributed vertical load), and lateral wind or seismic load.
(3) The so-called ground refers to natural or processed ground which guarantees vertical and lateral stability.
(4) The simplified calculation of loads bored by the columns or shear walls on the under layer is conducted with "footprint" analysis and later serves as significant basis for foundation design;
(5) The structure has sufficient rigidity to resist wind load and keep the figure intact;
(6) When earthquake strikes, lateral seismic load will be generated by the deadweight accumulated with the height of the structure. Besides, wind produces lateral load, which generates not only lateral shear force but also overturning moment at the foundation. These factors have direct impact on the design of the foundation and the ground beneath it and have to be taken into consideration in the overall design of the structure.

2. The model of overall analysis

With all the assumptions of overall analysis, model shown in Fig. 3.24 could be adopted in the early stage of structural design. Such a model should take into consideration the stress states and the interactions between the above-the-ground

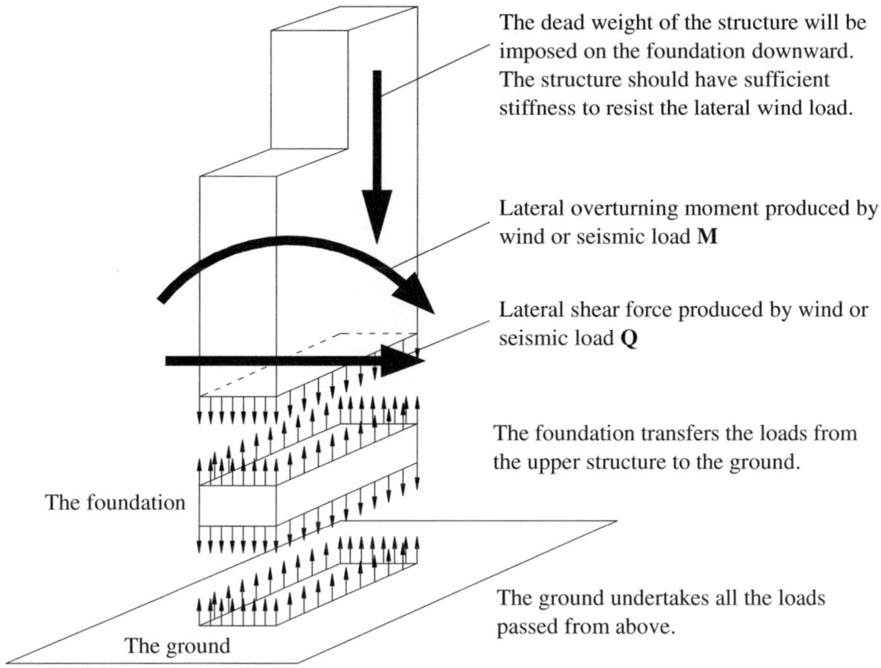

Fig. 3.24 Model of overall analysis of structure

structure, foundation, and the supporting ground underneath. The deadweight and vertical live loads of the upper structure will be imposed upon the foundation, which requires sufficient vertical strength and stiffness. Meanwhile, wind or seismic load will produce lateral shear force and overturning moment, which calls for sufficient lateral strength and stiffness. With the stress analysis of the foundation, we could see the stress, shear forces, and overturning moment generated by vertical and lateral loads being inflicted upon the foundation. The foundation transfers the loads from the upper structure to the ground, which undertakes all stress, shear forces, and overturning moment passed from the above.

The structure, being regarded as a rigid body, is under the action of deadweight, live loads, and lateral loads. Basic requirements for an overall analysis of the structure are as follows:

(1) Requirements of strength, stiffness, and stability of the structure are met at all load combinations;
(2) Requirements of strength and stability of the foundation and the supporting ground underneath are met at all load combinations.

In the early stage of design, the overall analysis of the structure could be conducted through basic static analysis. As for the stability and dynamics of the structure, they could also be simplified as equivalent static analysis accordingly.

3. "Footprint" analysis

"Footprint" is the plane layout of the foundation, which plays the key role in the force-bearing capacity of the structure. Loads are transferred along components of the structure (such as columns and walls) to the foundation. Similarly, the foundation, abiding by the principle of action–reaction, provides reaction force back to the structure in accordance with the way components arranged on the surface of the foundation and retains the balance of the structure itself. As we could easily see, when the plane arrangement of components on the foundation changes, the reaction force that the foundation provides changes in the way of distribution and in the magnitude of the force accordingly. This brings forward the design concept of reasonable allocation of components on the foundation and the optimization of "footprint."

Basic assumptions and the principle of "footprint" application are as follows:

(1) "Footprint" is the plane layout of the components on the foundation, which serves as the interface between the upper structure and the foundation;
(2) "Footprint" reflects the force-bearing capacity of the structure and has impact on the overall strength and stiffness of the structure;
(3) "Footprint" is closely related to the design of the foundation, for that it directly determines the loads and their distribution and therefore influences the magnitude and distribution of the reaction force that the supporting ground underneath provides;
(4) The optimization of "footprint" is the central point in the early stage of design when the upper structure and the foundation are being considered in an all-round perspective. Factors such as the quantity, the distribution, and the section selection of columns as well as the selection of foundation type and sectional area all have a direct impact on the total cost of construction;
(5) "Footprint" analysis assumes that under the action of vertical loads, every column or shear wall shares and undertakes corresponding vertical load in accordance with the tributary floor area. Since the force flow always takes the shortest path to the foundation, the midline of the floor between two columns becomes the dividing line of the force flow. As shown in Fig. 3.25, the floor area marked out by dotted lines is the shared area of columns or shear walls. Same analysis applies to columns, shear walls, and tubes;
(6) "Footprint" analysis assumes that under the action of lateral loads, the axial force of columns is linearly proportionate to the distance between the axis of the column and the neutral line in the "footprint" (Fig. 3.25). Thereby, we could estimate the axial forces of columns induced by lateral overturning moment. Note that the checking calculation of the axial forces of columns is generally conducted along the weak axis and there is no need to check along the strong axis.

As shown in Fig. 3.25a, as for evenly distributed columns, dotted line drawn along the midline between two rows of columns divides the tributary floor area for

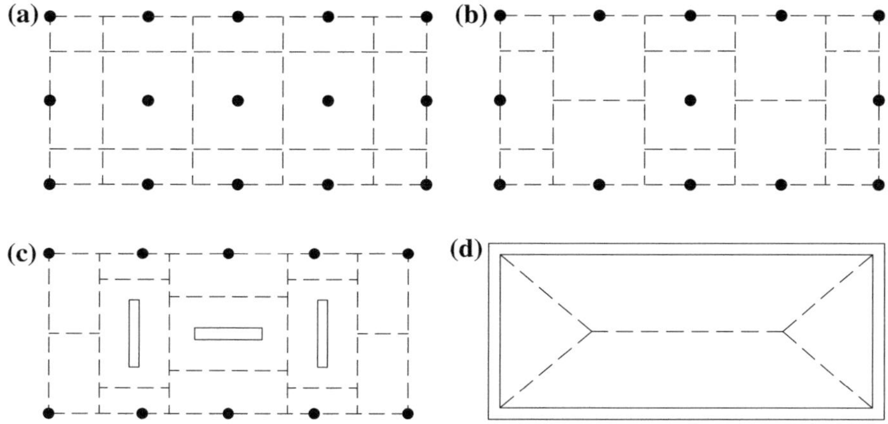

Fig. 3.25 Shared vertical loads suggested by "footprint" analysis. **a** Evenly distributed columns. **b** Non-evenly distributed columns. **c** Columns and shear walls. **d** Tube

each column. When it comes to non-evenly distributed column grid, we adopt the approach of equal portions, which means to partition in alignment with the principle that the force flow of the floor goes to the nearest column or wall, as shown in Fig. 3.25b. When the plane layout of column–shear wall combination is concerned, the principle of relative halving is still applicable, as shown in Fig. 3.25c. When tube structure shown in Fig. 3.25d is approached, the partition abides by the principle that the force flow takes the shortest path to the nearby tube wall, which principle is abided by the strand allocated on the floor under the action of evenly distributed load.

4. Example of "footprint" analysis

Next, with the help of an example of "footprint" analysis, we will demonstrate how to calculate the internal forces of columns or shear walls on the foundation, and estimate geometric dimensions of major components and the total cost of the construction.

Figure 3.26a shows an eight-story building, the plane layout of which displays an L in outline. The length is 42 m, the width is 20 m, and the height is 24 m. Vertical evenly distributed load on each floor is designed to be 10 kN/m^2, deadweight and live loads included. On the surface of the foundation, the total weight the structure has to bear is as follows:

Total weight = Vertical evenly distributed load × Floor area × Number of floors = 10 × (42 × 20 + 18 × 20) × 8 = 96,000 kN.

In the early stage of design, it is recommended to consider major components such as columns and shear walls in an overall way and make hypothetical distribution on the basis of that the operational requirements of the structure are met. For instance, we divide the length into 7 spans and the width into 2 spans. The columns are allocated as shown in Fig. 3.26b.

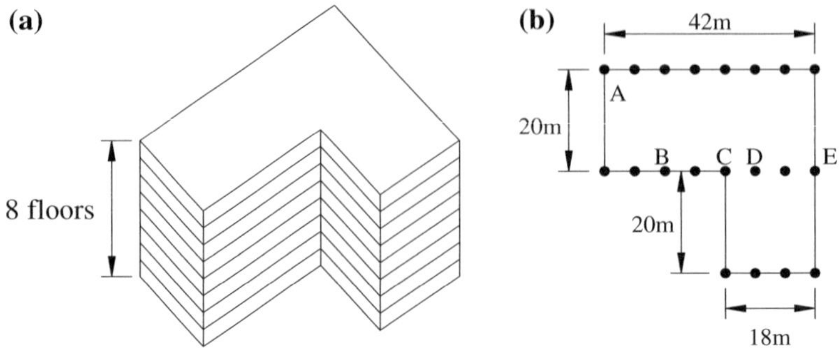

Fig. 3.26 Footprint method in vertical loading. **a** The diagram of floors. **b** The diagram of column allocation

The weight imposed upon unit area of the foundation is $q = 10\ \text{kN/m}^2 \times 8 = 80\ \text{kN/m}^2$. Shared floor area of each typical column is calculated as suggested by Fig. 3.25. Then, we have:

The weight corner pillar A bears is $q \times (3 \times 10) = 2400$ kN;

The weight side pillar B bears is $q \times (6 \times 10) = 4800$ kN;

The weight central corner pillar C bears is $q \times (6 \times 10 + 3 \times 10) = 7200$ kN;

The weight central pillar D bears is $q \times (6 \times 20) = 9600$ kN;

The weight side pillar E bears is $q \times (3 \times 20) = 4800$ kN.

When the structure is under the action of lateral loads, "footprint" analysis is also applicable to determine the axial force of columns resisting overturning moment. The structure shown in Fig. 3.27 has a height of 30 m, a length is of 48 m, and a width of 24 m. The wind load is designed to be 1 kN/m^2. The axial resistance of each column in the figure is linearly proportionate to the distance between the axis of the column and the neutral line in the "footprint." Figure 3.27 illustrates the calculation of the axial forces of column C1 to C5. This is a basic application of "footprint" analysis. Such a calculation principle is applicable to various occasions when there is variation in the figure of the structure, number of stories, and the condition of loads.

Since the wind is blowing toward the structure from different directions, dominant wind direction of structures from different regions might vary. Seeing that wind blowing from any direction could be broken into that along the two principal axes, the axial force of columns induced by the wind from the direction of two principal axes should be calculated, respectively, and then superimposed together in detailed stage of design. In the early stage of design, the axial force of columns induced by all wind loads might be conducted only in the direction of the weak axis of the structure for simplification. In the example as shown in Fig. 3.27, the axial force of columns induced by wind load in the direction of the weak axis is actually in need of calculation and readers could manually calculate in the same way as shown in Fig. 3.27 and compare results.

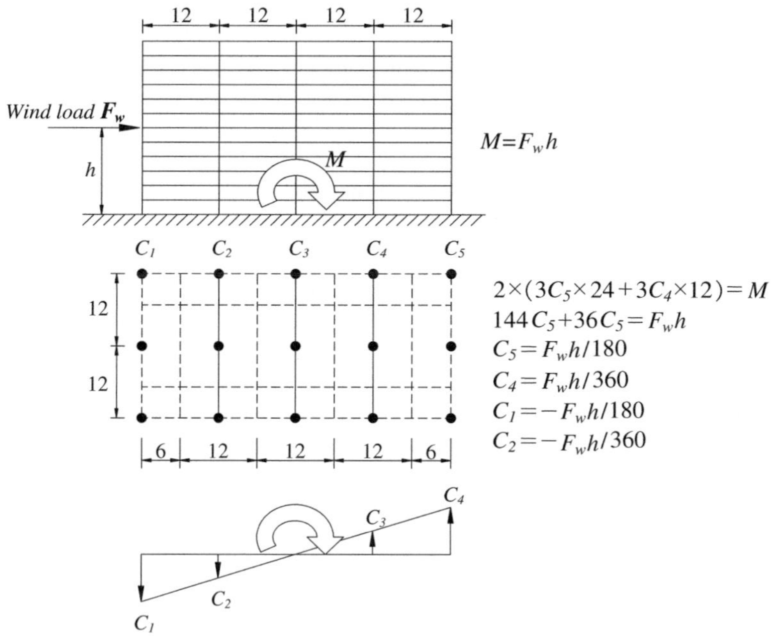

Fig. 3.27 "Footprint" analysis in the calculation of axial forces of columns under lateral wind load

Characteristics of "footprint" analysis could be summarized as follows:

(1) "Footprint" analysis enables a ready determination of load designed for columns (or major components) in a complex structure.
(2) The "footprint" could be optimized in light of varied allocations of columns in pursuit of a more reasonable load-bearing design for columns (or major components).
(3) Columns sections could be estimated on the basis of load analysis.
(4) As for the overturning moment induced by lateral load upon the foundation, "footprint" analysis could be applied to determine the allocated axial force of columns. In the early stage of design, the shear force generated by lateral wind load in the interface between columns and the foundation is often ignored. The checking of shear force is usually conducted in detailed design.
(5) "Footprint" analysis, when extended to other locales in structural design, shows us that it does not exist only in the base of the structure (the interface between the upper structure and the foundation) but also in each and every interface connecting the upper and lower structure (assuming that the lower structure has sufficient rigidity to support the upper structure), as shown in Fig. 3.36. It demonstrates that in conceptual stage of design, "footprint" analysis has special significance in the disassembly of structures of complexity, in the checking calculation when structure is assembled, as well as in layered or partitioned analysis of the structure.

5. Eccentric effects of wind and seismic loads

When designing for structures in seismic zone, we have to consider the seismic effect. Earthquakes induce the inertia force due to the mass of the structure. In design, all those inertia force is converted to a statically equivalent seismic force generated by the equivalent mass. For instance, high-rise structure could be simplified as a cantilever with conglomerated mass lump distributed along the height, as shown in Fig. 2.4.

If the resultant lateral force (wind or seismic) is H and the height from the action point to the foundation is h, the overturning moment imposed upon the foundation will be $M = H \cdot h$. If the resultant lateral force does not act on the centroid of the structure and the eccentricity is e, then the eccentric torque will be $T = H \cdot e$, as shown in Fig. 3.28, which will give rise to corresponding shear force in the foundation. The overturning moment M will be counterbalanced by the moment formed by the pair of reaction forces R_{fc} (compressive) and R_{ft} (tensile) from the foundation. That is $M = H \cdot h = R_{fc} \cdot d$, in which expression d is the arm of force between R_{fc} and R_{ft}. Under the assumption that normal stress at the bottom is linearly distributed, $d = 2B/3$, in which B is the width of the structure along H. Therefore, we see that the additional reaction R_{fc} and the shear force generated by lateral force H and torque T in the foundation will add to the complexity of stress, which should be considered in the early stage of design. Specific analysis will be discussed in Sect. 3.6 of this book.

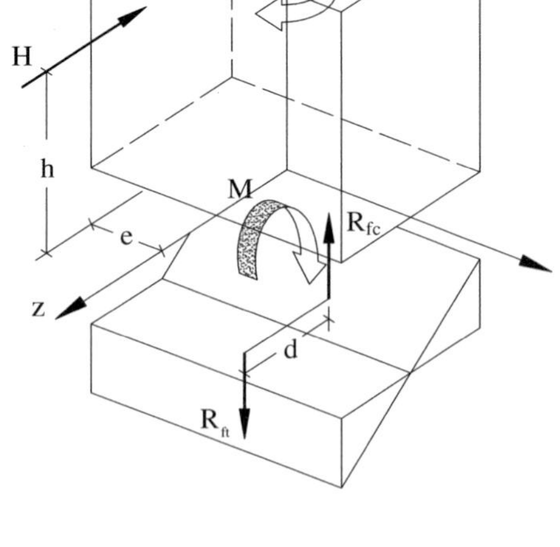

Fig. 3.28 Torque and overturning moment by eccentric lateral force

3.5 External Configuration and Balancing Design of Structures

The aspect ratio of the structure is of significance to the resistance to overturning moment as well as to the overall stability of the structure. Therefore, sufficient consideration should be given in the early stage of design. The fundamental principle of overall stability is the equilibrium requirement of the rigid body (Fig. 3.29):

$$H \cdot a = V \cdot d \tag{3.1}$$

$$V = H \cdot a/d = H \cdot c \cdot h/d \tag{3.2}$$

In formula (3.2), c, which equals a/h, is the coefficient of overturning arm. This coefficient determines the magnitude of the reaction force V. In addition, the larger d is, the smaller V becomes. Formulas (3.1) and (3.2) assume that the center of gravity acts upon the centroid of support. If total vertical load the structure bears is W, the reaction forces of the right and left column should be:

$$R_{右} = (W/2) + H \cdot a/d = (W/2) + H \cdot c \cdot h/d \tag{3.3}$$

$$R_{左} = (W/2) - H \cdot a/d = (W/2) - H \cdot c \cdot h/d \tag{3.4}$$

(d is the distance between columns).

If e is the eccentricity between the center of gravity and the centroid of support, the checking requirements of overturning resistance will be:

$$H \cdot a + W \cdot e = V \cdot d \tag{3.5}$$

$$R_{左} = (W/2) - (W \cdot e/d) - (H \cdot a/d) \geq 0 \tag{3.6}$$

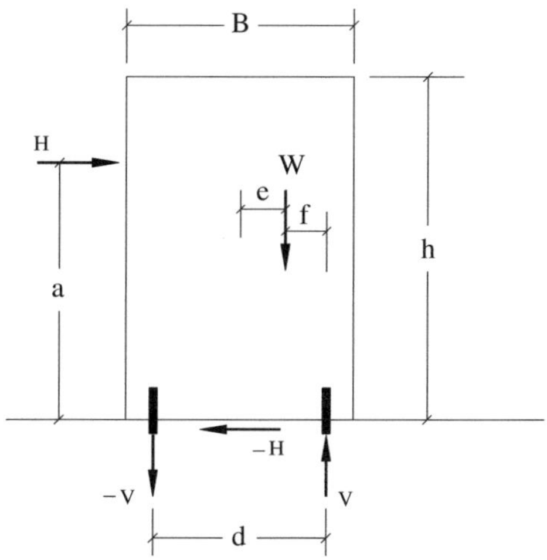

Fig. 3.29 Equilibrium requirement of a rigid body in its resistance to the overturning moment

To avoid tension in vertical components and foundation, it will not be allowed as the internal force of columns on the foundation ($R_{左} \geq 0$). This is the equilibrium requirement of foundation design. When requirement (3.6) is not met, the distance between columns *d* needs to be increased until Eq. (3.6) is met.

When the aforementioned equilibrium requirements are met,

$$W \cdot f > K(H \cdot a) \tag{3.7}$$

K is the safety coefficient and usually $K = 1.5–2.0$; *f* is the eccentric arm of force (Fig. 3.29).

1. Force-bearing capacity and stiffness

The force-bearing capacity of structures refers to the ability of the structure to undertake loads without failure. Stiffness is the property of the structure to prevent deformation. Structures have to be equipped with sufficient strength and stiffness to undertake loads and to restrict deformation. These deformation limits, when designing for high-rise structures, have to be met very strictly.

Due to elasticity, structures will display noticeable deformation when loaded. Figure 3.30 is derived from Fig. 3.29 when elastic deformation is considered. Bottom columns of the structure could be regarded as a group of springs. Under the

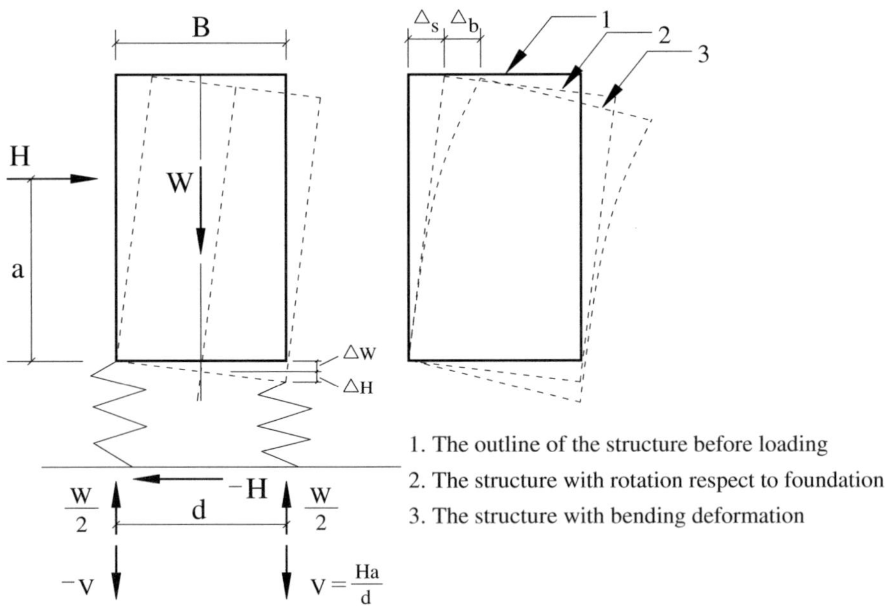

Fig. 3.30 Deformation of structures under loading

action of vertical load W, springs are compressed and the structure demonstrates vertical compressive deformation Δ_w. Under the action of lateral force H, the spring on the right is compressed, the spring on the left is stretched, and the structure displays an uneven deformation Δ_H as shown in Fig. 3.30. Due to the uneven deformation, the top of the structure displays lateral displacement Δ_s. Meanwhile, lateral force H will also give rise to bending deformation and lead to an additional displacement Δ_b on the top of the structure. The magnitude of Δ_b is closely related to the bending rigidity of the structure, which in turn is related with the width of structure B, the elastic modulus of the structure E, the shape of the section, and the height of the structure h.

Contour 1 in Fig. 3.30 represents the outline of the structure before loading. Contour 2 depicts the structure with rotation respect to foundation. Contour 3 represents the structure with bending deformation. From the figure above, we know that the higher the structure is the greater lateral force is, the smaller the moment of inertia is and the greater lateral displacement on the top of the structure is. Deformation is the significant issue in the design of high-rise structure. With extraordinary height and relatively small lateral rigidity, high-rise buildings undertake enormous lateral wind or seismic load and easily display excessive lateral displacement or even damage, which affects the functionality and safety of the structure.

Common practice is to regulate the bending stiffness of the high-rise to reduce lateral displacement under wind load and improve seismic response of the structure. This is self-contradictory. When structures have large lateral stiffness, they will not display excessive lateral displacement under the action of wind load. However, structures with large lateral stiffness will magnify seismic respond and do not favor seismic design. Thus, when the height of a high-rise is determined in conceptual design, designers generally will focus on the regulation and optimization of the overall stiffness of the structure. Usually, we will make changes to the "footprint" of the structure to meet the requirement of overall stiffness. High-rise structures, in their service condition, should have sufficient stiffness to resist wind load without producing excessive lateral displacement at the top of the structure. In the meantime, when earthquakes strike, high-rise structures should be ductile enough so that they will not drastically respond to the earthquakes (in general, the greater the stiffness of the structure is the heftier the seismic effects are) and ensure the safety of the structure.

Restriction upon lateral displacement is a major issue in the design of high-rise structures, and the magnitude of lateral displacement has direct impact on the functionality of the structure. Thus, several factors will need specific analysis as follows.

The "footprint" of the foundation could determine, to a large extent, the geometry of the cross section and the overall stiffness of the structure. The bending resistance of the structure is determined by the overall stiffness of the structure, which is related to the moment of inertia. Generally speaking, the load-bearing components are designed to be far away from the neutral axis in the "footprint," for that the contribution of components on the "footprint" to the overall moment of inertia of the structure is linearly proportionate to the distance between their axes and the neutral axis. Hence, each time we move one component on the cross section of the structure, we affect the overall moment of inertia. If the distance between the component and the neutral axis was d and that after moving is d', then the influence the movement has on the moment of inertia I is $(d'/d)^2$.

"Footprint" does not affect only the stiffness but also the distribution of reaction forces and supporting capacity of the foundation. Thus, it has considerable impact

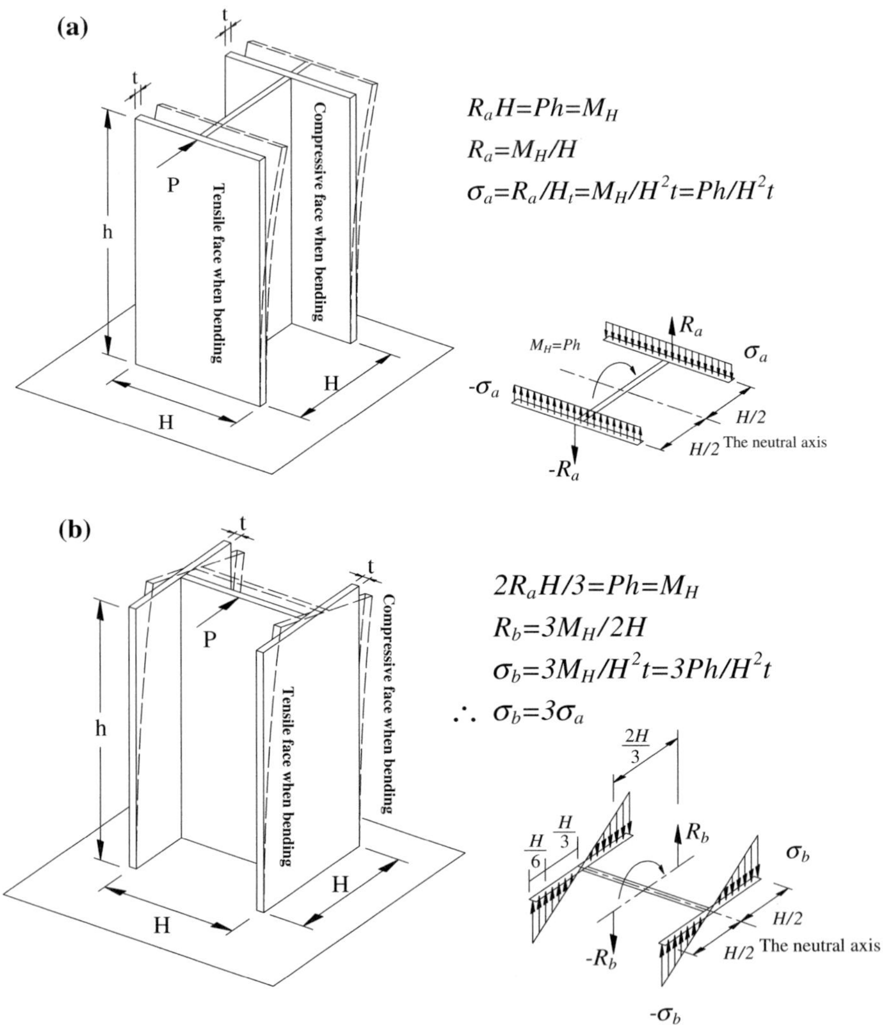

Fig. 3.31 Influence of the direction of force application on the stress distribution of I-shaped steel. **a** Direction of force application is perpendicular to the wings. **b** Direction of force application is parallel to the wings

on foundation design. Take symmetric I–shaped steel columns shown in Fig. 3.31 for instance. It is clear that under the same load, the distribution of reaction forces and supporting capacity of the foundation varies due to different arrangement of the column (different "footprint"). Figure 3.31a displays the situation when the two wings of this I-shaped steel is under the same tension (or compression), in which case the utilization of the material reaches its maximum. Meanwhile, the two wings of this I-shaped steel are at their utmost distance from the neutral axis, and the height of the section is sufficiently utilized. Since the coefficient of efficiency in the height of the section is at is maximum, assumed to be 1. Under the same lateral load with a slight change of direction of application, the stress distribution in two side wings of symmetric I-shaped steel displays as a triangle as shown in Fig. 3.31b and the utilization of material is only half as that of the previous example. Moreover, the distance between the resultant reaction force R_b and the neutral axis is no longer $H/2$ but $H/3$, and the coefficient of efficiency in the height of the section is only 2/3. When external load P is the same, the stress distribution of these two arrangements varies and one maximum stress is three times the other, as shown in Fig. 3.31b. In the meantime, the moment of inertia I_a as shown in Fig. 3.31a is three times the moment of inertia I_b as shown in Fig. 3.31b. That is $I_a = 3I_b$.

From the analysis above, we know that the "footprint" has a huge impact on the force-bearing capacity and stiffness of the structure. Figure 3.32 demonstrates the variances in bending rigidity or moment of inertia of the foundation when the shear walls are installed on the same cross section with different "footprint."

Figure 3.32 shows us that three different "footprint" (as shown in Fig. 3.32a–c) of two identical shear walls could lead to six times of difference in the moment of inertia, which means that with the arrangement of "footprint," we are able to effectively regulate the bending rigidity and bending deformation of the structure.

Therefore, in the overall design of the structure, engineers and architects, on the basis of meeting the requirements of the structure under all load combinations, are

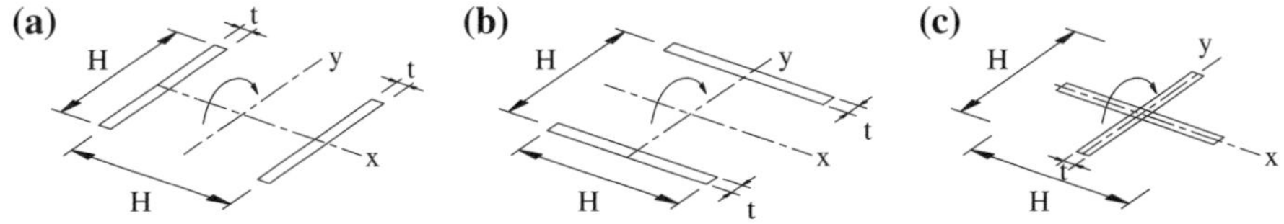

$$I_a = \frac{1}{12}(2t)H^3 \quad \text{If } t = \frac{H}{10} \quad I_a = \frac{H^4}{60}$$

$$I_a = 2tH\left(\frac{H}{2}\right)^2 = \frac{H^4}{20}$$

$$I_c = \frac{1}{2}I_a + \frac{H}{12}t^3 = \frac{H^4}{120} + \frac{H^4}{12000} \approx \frac{H^4}{120}$$

Ratio of inertial moments $I_a/I_b/I_c = \frac{H^4}{60} / \frac{H^4}{20} / \frac{H^4}{120} = 1/3/0.5$

Fig. 3.32 Influence of "Footprint" on the force-bearing capacity and rigidity of the structure welcome to utilize the concept and approach of "footprint" and reasonably arrange the way components are allocated on the foundation, making the sections of the components as far away from the neutral axis as possible and providing the structure with sufficient strength and stiffness.

2. Overall mechanical properties of the structure

In the early stage of design, designers, on the basis of meeting functional and operational requirements, have to take safety principles such as the requirements of strength, stiffness, and stability into consideration. Any violation of these principles will result in unacceptable consequences or even failure of the structure. In this stage, material mechanics could be simplified to solve problems mentioned above.

Take high-rise structures for instance. In the early stage of design, it is assumed to be a cantilever embedded to the ground. The height of the structure is H.

1. When lateral concentrated force P is applied to the top of the structure, the overall deformation will be

$$\text{Lateral displacement on the top is } \Delta = P \cdot H^3/3E \cdot I \tag{3.8}$$

$$\text{Flexibility of the structure is } \delta = H^3/3E \cdot I \tag{3.9}$$

$$\text{Stiffness of the structure is } K = 3E \cdot I/H^3 \tag{3.10}$$

2. When lateral load q is evenly distributed along the height of the structure,

$$\text{Lateral displacement on the top is } \Delta = q \cdot H^4/8E \cdot I \tag{3.11}$$

$$\text{Flexibility of the structure is } \delta = H^4/3E \cdot I \tag{3.12}$$

$$\text{Stiffness of the structure is } K = 8E \cdot I/H^4 \tag{3.13}$$

3. The greatest vertical load borne by the foundation of the structure is

$$N_{\max} = q_v \cdot A \cdot n \tag{3.14}$$

In Eq. (3.14), q_v is equivalent vertical evenly distributed load on each floor, which includes the deadweight of floor, walls, and pillars as well as in-service live loads on the floor. A is the surface area of each floor, and n is the number of stories in the structure.

4. The bending moment and shear force of high-rise structures
   Under the action of lateral evenly distributed load, the bending moment on any cross section is

$$M = qH \cdot h^2/2 \tag{3.15}$$

In Eq. (3.15), h is the distance between the cross section and top of the high rise; qH is lateral evenly distributed load. If H represents the overall height of the structure, then the greatest bending moment on the foundation will be

$$M_{max} = qH\,H^2/2 \quad (3.16)$$

Under the action of lateral evenly distributed load, the shear force on any cross section should be the sum of all lateral loads above.

$$Q = qH \cdot h \quad (3.17)$$

The greatest shear force on the foundation of this high-rise structure is

$$Q_{max} = qH \cdot H \quad (3.18)$$

5. "Footprint" is of utmost importance to mechanical properties of the structure, especially the force transference between the foundation and the supporting ground underneath. Figure 3.33 displays the difference in the bending rigidity of the structure (*EI*) when five "footprint" are proposed.

   Case 1: (as shown in Fig. 3.33a)
   Four 1 m × 1 m columns without connections in between.

$$I_1 = 4\left(1^4/12\right) = 4/12$$

   Case 2: (as shown in Fig. 3.33b)
   One 2 m × 2 m column allocated in the center

$$I_2 = 2^4/12 = 16/12 = 4(4/12) = 4I_1$$

   Case 3: (as shown in Fig. 3.33c)
   Four 5-m-wide 0.2-m-thick walls, derived from four 1 m × 1 m columns, are placed as if they were the four sides of a box only without any connection between. Out-of-plane rigidity of the wall boards is ignored for the slight magnitude. Then, we have

$$I_3 = 2\left(0.2 \times 5^3/12\right) = 50/12 = 12.5I_1$$

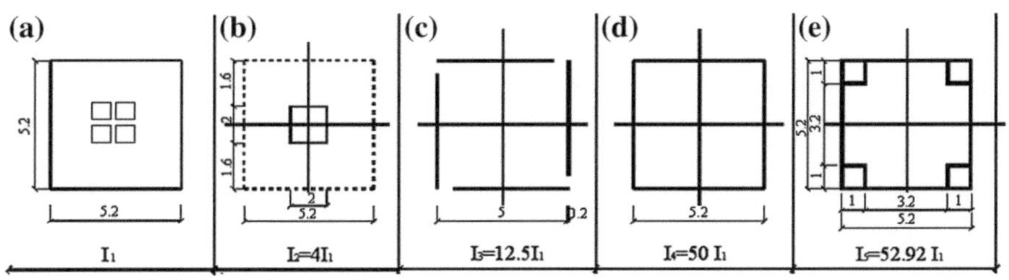

Fig. 3.33 Influence of "Footprint" on mechanical properties of the structures

Case 4: (as shown in Fig. 3.33d)
Four flat walls in case 3 are connected in the corner, in which case the cross section becomes a closed box. Then, we have

$$I_4 = 2\left(0.2 \times 5^3/12 + 0.2 \times 5 \times 2.5 \times 2.5\right) = 50\,(4/12) = 50I_1$$

Case 5: (as shown in Fig. 3.33e)
The four columns in case 1 are allocated in the corners. Then, we have

$$I_5 = 4 \times 1^2 \times ((5.2-1)/2)^2 = 4 \times 2.12 = 17.64 = 52.92I_1$$

From Fig. 3.33a–e, we could see that different "footprint" lead to a difference of 52.92 times in the moment of inertia.

3.6 Decomposition of Asymmetric Complex Structures

1. Symmetrical and asymmetrical structures

When the upper structure is vertically aligned with the foundation, asymmetrical stress induced by unevenly distributed live loads should affect the structure insignificantly. Under this circumstance, we still assume the stress is symmetrical in the early stage of design.

On account of requirements of function as well as aesthetics in structures, designers more often than not select unconventional figure and supporting system for the structure. For instance, an asymmetrical elevation and a correspondingly asymmetrical "footprint" constitute a typical asymmetrical structure. Asymmetrical issues raised from such selections have to be dealt with in the early stage of design. From the perspective of elevation, reasons for asymmetry or eccentricity of the structure are as follows:

(1) Asymmetrical arrangement of components and symmetrical supporting system
Eccentric deadweight of the structure results in eccentric stress in a symmetrically supported structure (Fig. 3.34a).
(2) Symmetrical arrangement of components and asymmetrical supporting system
Symmetrical deadweight of the structure results in eccentric stress in an asymmetrically supported structure (Fig. 3.34b).

Efforts should be made to minimize asymmetrical elevation or such supporting, for that these kinds of asymmetry considerably add to additional overturning moment and additional torque. When it is inevitable, the arrangement in the "footprint" should be optimized so that the eccentricity between the center of gravity and the centroid of support could be minimized.

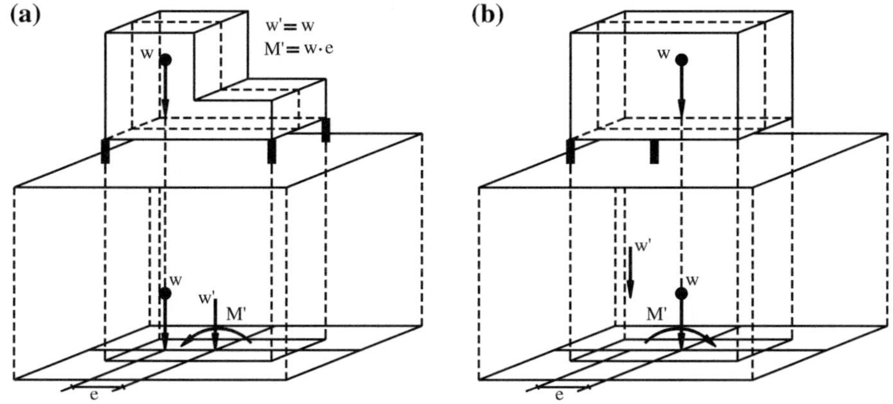

Fig. 3.34 Additional overturning moment of asymmetrical structures. **a** Asymmetrical elevation. **b** Asymmetrical support

As for asymmetrical structure, eccentricity gives occasion to additional overturning moment on the structure itself and the foundation. Additional forces imposed upon the foundation could be alleviated through the optimization of columns and shear walls on the foundation following the "footprint" principles. Generally speaking, the optimization of "footprint" is relatively simple and can be easily achieved by basic static calculation.

However, under the action of wind or seismic load, adverse effects upon asymmetrical structures might be aggravated, giving rise to additional torque T (Fig. 3.35). Two kinds of additional torque are possible in general term:

(1) When the "footprint" is symmetrical but the upper structure is asymmetrical, additional torque is generated by the lateral force H and the eccentricity e. That is $T' = H \cdot e$ (Fig. 3.35a).
(2) When the "footprint" is asymmetrical but the upper structure is symmetrical, additional torque is generated by the lateral force H and an H-induced eccentricity e. That is $T' = H \cdot e$ (Fig. 3.35b).

In the early stage of design, we have to take the influence of additional torque into consideration and often arrange paired shear walls to resist potential additional torque.

2. Disassembly of complex structures

Through the application of "footprint" analysis, complex structures could be disassembled and analyzed as simple substructures. For instance, substructure 3 in Fig. 3.36 is directly connected to the foundation; on top of substructure 3 is substructure 2, on the top of which lies substructure 1. The analysis usually starts from the highest substructure, in this case substructure 1, and goes downwards. Lower substructures serve as the foundation and provide support for the upper

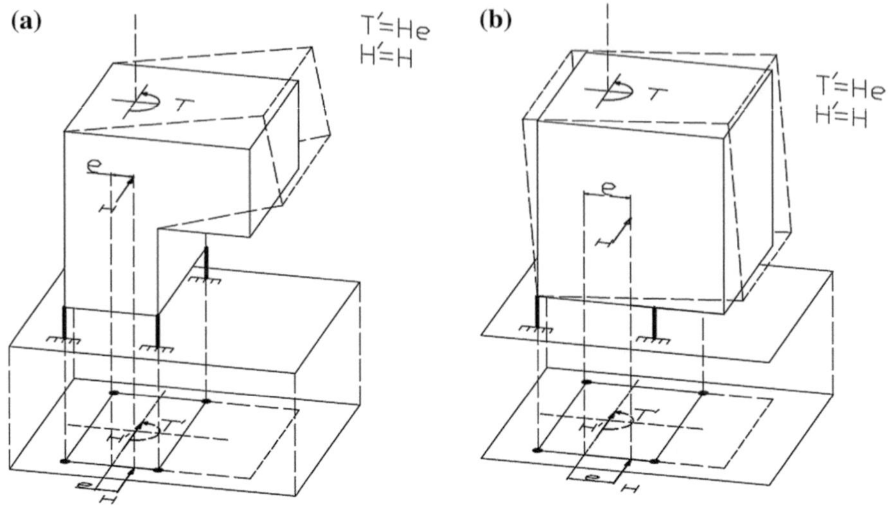

Fig. 3.35 Additional torque in asymmetrical structures. **a** Asymmetrical elevation. **b** Asymmetrical support

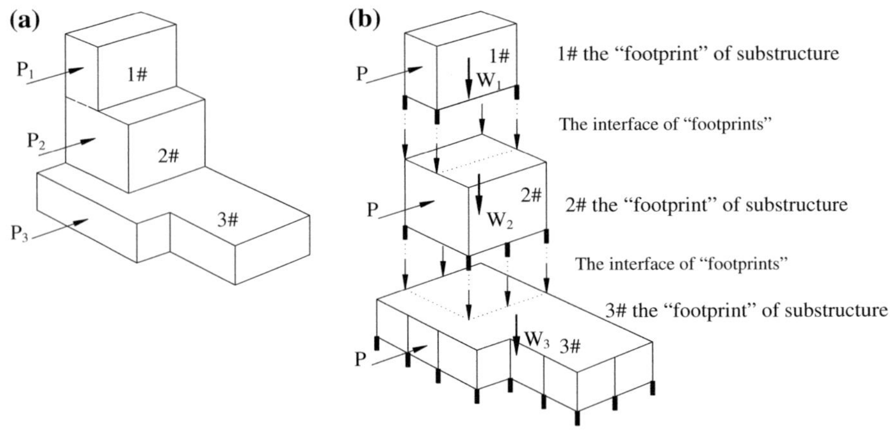

Fig. 3.36 Disassembly of complex structures

substructure. Calculation is simplified with "footprint" analysis and goes from one substructure downwards to another.

In the analysis of substructure 1, we know that it has lateral and vertical load P_1 and W_1 and its support comes from the "footprint" of substructure 1 on the upper surface of substructure 2. Meanwhile, the internal force of substructure 1 on the "footprint" is imposed upon substructure 2 and becomes a "load" for substructure 2. This "load" and lateral and vertical load P_2 and W_2 which have been acting upon substructure 2 from the start constitute the loads that substructure 2 has to undertake. So the analysis goes until the lowest substructure. In the analysis of substructure 3, loads it has to bear are comprised of the reaction forces of substructure 2 in its "footprint" and lateral and vertical load P_3 and W_3.

All complex structures could be disassembled and analyzed as a series of simple substructures. With the help of principles pertaining to the "footprint," complex structures are disintegrated as a series of substructures, which enables the arrangement of components in each substructure, the regulation and the optimization of sectional areas and an optimized design of the whole structure.

3. Example

This example will demonstrate how to disassemble a complex structure and conduct stress estimation and optimization of its substructures. We will first determine the internal forces of columns in the main frame with the help of the "footprint" analysis and further utilize the analysis for the regulation and optimization of structural arrangements.

The assumptions of "footprint" analysis are as follows: In the simplified calculation of the axial force of foundation when the frame structure is under the action of lateral load, we assume that there are only axial and shear force at the bottom of the column, and ignore the bending moment at the bottom of the column; the internal forces of the upper structure could be passed on to the lower corresponding structural components; the structure could be disassembled and calculated as substructures and the numbers of these substructures could be superimposed.

(1) Requirements of structural design

With reference to functional requirements of the structure, the overall dimensions, the number, and the height of stories are worked out as shown in Fig. 3.37, which provides a perspective view of the structure. The structure is disassembled to three substructures which further divided into four layers: The fourth layer above the ground is substructure 1; the third layer is substructure 2; and the first and second layer combined is substructure 3. Figure 3.37b is the "footprint" of A-A section and the elevation is 0.0 m; Fig. 3.37c is the "footprint" of B-B section and the elevation is 8.0 m; Fig. 3.37d is the "footprint" of C-C section and the elevation is 12.0 m; and Fig. 3.37e is the elevation of the structure along axis y.

The coding and the internal force of columns on each section are represented by C_{ijk} and N_{ijk}, and the subscript is explained in Fig. 3.37. Specifically, i is the identifier of the elevated section, j is the column number, and k is the row number. The coding and the internal force of columns on each section are shown in Table 3.1. For instance, $C_{\mathrm{A}14}$ represents the column at Column 1/Row 4 on A-A section (elevation 0.0 m); $N_{\mathrm{A}14}$ and $Q_{\mathrm{A}14}$ are the axial force and the shear force of

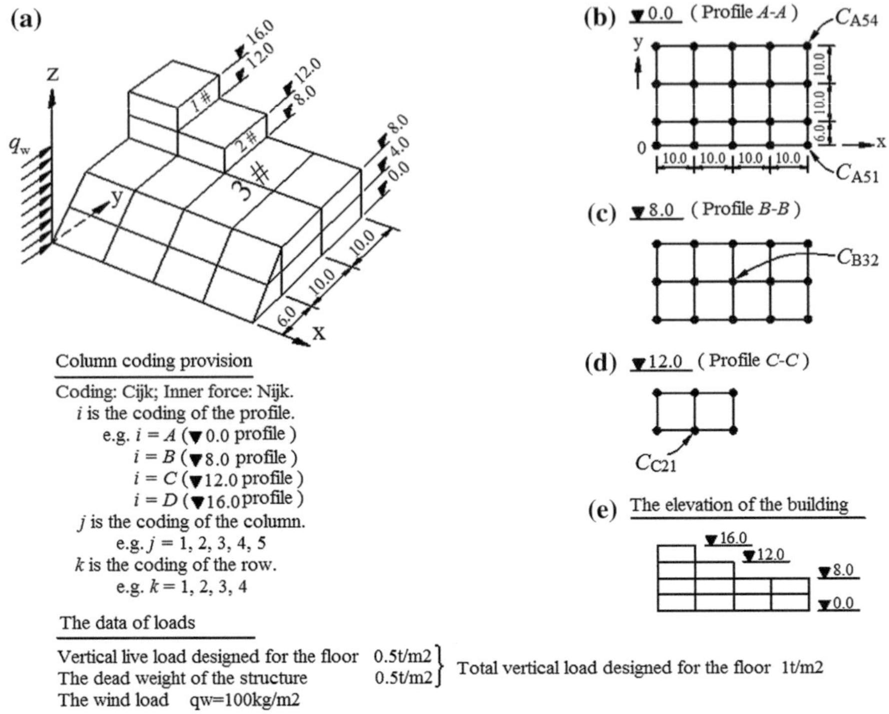

Fig. 3.37 Simplified calculation of the structure

this column. Likewise, C_{B34} represents the column at Column 3/Row 4 on B-B section (elevation 8.0 m), and N_{B34} and Q_{B34} are the axial force and the shear force of this column; C_{C23} represents the column at Column 2/Row 3 on C-C section (elevation 12.0 m), and N_{C23} and Q_{C23} are the axial force and shear force of this column.

(2) Estimation of loads

Estimation of design loads

Taking the magnitude of service load into consideration, we select reinforced concrete framework packed with brick masonry for the structure. Precast concrete floor is supported by 10 m × 10 m primary beam grid and 5-m-spaced secondary beams. The estimation of loads that the structure bears is as follows:

Vertical live load designed for the floor is 5 kN/m^2 (assume the coefficient of live load considered).

The deadweight of the structure (beams, columns, and floors are deduced into 0.2-m-thick boards) is 0.2 m × 25 kN/m^3 = 5 kN/m^2 (assume the coefficient of dead load considered).

Total vertical load designed for the structure is 5 + 5 = 10 kN/m^2.

Vertical load designed for substructure 1 is $W_1 = 10 \times 10 \times 10 = 1000$ kN (the coordinate of action point is (5, 21, 14)).

Vertical load designed for substructure 2 is $W_2 = 2 \times 10 \times 10 \times 10 = 2000$ kN (the coordinate of action point is (10, 21, 10)).

Vertical load designed for substructure 3 is $W_3 = 10 \times 40 \times (23 + 26) = 19{,}600$ kN (the coordinate of action point is (20, 12.25, 4)).

Estimation of wind load

With reference to regional wind load, wind load is designed as $q_w = 1$ kN/m^2 (assume the coefficient of wind load considered).

Lateral load designed for substructure 1 is $H_1 = 1 \times (4 \times 10) = 40$ kN (the coordinate of action point is (5, 16, 14)).

Lateral load designed for substructure 2 is $H_2 = 1 \times (4 \times 20) = 80$ kN (the coordinate of action point is (10, 16, 10)).

Lateral load designed for substructure 3 is $H_3 = 1 \times (40 \times 8) = 320$ kN (the coordinate of action point is (20, 3, 4)).

Estimation of seismic force

Approximately adopt a seismic load $H_{eq} \approx 0.2$ W. Then, we have:

Seismic force designed for substructure 1 is $H_{eq1} = 0.2 \times W_1 = 0.2 \times 1000 = 200$ kN (the coordinate of action point is (5, 21, 14)).

Seismic force designed for substructure 2 is $H_{eq2} = 0.2 \times W_2 = 0.2 \times 2000 = 400$ kN (the coordinate of action point is (10, 21, 10)).

Seismic force designed for substructure 3 is $H_{eq3} = 0.2 \times W_3 = 0.2 \times 19{,}600 = 3920$ kN (the coordinate of action point is (20, 12.25, 4)).

(3) "Footprint" analysis of load designed for substructures

"Footprint" analysis of load designed for substructure 1 (elevation 12.0 m)

Vertical load designed is 1000 kN.

Lateral load designed takes the seismic load as 200 kN (wind load 40 kN).

Overturning moment induced by lateral load is $M_{H1} = 200 \times 2 = 400$ kN m.

Shear force induced by lateral load is $Q_{H1} = 200$ kN.

"Footprint" analysis of load designed for substructure 2 (elevation 8.0 m)

Vertical load designed is 2000 kN.

Lateral load designed takes the seismic load as 400 kN (wind load 80 kN).

Overturning moment induced by lateral load is $M_{H2} = 400 \times 2 = 800$ kN m.

Shear force induced by lateral load is $Q_{H2} = 400$ kN.

"Footprint" analysis of load designed for substructure 3 (elevation 0.0 m)

Vertical load designed is 19,600 kN.

Lateral load designed takes the seismic load as 3920 kN (wind load 320 kN).

Overturning moment induced by lateral load is $M_{H3} = 3920 \times 4 = 16{,}080$ kN m.

Shear force induced by lateral load is $Q_{H3} = 3920$ kN.

(4) Determination of the internal forces of columns

From the "footprint" of foundation (elevation 0.0 m) as shown in Fig. 3.37b, we know that the moment of inertia respect to axis x I_x is smaller than that respect to

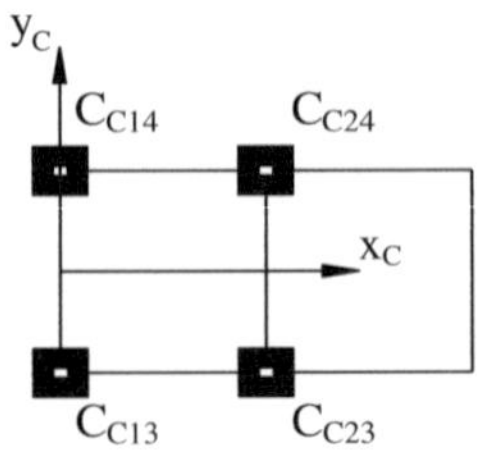

Fig. 3.38 Columns on C-C section

axis y I_y. Then, we only have to conduct estimative calculation of the internal forces of columns with the overturning moment respect to axis x.
Determine the internal force of columns on substructure 2 (elevation 12.0 m).
The internal forces N of columns in Fig. 3.38 are as follows:

$$N_{C14} = 1000/4 + 400/(2 \times 10) = 270 \text{ kN}$$
$$N_{C24} = 1000/4 + 400/(2 \times 10) = 270 \text{ kN}$$
$$N_{C13} = 1000/4 - 400/(2 \times 10) = 230 \text{ kN}$$
$$N_{C23} = 1000/4 - 400/(2 \times 10) = 230 \text{ kN}$$

Assume the sectional area of columns is the same. The shear forces Q of corresponding columns are as follows:

$$Q_{C14} = Q_{C24} = Q_{C13} = Q_{C23} = 200/4 = 50 \text{ kN}$$

An approximate section could be determined with the internal forces calculated above.
Determine the internal force of columns on substructure 2 (elevation 8.0 m).
The internal forces N of columns in Fig. 3.39 are as follows:

$$N_{B14} = 2000/6 + 800/(3 \times 10) + N_{C14} = 630 \text{ kN}$$
$$N_{B24} = 2000/6 + 800/(3 \times 10) + N_{C24} = 630 \text{ kN}$$
$$N_{B34} = 2000/6 + 800/(3 \times 10) = 360 \text{ kN}$$
$$N_{B13} = 2000/6 - 800/(3 \times 10) + N_{C13} = 536.7 \text{ kN}$$
$$N_{B23} = 2000/6 - 800/(3 \times 10) + N_{C23} = 536.7 \text{ kN}$$
$$N_{B33} = 2000/6 - 800/(3 \times 10) = 306.7 \text{ kN}$$

Assume the sectional area of columns is the same. The shear forces Q of columns in Fig. 3.39 are as follows:

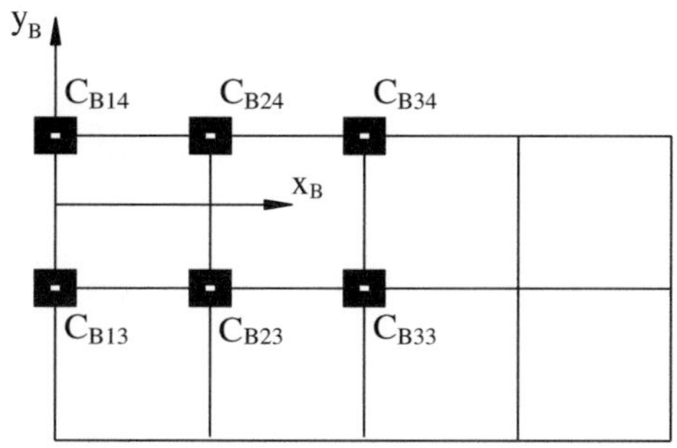

Fig. 3.39 Columns on B-B section

$$Q_{B34} = Q_{B33} = 400/6 = 66.7 \text{ kN}$$
$$Q_{B14} = Q_{B24} = Q_{B13} = Q_{B23} = 400/6 + Q_{C12} = 116.7 \text{ kN}$$

Determine the internal force of columns on substructure 3 (elevation 0.0 m). Diagrams of simplified calculation of the internal forces of columns on this substructure (elevation 0.0 m) are shown in Figs. 3.40 and 3.41. Readers could calculate the internal forces of columns the way we do with aforementioned two substructures, as shown in Table 3.2.

From Fig. 3.40, we know

$$\begin{aligned} M_{H3} &= 5(N_{A54} \times 13 + N_{A53} \times 3 + N_{A52} \times 7 + N_{A51} \times 13) \\ &= 5(N_{A54} \times 26 + (3/13)N_{A54} \times 3 + (7/13)N_{A54} \times 7) \\ &= 5((26 + (58/13))N_{A54}) = 16{,}080 \text{ kN m} \end{aligned}$$

The axial forces derived only from the bending moment M_{H3} are as follows:

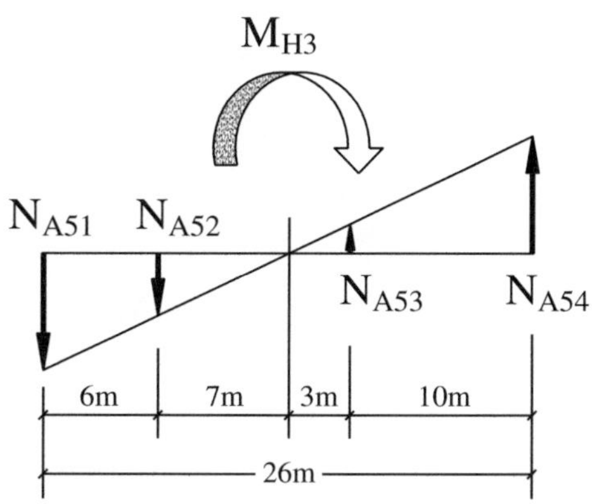

Fig. 3.40 Internal forces of columns

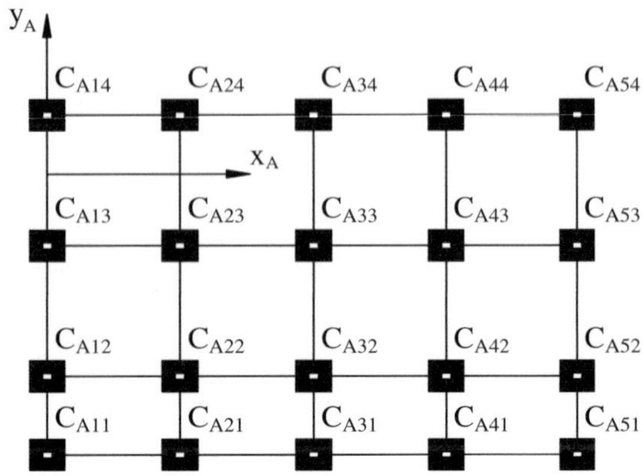

Fig. 3.41 Determination of the internal forces of columns

$$N'_{A54} = 16,080/152.3 = 105.6\ \text{kN}$$
$$N'_{A53} = (3/13)N_{A54} = 24.4\ \text{kN}$$
$$N'_{A52} = (7/13)N_{A51} = -56.9\ \text{kN}$$
$$N'_{A51} = 16,080/152.3 = -105.6\ \text{kN}$$

Correspondingly, total axial forces in Fig. 3.40 are as follows:

$$N_{A44} = N_{A54} = 19{,}600/20 + N'_{A54} = 980 + 105.6 = 1085.6\ \text{kN}$$
$$N_{A14} = 19{,}600/20 + N'_{A54} + N_{B14} = 980 + 105.6 + 630 = 1715.6\ \text{kN}$$
$$N_{A24} = 19{,}600/20 + N'_{A54} + N_{B24} = 980 + 105.6 + 630 = 1715.6\ \text{kN}$$
$$N_{A34} = 19{,}600/20 + N'_{A54} + N_{B34} = 980 + 105.6 + 360 = 1445.6\ \text{kN}$$
$$N_{A43} = N_{A53} = (19{,}600/20) + N'_{A53} = 980 + 24.4 = 1004.4\ \text{kN}$$
$$N_{A13} = (19{,}600/20) + N'_{A53} + N_{B13} = 980 + 24.4 + 536.7 = 1541.1\ \text{kN}$$
$$N_{A23} = (19{,}600/20) + N'_{A53} + N_{B23} = 980 + 24.4 + 536.7 = 1541.1\ \text{kN}$$
$$N_{A33} = (19{,}600/20) + N'_{A53} + N_{B33} = 980 + 24.4 + 306.7 = 1311.1\ \text{kN}$$
$$N_{A12} = N_{A22} = N_{A32} = N_{A42} = N_{A52} = (19{,}600/20) + N'_{A52} = 980 - 56.9 = 923.1\ \text{kN}$$
$$N_{A11} = N_{A21} = N_{A31} = N_{A41} = N_{A51} = (19{,}600/20) + N'_{A51} = 980 - 105.6 = 874.4\ \text{kN}$$

Since the first row of columns on substructure 3 in Fig. 3.41 is inclined, their internal forces have to be adjusted. The internal forces revised are:

$$N'_{A11} = N'_{A21} = N'_{A31} = N'_{A41} = N'_{A51} = 874.4 \times (5/4) = 1093\ \text{kN}$$

Basic shear force designed for the columns on 3# substructure is:

Table 3.2 Forces estimated for structural components (kN)

Substructure 3# along section A-A				Substructure 2# along section B-B				Substructure 1# along section C-C			
C_{ijk}	N_{ijk}(T)	Q_{ijk}(T)	Coordinate of stressed section	C_{ijk}	N_{ijk}(T)	Q_{ijk}(T)	Coordinate of stressed section	C_{ijk}	N_{ijk}(T)	Q_{ijk}(T)	Coordinate of stressed section
C_{A11}	1093	196	(0, 0, 0)								
C_{A12}	923	196	(0, 6, 0)								
C_{A13}	**1541.1**	**312.7**	(0, 16, 0)	C_{B13}	536.7	116.7	(0, 16, 8)	C_{C13}	230	50	(0, 16, 12)
C_{A14}	**1715.6**	**312.7**	(0, 26, 0)	C_{B14}	**630**	**116.7**	(0, 26, 8)	C_{C14}	**270**	**50**	(0, 26, 12)
C_{A21}	1093	196	(10, 0, 0)								
C_{A22}	923	196	(10, 6, 0)								
C_{A23}	**1541.1**	**312.7**	(10, 16, 0)	C_{B23}	536.7	116.7	(10, 16, 8)	C_{C23}	230	50	(10, 16, 12)
C_{A24}	**1715.6**	**312.7**	(10, 26, 0)	C_{B24}	**630**	**116.7**	(10, 26, 8)	C_{C24}	**270**	**50**	(10, 26, 12)
C_{A31}	1093	196	(20, 0, 0)								
C_{A32}	923	196	(20, 6, 0)								
C_{A33}	1311.1	262.7	(20, 16, 0)	C_{B33}	306.7	66.7	(20, 16, 8)				
C_{A34}	1445.6	262.7	(20, 26, 0)	C_{B34}	360	66.7	(20, 26, 8)				
C_{A41}	1093	196	(30, 0, 0)								
C_{A42}	923	196	(30, 6, 0)								
C_{A43}	1004.4	196	(30, 16, 0)								
C_{A44}	1085.6	196	(30, 26, 0)								
C_{A51}	1093	196	(40, 0, 0)								
C_{A52}	923	196	(40, 6, 0)								
C_{A53}	1004.4	196	(40, 16, 0)								
C_{A54}	1085.6	196	(40, 26, 0)								

$$Q_{H3}/20 = 3920/20 = 196 \text{ kN}$$

Shear forces passed from substructure 2 will have to be, respectively, superimposed upon six corresponding columns on substructure 3:

$$Q_{A33} = Q_{H3}/20 + Q_{B33} = 196 + 66.7 = 262.7 \text{ kN}$$
$$Q_{A34} = Q_{H3}/20 + Q_{B34} = 196 + +66.7 = 262.7 \text{ kN}$$
$$Q_{A23} = Q_{H3}/20 + Q_{B23} = 196 + 116.6 = 312.7 \text{ kN}$$
$$Q_{A24} = Q_{H3}/20 + Q_{B24} = 196 + 116.6 = 312.7 \text{ kN}$$
$$Q_{A13} = Q_{H3}/20 + Q_{B13} = 196 + 116.6 = 312.7 \text{ kN}$$
$$Q_{A14} = Q_{H3}/20 + Q_{B14} = 196 + 116.6 = 312.7 \text{ kN}$$

Table 3.2 is the table of summation of estimated forces bore by components, in which bold numbers indicate the greatest internal forces. Sectional area of columns could be estimated with these numbers.

4. Discussion

(1) As for complex structures, they could be disassembled as several substructures, the internal forces of which could be calculated, respectively, and then be superimposed. Calculated results are listed in Table 3.2.
(2) The load in this example is lateral wind or seismic load along axis Y in Fig. 3.41. Due to the small magnitude of I_x in the "footprint" of the foundation, relatively large internal force of columns would have to exist. Besides, vertical loads from the upper two substructures are superimposed and passed downwards. Thus, columns N_{A13}, N_{A14}, N_{A23}, and N_{A24} will have to undertake the greatest axial force and shear force. The greatest axial force estimated is up to 1720 kN, and the greatest shear force is estimated as 313 kN.
(3) If we consider another situation, when lateral wind or seismic load is along Y axis, columns N_{A11}, N_{A12}, N_{A13}, N_{A14}, and N_{A15} will have to undertake the greatest axial force and shear force. Moreover, since they are inclined columns, their axial forces will be increased to 1.25 times. However, since there is no superimposition from the two upper substructures, the greatest axial forces in this case still are smaller than the control axial forces of N_{A13}, N_{A14} in (2).
(4) If the load adopted is lateral wind or seismic load axis X in Fig. 3.41, for that the moment of inertia of the foundation $I_y > I_x$ in the "footprint," the internal forces generated will be smaller than the control axial forces of N_{A13}, N_{A14} in (2).

(5) Further analysis of data in Table 3.2 finds out that the internal forces of each substructure vary to a large extent. Hence, different preliminary internal forces are, respectively, planned for each substructure and the sectional area of columns on each of them. From Table 3.2:

For the columns on substructure 3: The axial force is designed to be 1715.6 kN, and the shear force is 312.7 kN;

For the columns on substructure 2: The axial force is designed to be 630 kN and the shear force is 116.7 kN;

For the columns on substructure 1: The axial force is designed to be 270 kN, and the shear force is 50 kN.

3.7 Overall Structural Performance

"Design for the overall performance of the structure," currently being promoted in structural engineering, is considered an advanced design concept. Such a design enables the structure to better perform in its service condition by way of particular consideration of certain structural system or locale. Consequently, the structure would be safer, more economical, and more user-friendly.

Modern structural design does not only require of sufficient strength and stiffness from the structure, but also calls for a performance-based structural design, which asks for excellent performance from in-service structures. For instance, prestressed structures could improve overall deformation and the force-bearing capacity of in-service structures; when integrated with smart materials, structures would be able to regulate dynamic and static response to external loads and become structures with "vibration control"; when design of a "structural fuse" is added to the structure, failure of the main structure due to extreme loads could be prevented and the safety of personnel and property is guaranteed; when concepts such as "robustness" and "vulnerability" are considered in the design, structures, when struck by impact loads, would not easily yield extensive or complete failure; and when "life circle" is considered in the design, the total cost of a structure in its service life would be brought to the minimum, which contributes to the sustainable development of the society. All these performance-based structural designs are to be introduced in this section.

1. Concept of "prestressed structure" in design

Basic principles of prestress have been applied to construction for centuries [6]. For instance, ropes or metal straps were used to bind the boards together to make buckets, as shown in Fig. 3.42. When tightened, the hoop is prestressed and in turn generates prestress between boards of the bucket. And this prestress would help the bucket counterbalance circumferential tension generated by the internal hydraulic pressure. In other words, hoops and boards are prestressed before the application of loads. The same principle applies to structures. Prestress could significantly improve the force-bearing capacity of the entire structure or the locales. It has been

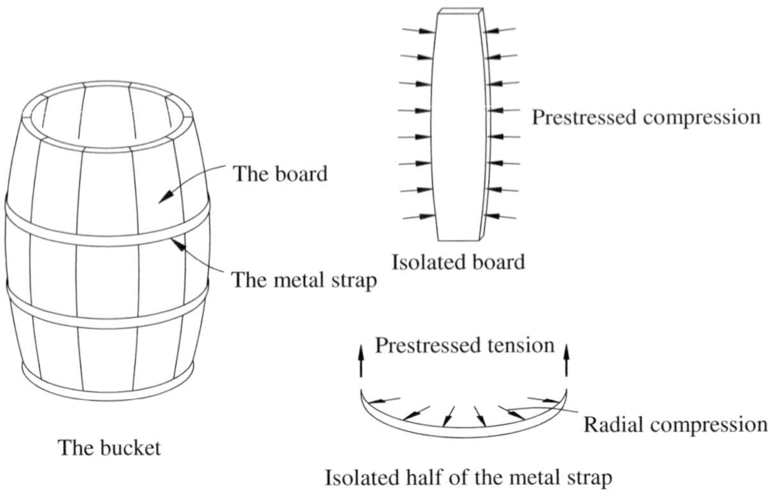

Fig. 3.42 Principle of prestress in bucket making

widely applied to prestressed reinforced concrete structures. Modern structural design learns from the principle of prestress and applies it to the whole structure. For instance, the application of prestress to spatial steel structure could regulate overall stress deformation of the structure.

Prestressed concrete was not a success at first, for that small-magnitude prestress in the reinforcements was soon lost due to concrete shrinkage and creep. The success of modern prestressed concrete began with the application of high-strength reinforcement steels, for that considerable tension could be applied, which in turn would compensate the prestress loss, and provide in-service concrete with effective prestress. It should also be noted that the success in the popularization of prestressed concrete has been closely related to the development of materials and methods of construction.

Prestressed concrete structure refers to the concrete structure with internal compressive stress, the magnitude and distribution of which could offset certain amount of tensile stress and deformation induced by external loads. Pre-established internal compressive stress is usually delivered by tensioned steel.

The mechanism of prestressed reinforced concrete could be explained from three aspects as follows:

(1) Prestress turns concrete into elastic material. Concrete, being weak in tension and strong in compression, is provided with compressive prestress and remains in an elastic state. The compression comes from tensioned high-strength reinforcement steels when trying to shrinkage but prevented by concrete. This compressive state allows concrete be able to undertake certain amount of external tensile force. In this way, concrete has transformed from tensile brittle material to compressive elastic material.

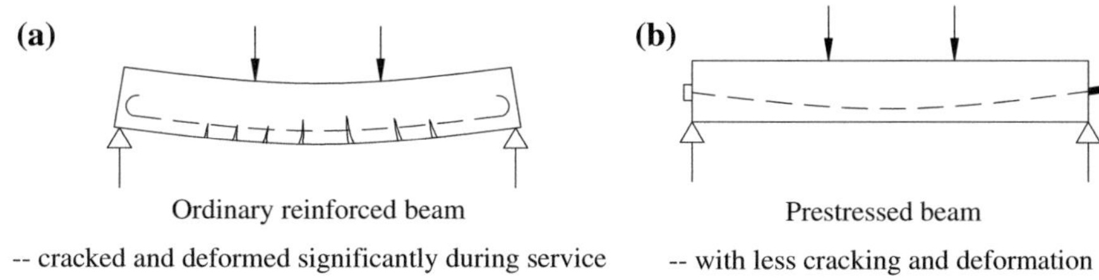

Fig. 3.43 Comparison of stress and deformation between ordinary RC beam and prestressed RC beam [6]. **A** Passive design. **b** Active design

(2) Prestress helps high-strength reinforcements better cooperate with concrete material. Same as ordinary reinforced concrete, reinforcements within prestressed concrete are pulled and the concrete is compressed, the pair of which constitutes a couple to resist external bending moment. Reinforcements in ordinary reinforced concrete usually undertake the tension after the cracking of the concrete and provide relatively small restraint on the cracking and deformation for the concrete. As a result, the concrete, before the strength of the reinforcements was utilized, was already cracked and deformed significantly. And this is a "passive design" (Fig. 3.43a). As for prestressed concrete, compressive stress and strain have been pre-applied to the concrete, which in turn generates tensile stress and strain in high-strength reinforcements. In this case, advantages such as the tension resistance of reinforcements and the compression resistance of concrete could be better utilized as these two materials better cooperate with each other and cracking and deformation of concrete could be lessened after the application of external loads (Fig. 3.43b). And this is an "active design."

(3) Prestress counterbalances external loads as well as cracking and deformation induced by the deadweight of the structure. This is a concept of "balanced design." When the prestress applied could counterbalance the bending moment and deformation given rise to by the deadweight of the structure, such prestress would be considered as prestress of "balanced design." Such prestress is frequently employed in the calculation of prestressed concrete structures, especially in the overall design for large-span structures, in which case "balanced design" has multiplied applications. The reason goes to the enormity of the deadweight of concrete, for such a dead load alone might generate considerable bending moment and cracking for large-span structures.

In sum, prestressed reinforced concrete serves as an effective solution in structural engineering. It effectively combines prestressed high-strength reinforcements with concrete, gives full play to the advantage of each material, and lessens the deformation of the structure. Prestressed reinforcements could transfer the majority of loads to the support and enhance the shear resistance of concrete beams or slabs at the same time. Normal reinforced concrete beam or slab design uses a passive structural design methodology. Beams and slabs under the action of the deadweight and external loads would generate relatively large bending moment and deformation, in which case cracking is unavoidable. Prestressed reinforced concrete beams or slabs use active design approach. Certain prestress is applied to generate upward bending moment and deformation, which would in turn offset a majority of bending moment and deformation induced by the deadweight and external loads. Then, better performance could be achieved with less or no cracking of concrete.

Prestressed concrete structures could be classified by prestress design and construction methods:

(1) Externally and internally prestressed concrete: Prestressed concrete structures generally refer to those with high-strength reinforcements providing prestress from the inside. Prestress could also be applied through the regulation of external supports or reaction forces. For instance, jack could be used to adjust the height of supports so that continuous beams are prestressed (Fig. 3.44).
(2) Line prestress and ring prestress: Ring prestress generally refers to that is applied to circular structures, such as cylindrical tank structures and latticed shell structures. Line prestress refers to such ways that prestressed reinforcements are arranged as in straight lines, folded lines, or curved lines.
(3) Pretensioning method and posttensioning method: Pretensioning method refers to the method that stretches the reinforcements first, then places the concrete. After concrete harden, the prestress transfers through the bonding between reinforcements and concrete. Such method is more often adopted in mass production by precast plants or in laboratory. For posttensioning method, first leave unstressed reinforcements in tubes, then cast the concrete, and finally apply tension to reinforcements after the concrete is hardened. Such method is frequently used in cast-in-place components and prefabricated components.
(4) Prestressed reinforcements with or without end anchor: In posttensioning method, end anchors could be used to fixate reinforcements so that prestress is passed onto the concrete. This is called posttensioning with end anchorage. In pretensioning method, reinforcements pass the compression onto the concrete through the bonding between reinforcements and concrete. This is called pretensioning without end anchorage.
(5) Prestress with or without bonding: Prestress with bonding means that prestressed reinforcements are cemented with concrete, and prestress is transferred through this bonding. Pretensioning without end anchorage must rely on bonding. Prestressed reinforcements in posttensioning with end anchorage could produce prestress with bonding through grout injection or prestress without bonding with no grout injected. Prestressed reinforcements without

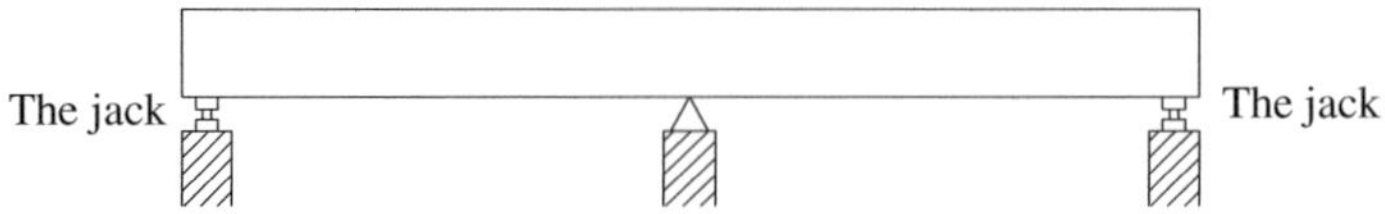

Fig. 3.44 External prestress applied through the regulation of external supports or reaction forces

bonding could go through oiling or galvanizing to prevent rust. Advantages are excellent rust prevention, compensable prestress loss, and replaceable prestressed reinforcements.

Prestressed methods mentioned above have their own distinct features and should be selected in accordance with requirements of design and construction. Generally, several aspects need to be taken into consideration: duct installation approach, component dimension, reinforcement layout, steel stretching and anchorage details, prestress applying equipment and method, etc. Final decision also has to take safety, economy, time, and other factors into consideration.

Comparison of applicability, safety, and economic efficiency between prestressed concrete and normal reinforced concrete is as follows:

(1) Applicability: Prestressed concrete is more suitable for large-span and heavy-loaded structures, for that the deformation of in-service prestressed concrete is quite small which makes it easy to meet the service requirement of the structure. Prestressed concrete structure is light in weight and therefore eligible for structures with the need for aesthetic expression. Moreover, it provides larger net structural space with thinner cross sections. Figure 3.45 shows the deformation comparison between normal reinforced concrete structure and prestressed concrete structure. Flexural deformation of prestressed concrete structure is smaller under the action of constant load. When prestress of "balanced design" is adopted, the structure under the constant load of deadweight nearly displays no deformation, for that the inverted arch effect generated by prestress counterbalances deformation induced by constant load of deadweight. Under the action of live loads, deformation of prestressed concrete structure is also relatively small. Because there is no cracking on the section of prestressed concrete, the moment of inertia to resist external force is 2–3 times greater than that of ordinary reinforced concrete after it is cracked.
(2) Safety: It is unfair to say that one structural form is safer than another. Safety depends more on the whole process of design and construction instead of the form of the structure. On the one hand, prestressed concrete structures do have

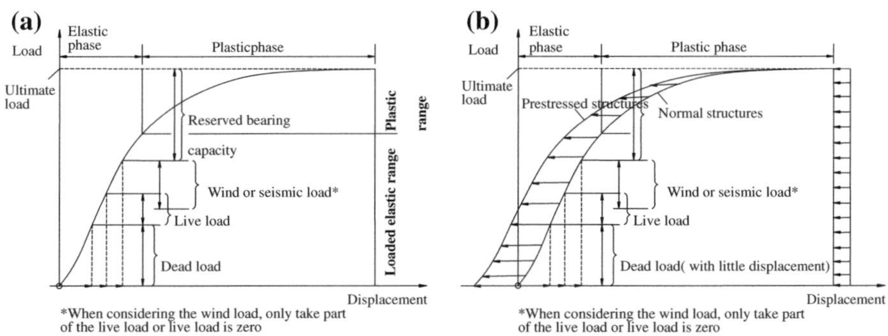

Fig. 3.45 Comparison of load-deformation curves between normal and prestressed structures [6]. **a** Normal structure. **b** Prestressed structure

less cracking, small deformation, less reinforcement corrosion, and longer life span. One the other hand, during the process of prestress application, reinforcements and concrete undertake relatively large internal stress, which also poses challenge to materials. Generally speaking, prestressed concrete and normal concrete have almost the same service life. However, prestressed concrete without bonding facilitates rust prevention and replacement of reinforcements, and therefore, no-bonding prestressed concrete structures would have a longer service life.

(3) Economic efficiency: The strength of both reinforcements and concrete in prestressed concrete structures is high, and hence, prestressed concrete structures are light in the deadweight, which does not only save construction material but also reduces cost in transportation and construction. However, since the unit price of high-strength reinforcements and concrete is higher and relevant equipment and anchorage devices are necessary in the application of prestress, prestressed concrete is not advisable for all structures. Optimal choice should be made after a comprehensive evaluation of economic efficiency and applicability.

Prestress is also widely applied to modern novel spatial steel structures. For instance, ring prestress when applied to large-span latticed shell steel structures could help reduce vertical deformation. Prestress is also indispensable for beam string structures (Fig. 3.17) in the making of originative spatial structural systems.

2. The concept of "vibration control" in structural design

Vibration control is to install certain control device (equipment, substructure, external force application, etc.) in specific location to regulate dynamic performance. In this way, under the action of external dynamic loads, response of the structure is reasonably controlled. Safety of the structure, personnel, appliances, and equipment is better protected.

In the early 1950s, former Soviet Union installed four pendulums in a 100-m-high-steel TV tower, increasing logarithmic decrement from 0.12 to 0.56. Thus, wind-induced vibration response of the tower was considerably reduced. This is the earliest success in dynamic response reduction through vibration control device.

Vibration control is new in structural engineering and still is in the early stage of development and application. Table 3.3 lists vibration control systems and their features. Table 3.4 sums the features and advantages of vibration control.

Another classification of vibration control is to identify with the existence/nonexistence of external input energy, as shown in Fig. 3.46. In line with such a classification, vibration control is categorized as active control, semi-active control, and mixed control [12].

(1) Passive control

Passive control does not need external input energy. Usually, the dynamic features of the structure are improved by adding a subsystem to certain part of the structure, or modifying certain local components.

Table 3.3 Vibration control systems and their features

Vibration control systems		Introduction	Classification	Range of application
Vibration isolation		Lateral soft support is installed between the structure and foundation to prolong vibration cycle of the system and keep distance from preferred period distribution of seismic effect, in which case resonance would not take place	Rubber bearings	Buildings and bridges that need high-level security in earthquakes
			Sliding-frictional bearings	
			Suspension and sliding isolation	
Energy dissipation	Dampers	Dampers are installed in structures to consume seismic input energy	Friction or metallic damper	Multi-storied, high-rise, ultra-high-rise structures with small lateral stiffness, towers, large-span bridges, etc.
			Tuned mass damper	
			Viscous, viscoelastic damper	
	Energy dissipation components	Energy dissipation apparatus are installed in structures to dissipate seismic input energy	Energy dissipation supports	
			Energy dissipation shear walls	
Vibration control		Sensors are installed in structures to collect and transmit data of vibration to the control system which would react correspondingly	Active control	Structures that need higher level of protection against wind and earthquakes
			Semi-active control	
			Mixed control	

Vibration isolation system installs certain vibration isolation apparatus between the upper structure and the foundation so as to reduce seismic energy that is being transferred upward and ultimately minimizes vibration. Vibration isolation apparatus has to have features as follows: considerable deformation capacity, sufficient initial stiffness, and strength, providing substantial damping and excellent energy-consumption capacity. Vibration isolation above foundation significantly lowers natural frequency of the structure and is fit for low-/medium-rise structures and rigid structures. It is not applicable to high/ultra-high-rise structures.

Table 3.4 Features and advantages of vibration control

	Comparison between old and new vibration control system		Advantage of vibration control
	Conventional vibration control	Vibration control	
Meaning and means	Constitutionally strengthen the structure through the enhancement of local components	Modify dynamic features of the structure via increment of flexibility through isolation, energy dissipation, or control	Effective vibration absorption, fewer restrictions on design, and easy maintenance
Basis of design	Fortification intensity	Sudden high-intensity seismic force is considered	Safety is ensured
Target of protection	The structure alone	The structure, equipment, decoration, etc.	Social requirements are met
Range of application	Optional for structures of novel design	Eligible for new structures, anti-seismic improvement of old structures, general structures, as well as those of significance	Much wider range

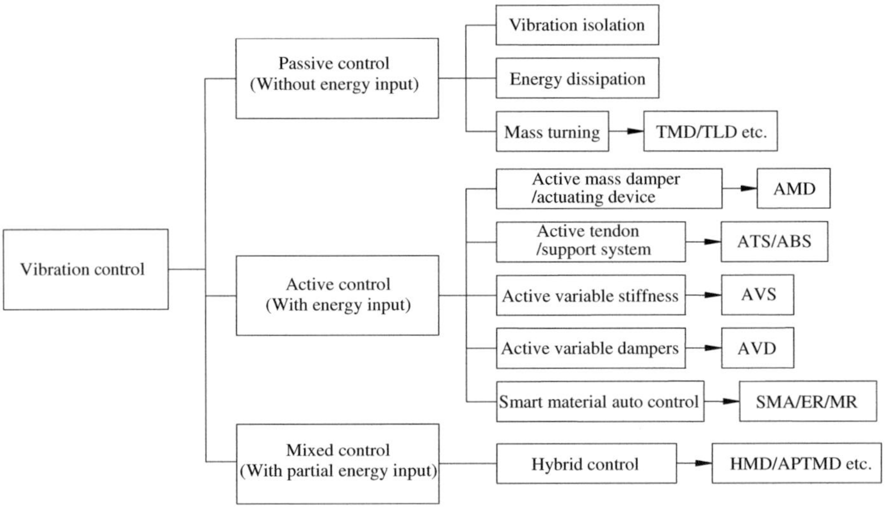

Fig. 3.46 Classification of vibration control methods [12]

Energy dissipation system designs some non-structural components or dampers to be energy dissipation units. Under the action of wind load or small earthquake, energy dissipation units (or dampers) are in elastic state and the structural system still has sufficient lateral stiffness to meet the requirements of structure in its service condition. Under the action of strong wind or severe earthquake, energy dissipation units (or dampers) will first enter a non-elastic state, producing large damping and dissipating energy, and reduce dynamic response of the structure. With reference to the difference in energy consumption units, energy dissipation system is divided into the following:

Energy dissipation components, such as energy dissipation supports, energy dissipation shear walls, and energy dissipation joints.

Dampers are subdivided into:

- speed-dependent dampers, such as linear viscous damper or viscoelastic damper;
- displacement-dependent dampers, such as metallic yielding damper or friction damper;
- tuned vibration-absorbing dampers, such as tuned mass damper or tuned liquid damper. Tuned mass damper is inertial mass added to the top of the structure and connected with the main structure, regulating dominant vibration mode through mass shifting. The inertial mass added usually is the water tank, mechanic room, or rotating restaurant in high-rise structures.

Passive control does not need external energy. Thanks to the simple mechanism, easy installation, economic efficiency, reliability, and satisfactory inhibiting effect on multiple vibration modes, it has been applied in many structures. Nonetheless, passive control only has damping effect on certain given dynamic features and is devoid of capabilities such as tracking and regulating, in which case its control effects are obviously dependent upon the characteristics of frequency spectrum of in-put stimulation and dynamic features of the structure. Besides, durability research, installation, maintenance, and replacement of apparatus are issues remaining to require more research work.

(2) Active control

Active control is a vibration control system with external energy input. The mechanism is given as follows: (1) Sensors monitor external disturbance and dynamic response of the structure; (2) based on data collected, computer system (controller) calculates the magnitude of control force needed to be applied, based on precise vibration model of the structure and active control algorithm; and (3) optimal control force is finally applied to the structure by external energy input. The whole process is shown in Fig. 3.47. The active control based on feedback observation is called feedback control or closed-loop control. The active control-based environmental interference observation, e.g., ground movement is called feed-forward control or open-loop control. If external interference and structural response are observed simultaneously, we call it open-/closed-loop control. Open-loop control requires less real-time calculation, but needs high-precision units and has low signal-to-noise ratio. Closed-loop control demands much more real-time computation capacity, but it does not require as-precise sensor. In addition, it has high signal-to-noise ratio and continuous monitoring capability.

Active control has continuously variable control force and adapts for a vast range of frequency. Satisfactory effects could be achieved using modern control

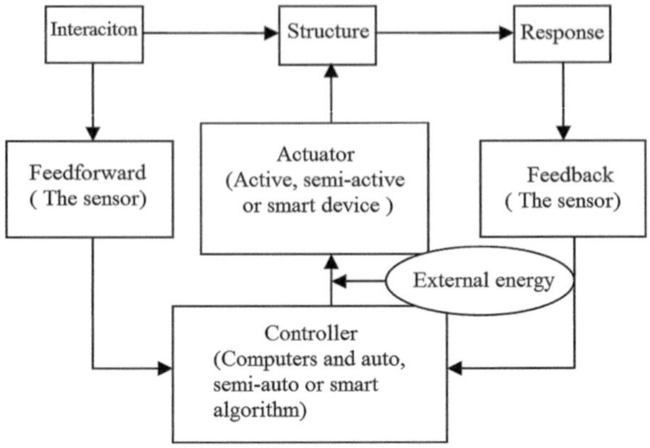

Fig. 3.47 Principle of active control

theory, which has been a very active research area [13–15]. However, its disadvantage is also obvious. Since structures which require vibration control are usually vast in volume and heavy in mass, high-power actuator and energy storage equipment are necessary to conduct control actions. Moreover, accurate active control algorithm is built upon precise structural vibration model. However, the question is how precise can a structural model can be. Compared to force-bearing components, effects of non-force-bearing components are usually ignored in design calculations. Nevertheless, vibration control targets at existing structures, in which case non-force-bearing components and mass change have major influence upon the vibration models, and as a result, significant changes in dynamic features might be possible. For the moment, a majority of active control still employ simplified dynamic model of the structure with only major force-bearing components included, which will undoubtedly affect the effectiveness of active control and in extreme cases might lead to the instability or even magnify the dynamic structural response. And this is the major reason behind the rare application of active control in actual structures. In addition, how to offset delay in control force application remains one of several issues that need to be solved.

(3) Semi-active control

Semi-active control has both adjustable parameters of active control and reliability of passive control. This system usually takes passive control as the primary system and monitors vibration response as well as live load information. Then, it regulates parameters (such as damping and stiffness) with minimum energy to achieve optimal-state-oriented dynamic performance. Thus, energy needed by semi-control is far less than that needed by active control.

Semi-active control, in essence, is a parameter control method. It reduces the vibration of the structure through the modification of stiffness or damping coefficient. Therefore, it has two ways of control: active variable stiffness system (AVS) and active variable damper (AVD).

(4) Mixed control

Mixed control installs two or more control devices in one subject (a structure or a structural system comprised of numerous substructures or adjoining structures) so as to utilize the advantages of different control system. According to the role control devices play, there are two typical combinations:

(i) Principal–subordinate combination. It has one principal control through which the control of the structure by other controls is realized. For instance, if passive control serves as the principal control in control system, active control will ease its displacement control on passive control and provide necessary resilience for it.
(ii) Parallel combination. Two controls work independently to regulate the structure. Passive control provides protection against frequent load to the structure, while active control provides protection against rare seismic load and serves as the last line of defense before destruction.

Judging from the feasibility and performance of employing two or more control systems, mixed control outweighs active control and passive control. Nonetheless, the installation of two control systems on one structure might not be economic. However, there is not a lot real-world application. This is due to that more research is needed in the area of system selection, design, and cooperation to achieve optimal vibration control performance.

3. The concept of "structural fuse"

As a common sense, electrical appliances have to integrate fuses in their design so that the appliances are protected from strong currents. This concept is transferable to structural design. Modern structural design does not only have to meet requirements of strength and stiffness but also have to take performance of the structure into consideration. In particular when the structure is under the action of strong earthquakes or storms, the concept of "fuse" in electrical appliance design could be applied to structures. In this way, the main structure is protected against strong earthquakes or storms. When the structure is in its service condition, structural fuse should be intact and make sure the overall force bearing and operation of the structure are up to the standard. When strong earthquakes or storms strike, structural fuse will absorb the energy, and fail and alter damping coefficient and stiffness of the structure. Then, dynamic response of the structure is altered, and major force-bearing components remain intact or only suffer a minimum damage. In this way, the safety of personnel and property within the structure is protected, and the restoration effort afterward is minimized.

Designs of common structural fuses are as follows:

(1) Weakened column-beam connections. This is to use a dog bone connection to reduce the sectional area of a beam connected to the column in a frame structure (as shown in Fig. 3.48). And this purposely weakened area will first transform into a plastic hinge, which will absorb much more energy when deforming under the action of strong seismic event. It attracts energy from

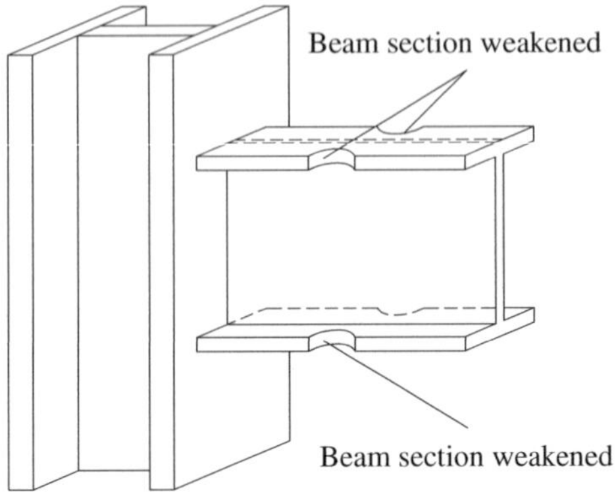

Fig. 3.48 "Dog bone" connection at joints

external loads and changes the damping, stiffness, and ultimately dynamic response of the structure. In this case, those weakened connections are the "fuse" of the structure. Major force-bearing components (columns) remain intact or suffer a minimum damage. After the event, the damage could be restored with less effort.

(2) Variable dynamic feature. Connection beams with relatively small magnitude of strength and stiffness are often installed between the inner and outer tube in tube-in-tube structures. These beams serve as the "fuse" of the structure (Fig. 3.49). When strong earthquakes or storms strike, connection beams fail and absorb a great amount of energy. In addition, outer tube separates from the inner tube. As a result, great changes in damping coefficient and stiffness of the structure take place, which alter the dynamic response. Then major force-bearing components (the inner and outer tube) remain intact or suffer minimum damage. The most well-known example with such a "fuse" installed is the Bank of America Building designed by T.Y. Lin, which survived the Nicaragua earthquake, as shown in Fig. 3.50. On December 23, 1972, when the earthquake struck Managua, Nicaragua, most structures collapsed while

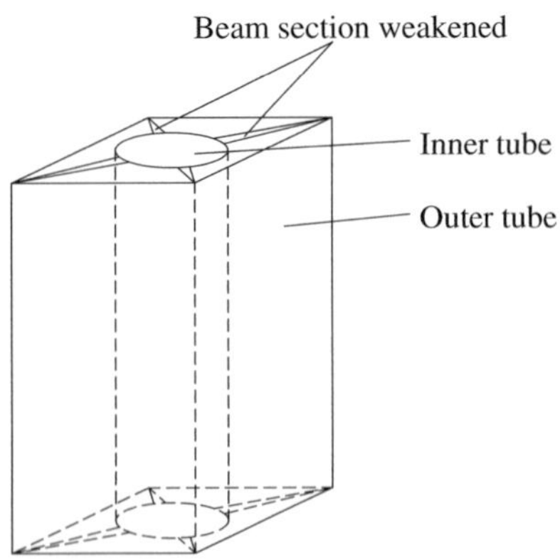

Fig. 3.49 Tube-in-tube structure with connecting beams

Fig. 3.50 Bank of America survive in the epicenter [6]

Bank of America building designed by T.Y. Lin International remained structurally sound. During the earthquake, a peak ground acceleration of 0.35 g was recorded, which is six times than the designed ground acceleration (0.06 g). After the earthquake, connecting beams designed with reduced section were found failed and allowed the inner and outer tube to vibrate independently. Thus, lateral displacement was magnified, seismic response was alleviated, and the structure maintained its structural soundness.

(3) Energy absorbing shear wall. Specially designed "shear wall with slits" could be used as the "fuse" of shear wall structural system (Fig. 3.51). Shear wall structural system is designed to resist lateral force and to increase lateral stiffness. Slits are engineered in the shear wall so that lateral stiffness of the shear wall is reduced. In this way, when the structure is under the action of extreme lateral loads (major earthquakes or severe storms), shear failure would first happen to those shear walls with slits; considerable shear deformation

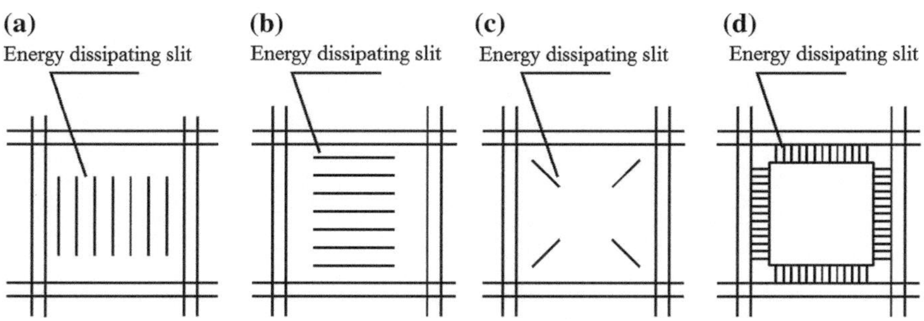

Fig. 3.51 Shear wall with slits. **a** Shear wall with vertical slit, **b** shear wall with horizontal slit, **c** shear wall with skewed slit, **d** shear wall with periphery slit

would take place, absorbing energy produced by earthquakes or storms, leaving major force-bearing components intact or minimum damaged. In comparison, sudden brittle failure is prone to happen to solid shear wall when massive earthquake hits due to its large lateral stiffness and small deformation, which easily leads to devastating structure destruction of the structure and personnel casualties.

(4) Knee bracing support. Lateral resistance of a steel frame could come from central support, eccentric support, or knee bracing support (Fig. 3.52). Knee bracing support is a typical "fuse" in structural design [16, 17]. Under the action of major earthquakes or storms, the wreckage of central support or eccentric support of a steel frame would result in severe destruction of the whole structure, leaving the safety of personnel and property in peril and adding difficulty to restoration. In comparison, under the same situation, knee bracing support, having lower strength and stiffness than the major components, would first transform into plastic hinge so as to absorb seismic energy and change damping and stiffness of the structure, in which case dynamic response of the structure is alleviated and major components (columns and beams) remain intact or suffer a minimum damage. Knee bracing support comes in several styles, as shown in Fig. 3.53.

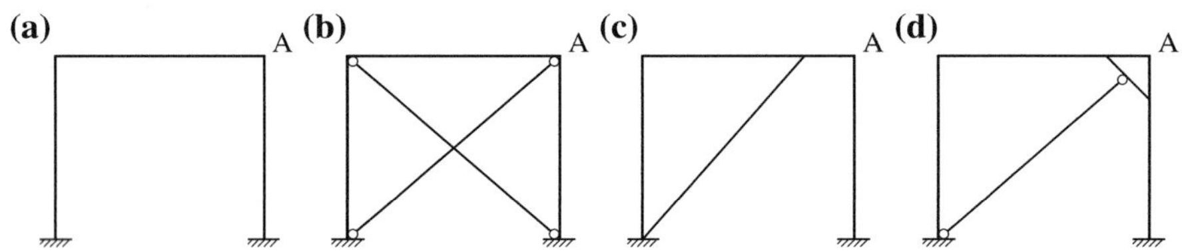

Fig. 3.52 Conventional supporting systems for steel frames. **a** Unbraced frame. **b** Centrically braced frame. **c** Eccentrically braced frame. **d** Knee-braced frame

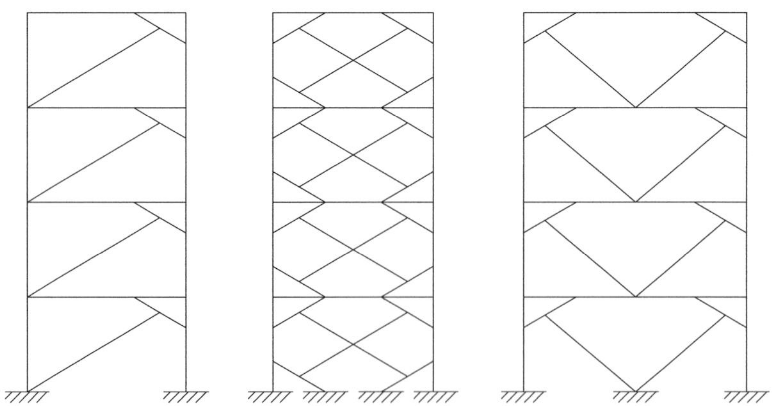

Fig. 3.53 Several styles of knee-braced support

In sum, the concept of "fuse" in structural design is in essence a measure of vibration control, "energy dissipation" passive control to be exact.

4. The concept of "robustness" and "vulnerability" in structural design

On September 11, 2001, twin towers of New York World Trade Center collapsed cast catastrophically. The incident left much for engineers to ponder over.

Why did World Trade Center (WTC) collapsed after the collision with the plane, while Pentagon was only partially damaged by the other hijacked plane the same day? Judging from the consequences of these two incidents, the latter was only damaged, while the former was, of quite another level of severity, catastrophically destroyed. Evidently, devastating catastrophe would take a heavy toll on lives and properties: About 3000 lives were taken and more than 6000 injured.

Structural "vulnerability" refers to the high sensitivity of the structure to the damage, meaning under the action of sudden external factors failure of some local components gives rise to disproportionate (magnified) destruction or complete collapse of the structure. Vulnerable structures easily give occasion to calamity or even total destruction. As for structural "robustness," it means under the action of sudden external factors structures are not quite responsive to the damage since only local components are failed, in which case the restoration is likely facilitated. Structures with sound robustness hardly respond to sudden external damage and would not lead to fatal destruction. Pentagon suffered limited loss from its collision with hijacked plane. From the ready collapse of twin towers of WTC, we know that they were structurally vulnerable (refer to the example and discussion in Sect. 5.6 to see the WTC analysis).

Figure 3.54 is a diagram of New York WTC. The inner and outer tubes serve as the main force-bearing component. Loads on the floor are transferred to secondary truss beam and then primary truss beam, from the ends of which loads are passed on

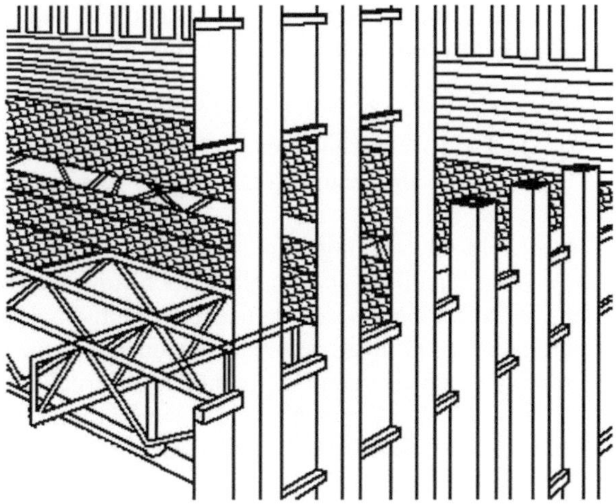

Fig. 3.54 Diagram of New York World Trade Center [9]

to the inner and outer tubes. Since the ends of primary beams on the upper floor are supported by columns underneath in the outer tube, when one floor is overloaded and suffers damage of floor slabs and primary beams induced by sudden occurrence, joints connecting the end of primary beams and columns in the outer tube are wrecked, and as a result, columns underneath are destroyed (buckling failure of slender columns). Meanwhile, loads on this floor fall upon the floor underneath, which results in extra overload and leads to the buckling failure of columns underneath and so it goes. Thus, we know that New York WTC does have this "implicit damage chain." However, structures with such "chains," without triggering incidence, could function quite normally. Note that the concept of implicit damage chain is put forward for the very first time by the author so that it could be used in qualitative analysis of "robustness" and "vulnerability" of the structure.

"Implicit damage chain" refers to the defects within the internal constitution of the structure, which would set off a chain reaction once damage of local components occurs and lead to the wreckage of next force-transferring component, until the load reaches the foundation [9]. If the wreckage of one component would result in that of the next one and this chain-like damage has the potential to produce massive destruction or even total collapse, then there exists an implicit damage chain. We call it "implicit" for that when the structure is in its service condition components on this chain could normally transfer forces and guarantee a safe function of the structure. Therefore, whether the "implicit damage chain" exists and whether it would be triggered by some destructive factors are the key issues. Making the wreckage of component i the destroyer of the next-in-line component $i + 1$, the wreckage pattern of components along this "implicit damage chain" is:

$$[\text{Structural component}]_i \rightarrow [\text{structural component}]_{i+1} \rightarrow \text{foundation}$$

If the wreckage of component i would lead to the destruction of the next component $i + 1$ in the same direction, then "implicit damage chain" exists within the structure, and the structure is regarded as "vulnerable."

Discuss the constitution of cable-stayed bridge. Loads are transferred in such a direction:

$$\begin{aligned}\text{Deck} &\rightarrow \text{transverse beams} \rightarrow \text{longitudinal beams} \rightarrow \text{stay cables}\\ &\rightarrow \text{the main tower} \rightarrow \text{pier foundation}\end{aligned}$$

In a cable-stayed bridge, lateral beams provide support and prevent longitudinal beams from buckling, and longitudinal beams transfer loads to stay cables. Thus, once transverse beams are damaged, longitudinal beams are prone to buckling, which might result in the failure of stay cables on the one side, flexural failure of the main tower, and eventually a complete destruction of the bridge. Therefore, we know that "implicit damage chain" exists within the bridge.

Take suspension bridge shown in Fig. 3.55 as an example. Loads are transferred in such a direction:

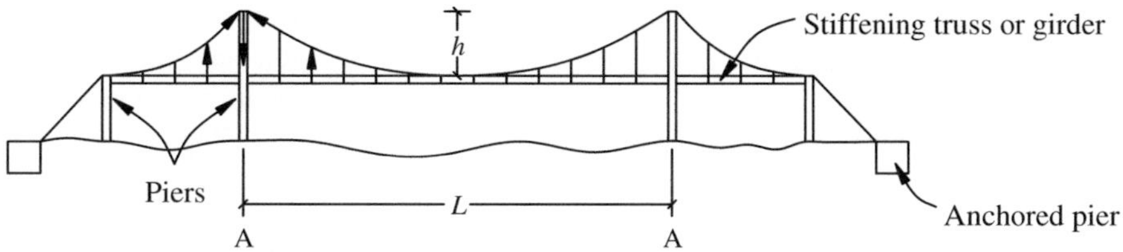

Fig. 3.55 Suspension bridges

$$\text{Deck} \rightarrow \text{transverse beams} \rightarrow \text{longitudinal beams} \rightarrow \text{hangers} \rightarrow \text{suspenders} \rightarrow \text{main towers} \rightarrow \text{pier foundation}$$

On the other hand, failure of one component in force-transference direction would not necessarily give occasion to that of the next component and damage might only be inflicted upon components before the failed one. In this case, components after the failed one in this force-transference direction would constitute "the main structure" relative to the destroyed parts ahead and some parts of the structure get to remain intact. Hence, "implicit damage chain" does not exist in this structure.

Generally speaking, the failure of one component will be accompanied by damage of different range. Compare the constitution between suspension bridge and cable-stayed bridge. When some of the transverse beams fail, thanks to the nonexistence of "implicit damage chain," suspension bridge suffers quite limited damage, which is confined to the adjoining transverse beams. The safety of the whole bridge would not be endangered, and repair should not be difficult. In comparison, when some of the transverse beams fail in cable-stayed bridge, due to the existence of "implicit damage chain," chain destruction of the entire bridge might take place, against which we should always be on guard.

Authors of this book believe that the existence/nonexistence of "implicit damage chain," whether the "chain" could be triggered, and whether effective interrupt mechanism exists have direct influence upon the "robustness" and "vulnerability" of the structure.

Accordingly, "implicit damage chain" should be avoided in structural design. When such a "chain" is unavoidable, it is of great significance to stop the "implicit damage chain" from being converted into "explicit damage chain." The importance of each component to the constitution of the structure could be analyzed, and the importance of coefficient be calculated. Protection of components of more importance or improvement on reliability designed for such components could effectively prevent "implicit damage chain" from being triggered. More research could be conducted over the interrupt mechanism of "damage chain" so as to confine the damage to a limited extent. For instance, isolation of potentially damageable components and installation of settlement cracks or induced cracks would restrain the collapse or damage of the structure within certain scope, which is also being the concept of interrupt mechanism of "damage chain" in structural design. Further discussion over interrupt mechanism of "damage chain" is in Sect. 6.7.

In sum, concepts such as "robustness" and "vulnerability" of the structure should be included in structural design [9]; investigation and research on structural constitution could help to design "robust" structures and avoid chain destruction.

As for conventional structures, structural measures could be adopted to reduce vulnerability. However, when structures of more significance are concerned, structural measures alone might not be sufficient and other measures should be explored. Measures of such kind are summed up as follows: alternative path of load, enhanced key components, installation of energy-absorbing apparatus, and event control, which could also be adopted in combination, when necessary [18, 19].

(1) Alternative path of load. To avoid sudden wreckage of key components, alternative path of load is often engineered, which is quite simple and efficient. This approach will first "conceptually remove" key components and equip the structure with an alternative path of load. In other words, redundant connections are constructed and when key components fail alternative path of load would step in.
(2) Enhanced key components. Through analysis of contingencies, key components are determined and strengthened by design. Load designed in this case is not the conventional type but rather potential load that could take place in contingencies such as explosion, collision, and fire. As it should be, economic efficiency has to be taken into consideration in the process.
(3) Installation of energy-absorbing apparatus. To strengthen robustness of the structure from the perspective of energy dissipation is another effective approach. The installation of energy-absorbing apparatus in the structure could absorb or dissipate energy induced by contingent load and then prevent the structure from premature collapse. Concepts of "vibration control" and "fuse" of the structure, mentioned in the previous section, are classified in this category.
(4) Event control. Above-mentioned are all structural approaches to reduce vulnerability. An alternative approach would be the event control. Measures of prevention against direct application of contingent load to the structure have their difficulties in construction. Nonetheless, thanks to their outstanding efficacy, measures of such kind have been much successfully applied in practical constructions. For instance, isolated islands could be constructed so as to prevent pier foundation of bridge from accidental collision with cargo ship; explosion-proof wall or water pool could be constructed around structures of significance to isolate and protect; anti-bird net has been installed above the "Water Cube" National Aquatics Center to prevent inflatable membrane from being clawed by birds, etc.

For the moment, quantitative design concerning "robustness" and "vulnerability" of the structure is still in research with some achievements [19, 20], while qualitative measures have been much applied in construction. Furthermore, the concepts deserve more attention from industry and academia.

5. Concept of "whole life circle" in structural design

General objective of construction is to achieve economically advisable function. Construction cost of a structure alone is not sufficient to demonstrate economic rationality. An extensive performance of all parts in whole life circle should be systematically taken into consideration. In other words, economic efficiency of initial construction should not be the only issue; "management of reliability in whole life circle" for construction projects should be concerned. "Life circle" design for a structure is only part of "management of reliability in whole life circle" of construction projects.

(1) "Management of reliability in whole life circle" of construction projects

"Management of reliability in whole life circle" refers to management of reliability of all construction processes in every state of the structure. Take a project of highway network as an example. Seven states are included: survey, design, construction, inspection, operation management, maintenance, demolition.

"Survey" means to conduct feasibility research for the establishment of highway project and collect necessary original data of economy, society, topography and morphology, geology investigation, etc.

"Design" includes feasibility research for the establishment of highway project, conceptual design, preliminary design, and detail design of each branch of this project (such as highway, bridge, tunnel, and communication in highway network system). Design of "whole life circle" is emphasized here.

"Construction" is the entire process of execution after the stage of comprehensive design.

"Inspection" means to observe and inspect regularly through the life circle of projects (structures) to make sure they work in a safe and normal way. When traffic engineering is concerned, such as in railway, highway, harbors, and airports, inspections could be quite strenuous and they are actually the majority of routine management.

"Operation management" refers to the routine management that maintains a normal operation of project after the completion of construction.

"Maintenance" is based on "inspection." When damages or problems that might affect normal function of the project (structure) within its life circle are identified, maintenance has to step into retain or restore a normal working condition.

"Demolition" refers to the work pertaining to waste transportation and environment restoration entailed by on-site dismantling of the structure due to the expiration of its life circle or the alteration of function. Previously, some structures that had not taken demolition into consideration in design spent large amount of money on it when the time came, which serves as a lesson for us to learn.

"Management of reliability in whole life circle" of construction project will consider not only the cost of construction but also the reliability, economic efficiency, and rationality of the project in its whole life circle and the basic principles of such a management are as follows:

(i) Overall requirements are met in the respect of operational functions;

Social benefit of the project is maximized;

(ii) Total cost of construction, installation, and demolition is minimized;
(iii) The project is environmentally friendly and facilitates sustainable development of society.

"Management of reliability in whole life circle" of construction project starts from feasibility research and major factors that might affect the establishment of the project are as follows:

(i) Function and nature of the project;
(ii) Scale of investment, stage of construction, objective, and corresponding investment in each stage;
(iii) Site condition, geology, geomorphology, geology, ecology, resource, labor, and other conditions of the project.

On the basis of project feasibility research, subsystems and their corresponding facilities could be arranged. Specifically speaking, structural constitution, general layout, transportation, power supply, and on-site management are all included, and they are the basis upon which systematic model of the project will be built. Theoretically speaking, mathematical methods could be applied in this stage to search for an optimized solution for this systematic model. However, large-scale systematic model in this way is too complicated, and for the moment, it only has theoretical significance. Many more cases of optimization of practical constructions began with feasibility research of the project, took into consideration the opinions of proprietors (investors) and designers, and derived a comparatively "satisfactory solution" (instead of the theoretically optimized solution) as the optimal plan.

(2) "life circle" design of structures

"Whole life circle" design of structures is only part of "management of reliability in whole life circle" of construction projects. In the early stage of design, feasibility and rationality regarding the whole process of design, construction, maintenance, and demolition of the structure have to be analyzed. Thanks to the clear-cut objectives, "whole life circle" design of structures is more operable than "management of reliability in whole life circle" of construction project.

The following principles should be abided by during the implementation of "whole life circle" design of structures:

(i) When selection of structural type, equipment, and material is concerned, priority is to be given to the principle of sustainable development on the basis of economic efficiency and rationality of "whole life circle" process of the structure. Take high-rise structural design for instance. Multiple designs are available for adoption, such as frame structure, shear wall structure, frame-shear wall structure, frame-tube structure, and tube-in-tube structure. Meanwhile, various materials are eligible to be construction material, such as reinforced concrete or steel. Selection of both type of structure and construction material should be based upon principle of sustainable development, which emphasizes practicality of functions and rationality of economy.

(ii) Selection of structural type is to be conducted with synthetic consideration of requirements of functions and economic efficiency, terms of construction, cost of material, and maintenance. For instance, when selecting for a large-span spatial structure, comparison could be made from the aspects mentioned above between arch structure, thin-shell structure, suspension structure, membrane structure, and inflatable structure.

(iii) During the selection of a structural type, engineers need to cooperate closely with architects and endeavor to meet functional requirements of the structure in the aspects of aesthetics, green environment, and ecological energy-saving.

(iv) In this stage, techniques and conditions of construction are factors of concern, the decision of which has to be made upon consultation of professionals from structural engineering, architecture, construction techniques, and construction management.

(v) For the moment, when installation is concerned, priority should be given to energy saving and consumption reducing. In material selection, thermal insulation material could be a perfect choice to reduce long-term energy consumption from air-conditioning.

(vi) Structural type selected is supposed to meet requirements of technique and economic efficiency, having the structure equipped with best operational functions so as to achieve best possible social and economic benefits. Note that economically optimized design might not be equally rationalized, for that it could only provide a norm-abiding and least costly low-level plan. Such a design might increase the cost of maintenance within the service life of the structure, lower anti-risk capacity of the structure, and shorten its service life. The cost of secondary disasters induced by failure of the structure might be much higher than that was saved by material economization, which is attention-worthy in "whole life circle" design of structures.

3.8 Conclusion

In this chapter, classification of structural systems is first introduced so that readers could have an all-round grasp of structural types and their corresponding force-bearing characteristics. Composition characteristics of various vertical high-rise structural systems, lateral large-span structural systems, and spatial structural systems, together with the force bearing and force transfer by their corresponding substructures, have been introduced. Control of lateral stiffness and strength is stressed in vertical high-rise structural system; that of vertical stiffness and strength in lateral large-span structural system; and that of lateral and vertical stiffness and strength in spatial structural system.

Readers get to know specific structural composition and layout of all kinds of structural systems in this chapter, which includes:

Vertical structural system mainly refers to frame structure, shear wall structure, tube structure, and vertical mixed structure (frame-shear wall structure, frame-tube structure, and shear wall-tube structure), etc.;

Lateral structural system mainly refers to simply supported beam, continuous beam, truss structure, arch and portal-rigid frame structure, suspension and cable-stayed structure, lateral mixed structure (arch and suspension structure, arch, and cable-stayed structure, truss and suspension structure), etc.

Spatial structural system mainly refers to spatial truss structure, space latticed shell structure, membrane structure, and other new spatial mixed structure (beam string structure, retractable roof structure).

The introduction of overall force analysis is the highlight of this chapter. "Force flow" analysis could analyze vertical transfer of internal forces in a visually direct way, and "force flow" has its impact on elevation arrangement of the structure. "Footprint" is used to analyze plane arrangement of the structure, for that it reveals relation between structure and substructures as well as that among substructures. Concepts such as aspect ratio and balanced design touch upon factors and approaches that need to be taken into consideration when a rationalized overall design of the structure is concerned.

Through the demonstration of example in which complex structure is dissembled and calculated, respectively, readers are provided with the means to gradually reach deep into the nature of the problem and grasp the approach to dissemble a complex structure and that to assemble simple structures or substructures into a complex structure.

In-depth discussion has been conducted over several influential concepts regarding modern performance-based design, such as "prestress," "fuse" of the structure, "vibration control," "robustness," and "vulnerability" of the structure, and "whole life circle" design of structure. Concepts and influencing factors of "robustness" and "vulnerability" of the structure have been elaborately introduced with the impact of existence, trigger, and interrupt mechanism of "implicit damage chain" included, for these two properties of structure have received universal attention from structural engineers after the collapse of WTC in "911" attack.

Chapter 4
Systematic Analyses of Substructures

4.1 Analyses of Structural Components

4.1.1 Analysis of Horizontal Components

1. **Shape and size of cross section**

Substructure system is composed of horizontal and vertical components. The former mainly refers to slabs, beams, secondary beams, primary beams, and trusses. Slabs and trusses could be treated as beams in approximate calculation. Generally, horizontal components bear distributed or concentrated loads along their length within the plane. Concentrated reactions at supports, along with vertical loads, generate shear forces, bending moments, and sometimes torques or axial forces in horizontal components. In the early stage of design, since torques or axial forces that horizontal components undertake are of small magnitude, the influence of these forces could be ignored. Major attention should be paid to structural resistance to shear forces and bending moments. Horizontal components made of steel commonly have cross sections shaped as I, H, L, T, round, or square tube. As for concrete horizontal components, rectangle, T, H, round, or square are popular choices when it comes to the selection of shapes of cross section.

First step in preliminary design of horizontal components is to estimate span-depth ratio; the second is to calculate bending moments and shear forces that horizontal components undertake; the third is to estimate actual dimensions of cross sections so that the requirement of strength could be met on the basis of estimated span-depth ratio and deduced bending moments and shear forces; and the fourth is to check whether the deformation restrictions of components are met and to increase the sectional area if requirements are not met.

In the early stage of design, span-depth ratio of horizontal components has to be estimated by experience to achieve a reasonable design of sectional area. Span-depth ratio varies according to type of material, type of load, shape of

S. Lin and Z. Huang, *Comparative Design of Structures*,
DOI 10.1007/978-3-662-48044-1_4

sectional area, supports, and span. For instance, span-depth ratio of a steel beam is greater than that of a concrete beam. When the load on the beam is of large magnitude, a relatively small number is taken for span-depth ratio. When the section is in rectangle shape, a relatively large number is selected for span-depth ratio. When the supports of a beam are rigidly constrained, its span-depth ratio could be greater than that of a simply supported beam without constraints. When the span of beam is of great length, a relatively small number is taken for its span-depth ratio. Span-depth ratio of a continuous beam is greater than that of a simply supported beam. Span-depth ratio of a cantilever beam could be taken as 1/3 of that of the simple beam.

Conventional span-depth ratios of horizontal components are listed in Table 4.1. When it comes to actual application, preliminary selection of span-depth ratio should be conducted in accordance with the specific conditions of components so that the story height could be estimated and feasibility of early structural plan was analyzed.

Proper selection of span-depth ratio is determined by various factors, including stress state of structure, deformation limitations, vibration control, spatial requirements, and feasibility of construction. When the span-depth ratio is of small magnitude, height of beam is relatively large, which should benefit structural resistance to the bending moment and shear force along with deformation control; meanwhile, it has adverse effects upon reduction of thickness of story height and adds to the overall height of structure as well as the cost of construction. Therefore, a relatively large span-depth ratio should be selected when requirements of structural resistance to the bending moment and shear force as well as deformation control are met. Contradictory as they are, a solution could be derived from a synthetic consideration of all aspects.

2. **Diagrams of bending moment and shear force**

Quick and accurate calculation of maximum bending moment and shear force, as well as drawing diagrams of bending moment and shear force, serves as an indispensable expertise of a civil engineer. This section explains some fundamental rules regarding diagram of bending moment and shear force.

Table 4.1 Conventional span-depth ratios of horizontal components

Component	Span-depth ratio (L/d)
Concrete primary beam	10–15
Concrete secondary beam	20–25
Concrete solid slab	25–30
Prestressed concrete primary beam	15–25
Prestressed concrete secondary beam	30–35
Prestressed concrete slab	35–45
Steel I-shaped primary beam	15–25
Steel I-shaped secondary beam	25–35

Figure 4.1a shows the diagram of bending moment and shear force of a simply supported beam without end constraints under the action of distributed load; as for the former, it displays a parabola with a ($ql^2/8$) mid-span maximum bending moment. Figure 4.1b shows the diagram of bending moment and shear force of a continuous beam with rigid end constraints under the action of distributed load (it has the same diagram of shear force with simply supported beam); as for the diagram of bending moment, it is still parabolic with a ($ql^2/24$) mid-span maximum bending moment and a ($ql^2/12$) maximum negative bending moment at supports. Note that the sum of mid-span maximum bending moment and absolute value of maximum negative bending moment at supports is ($ql^2/8$) and it is equal to mid-span maximum bending moment of a simply supported beam. In fact, the diagram of bending moment of a continuous beam is derived from moving that of a simply supported beam upward by ($ql^2/12$) due to the rigid end constraints. Figure 4.1c shows the diagram of bending moment of a continuous beam with semi-fixed-end constraints under the action of distributed load with a ($ql^2/16$) maximum mid-span bending moment and an identical maximum negative bending moment at supports. Note that the sum of maximum mid-span bending moment and absolute value of maximum negative bending moment at supports is also ($ql^2/8$) and it is equal to maximum mid-span bending moment of a simply supported beam. Diagram of bending moment of a continuous beam with semi-fixed-end constraints as a matter of fact is derived from moving that of a simply supported beam by

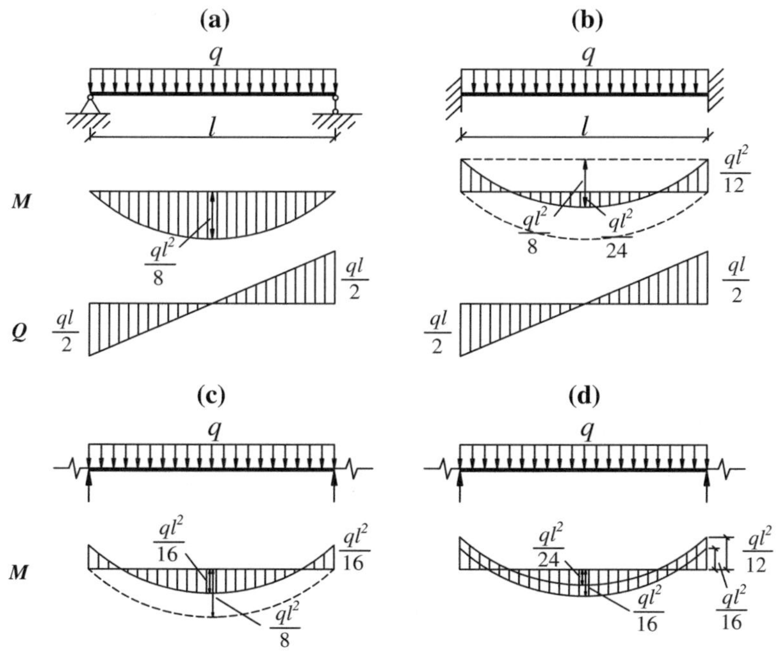

Fig. 4.1 Diagram of bending moment and shear of beams with different support constraints

($ql^2/16$) upward due to the partial constraints at both ends of the beam. Since the actual load that horizontal components bear and end constrains might be subject to change, an unfavorable combination of Fig. 4.1b, c is generally taken as the envelop diagram of bending moment in the early stage of design (as shown in Fig. 4.1b), in which case maximum mid-span positive bending moment ($ql^2/16$) and maximum negative bending moment at supports ($ql^2/12$) are employed to conduct preliminary design and analysis of components.

When maximum bending moment along cross section of the beam diminishes, sectional area downsizes with it and so does the cost. From the comparison above, we know that maximum bending moment of a continuous beam is much smaller than that of a simply supported beam; therefore, compared with a simply supported beam under the same load, continuous beam could have a smaller sectional area therefore is more efficient and economical. Figure 4.2 illustrates the distribution of bending moment of a three-span continuous beam and its maximum mid-span and support bending moment. It shows always smaller values than a simply supported beam with the same span. When cantilever beams of proper length (account for about 40 % of the entire span) are connected with continuous beams, maximum bending moment along the cross section could be maintained at a relatively small number and bring about better efficiency, as shown in Fig. 4.2c. In this case, maximum bending moment of cantilever beam is 0.5 q (0.4 l) 2 = 0.08 ql^2, which approximately equals to $ql^2/12$.

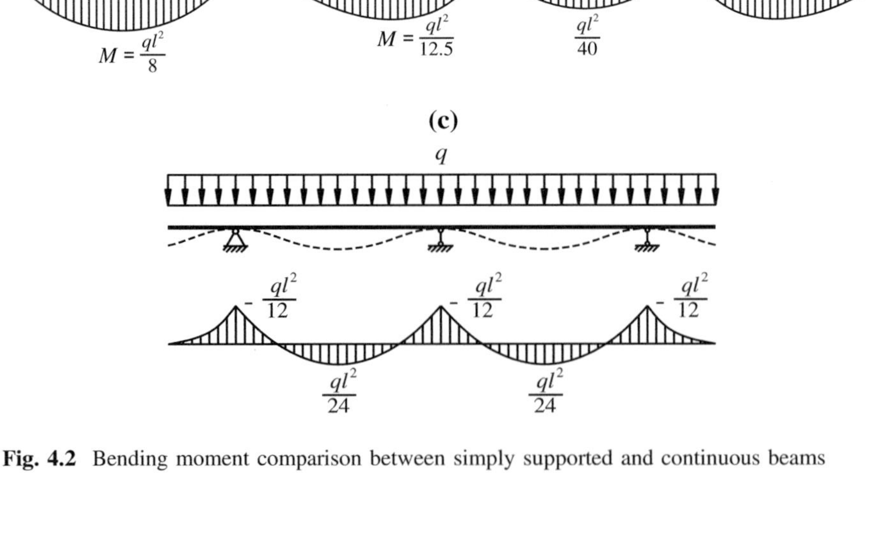

Fig. 4.2 Bending moment comparison between simply supported and continuous beams

When it comes to the inflection point (it is at 20–30 % of span starting from support) of a continuous beam, no bending moment exists (as shown in Fig. 4.2b), and this inflection point could always be employed as a connecting joint between beams. Note that shear force exists at this point and joint of beams installed at this location has to be able to transfer shear force.

From Fig. 4.2b, we could see that when each span is of the same length, maximum bending moment of the side span of a continuous beam is greater than that of the inner span. When the convenience of construction calls for an identical sectional area in each span, the length of side span is often reduced to be about 80 % of that of central span. In this way, maximum bending moment of the side span approximates that of the inner span, and identical design of sectional area is achieved. Readers could calculate the optimum length of the side span.

3. **Resisting moment of internal force**

Bending moment induced by load upon the beam should be balanced by resisting moment generated by the internal force within section of the beam. This resisting moment of internal force varies with the shape of section of horizontal component (as shown in Fig. 4.3). Beam with rectangular cross section as shown in Fig. 4.3a undertakes bending moment and shear force under the action of distributed load; as for the bending moment, it generates a triangle stress distribution within the section of the beam, in which case the upper part of the beam is compressed and the lower part is stretched; resultant tensile force and resultant compressive force pairs up a resisting moment to balance load-induced bending moment. Within I-shaped section of the beam as shown in Fig. 4.3b, the resisting moment is formed by resultant force of tension and that of compression within the upper and lower flange plates. As for the truss beam as shown in Fig. 4.3c, its resisting moment is mainly composed of resultant force of tension and compression from the upper and lower chords. When it comes to the triangle truss beam as shown in Fig. 4.3d, the resisting moment comes from resultant force of tension and compression within the upper and lower chords.

When material serves within the elasticity range, in accordance with plane-section assumption from elasticity mechanics, resisting moment of internal force of a rectangular-reinforced concrete beam is in balance with bending moment induced by external loads, as shown in Fig. 4.4. According to this equilibrium, the maximum stress $\sigma_{\max}$ could be calculated. If this maximum stress is smaller than the ultimate strength designed for concrete, components remain intact, which indicates that the estimated dimensions of cross section meet the requirement of strength design. On this basis, further checking of deformation requirements is needed. Strength design requirements within the range of elasticity are as follows:

$$\sigma_{\max} = Mr/I \tag{4.1}$$

$$W = I/r \tag{4.2}$$

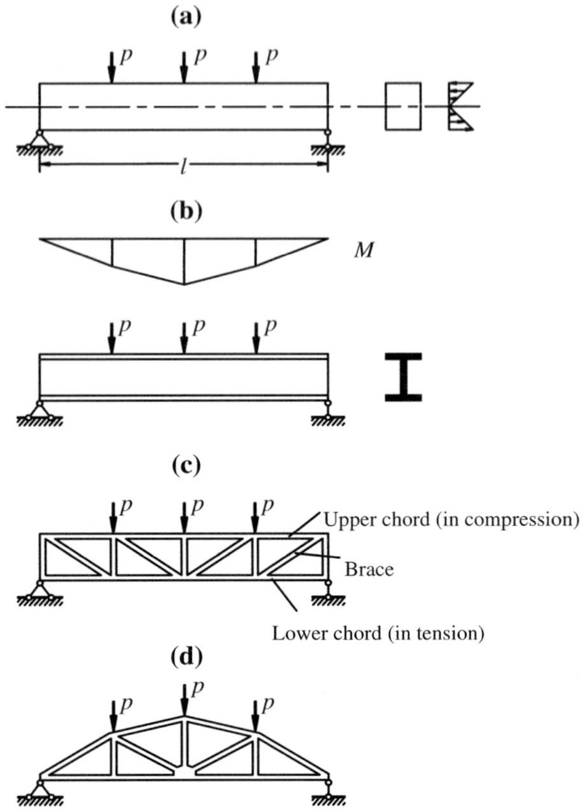

Fig. 4.3 Resistance of internal force from horizontal components of structure

$$\sigma_{\max} < [\sigma] \tag{4.3}$$

$$W > M/[\sigma] \tag{4.4}$$

where $[\sigma]$ is the designed strength of material; I refers to the sectional moment of inertia; r is the sectional radius of gyration; W is the sectional moment of area; and M is the maximum sectional bending moment.

Sectional moment of area W could be calculated by Eq. (4.4), so it is with estimated sectional area and shape.

As for steel structures, assumption could be made that they work within the range of elasticity. Equations (4.1) and (4.3) could be employed to determine whether the structure meets the strength requirement. In this case, yield strength of steel is taken as $[\sigma]$. When structure enters its plastic working state, resisting moment should be largely increased. For instance, the resultant force of I-shaped steel beam, when in an elastic state, is C as shown in Fig. 4.5a; and that of I-shaped steel beam, when it works in plastic state, should be $C' \approx 2C$ as shown in Fig. 4.5b. Assume that resisting moment of this beam in its elastic state is $M_1 = 0.9dC$, that in

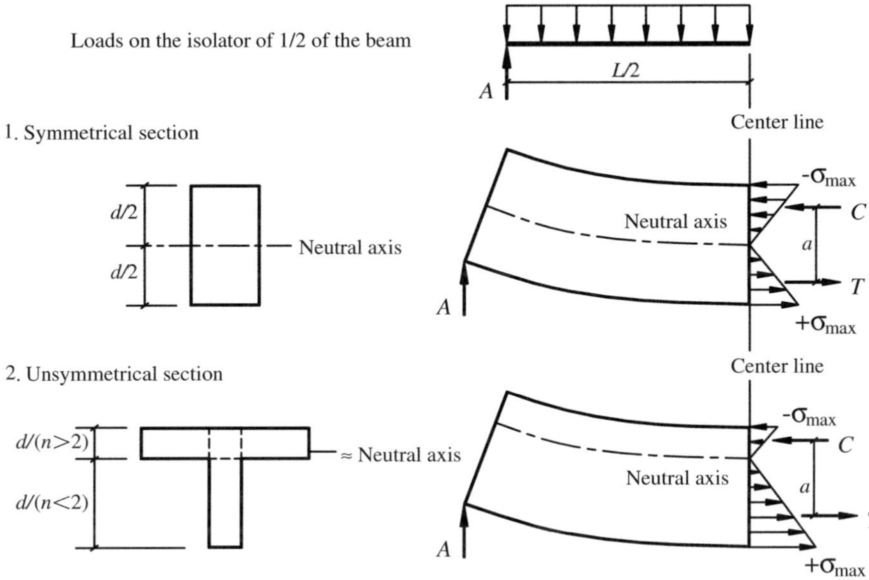

Fig. 4.4 Resisting moment of internal force balances with load-induced bending moment

its plastic state is $M_2 = 2 \times 0.8dC = 1.78M_1$. Compared with structure in its elastic state, bearing capacity of structure in its plastic state is enhanced by about 78 %.

When it comes to RC structure, it generally works in plastic state; for that concrete has a low tensile strength and is prone to cracking. In an ideal ultimate state, reinforcements are pulled into yield state while compressed concrete gets to its compressive strength. When this happens, resisting moment of internal force transforms as shown in Fig. 4.6b. Stress distribution in compressed area of concrete is determined by constitutive relation between elasticity and plasticity of concrete. Similarly, calculation of plasticity of steel structure is determined by simplified model of constitutive relation between elasticity and plasticity of steel (as shown in Fig. 4.7). Resisting moment of internal force could be calculated by distribution of elasticity and plasticity of internal force as shown in Fig. 4.6, and it should be equal to load-induced bending moment, in which case checking of strength reinforcements should be carried out.

$$M = 0.9d'T' = 0.9d'A_s[\sigma_s] \tag{4.5}$$

$$A_s \geq M/(0.9d'[\sigma_s]) \tag{4.6}$$

where $[\sigma_s]$ is the designed strength of reinforcement; A_s is the sectional area of reinforcement; d' is the computed height of cross section; and M is the maximum bending moment upon cross section.

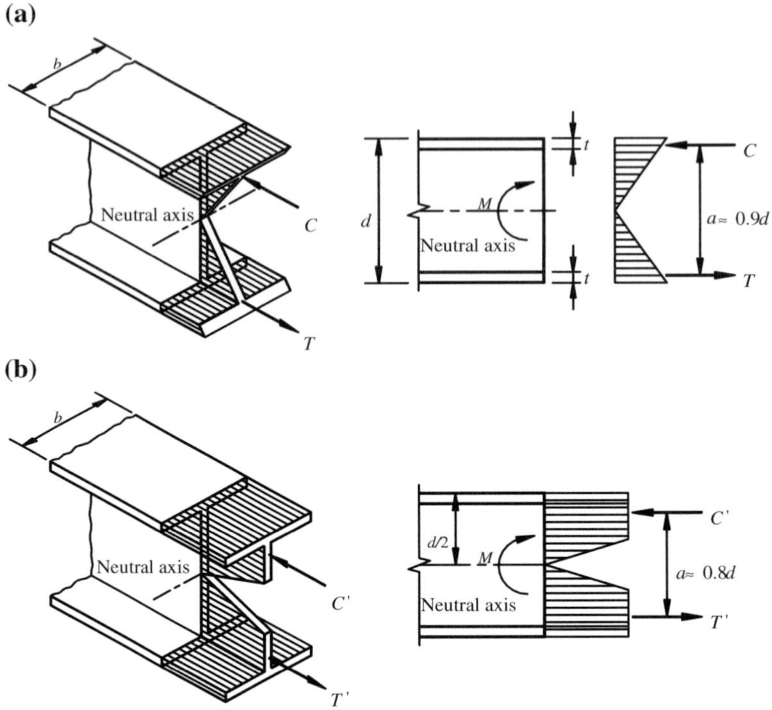

Fig. 4.5 Resisting moment of internal force of steel beam when it works in elastic and elastic–plastic state

Sectional area of reinforcement that meets requirement of strength could be calculated by Eq. (4.6), and so it goes with the estimation of quantity of steel consumption and cost of construction.

4. **Analysis of shear resistance of horizontal components**

To undertake external loads, horizontal components should be able to resist shear force. Capacity of horizontal components to resist shear forces ensures that they were equipped with overall bending resistance. And beams should be able to transfer transverse shear stress to make an entity of structure. Otherwise, each independent beam zone will end up with considerable deformation only bearing a small amount of bending.

From the feature of shear stress, we know that transverse shear stress exists wherever there is same vertical shear stress within the section in quantity. Figure 4.8 shows the distribution of shear flow within the beam. As for the components with rectangular cross section, the distribution of shear flow is like a parabola with maximum shear stress, about 1.5 times of average shear stress, at neutral axis of the beam. To simplify the calculation, average shear stress and maximum shear stress of the cross section could be calculated as follows:

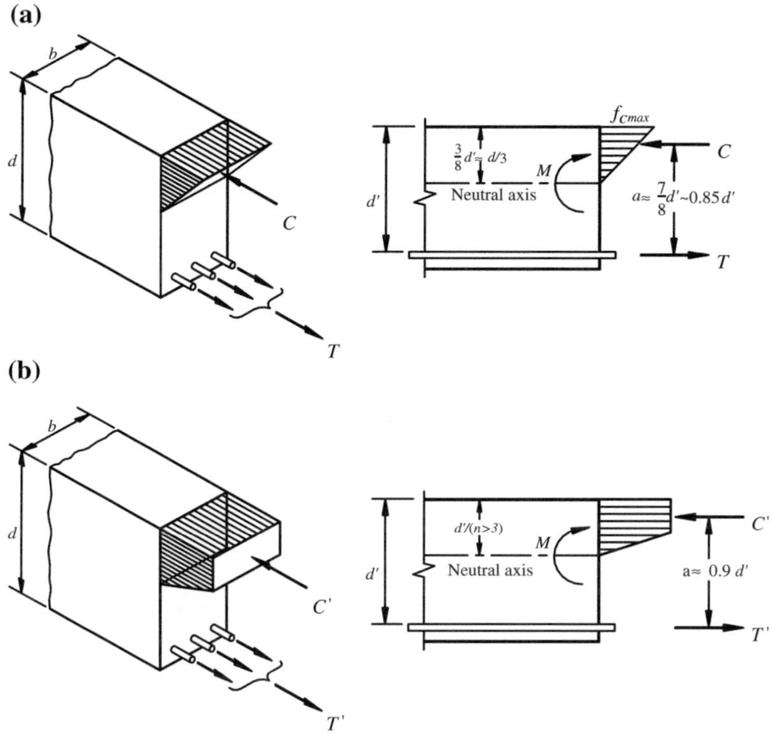

Fig. 4.6 Resisting moment of internal force of concrete beam when it works in elastic and elastic–plastic state

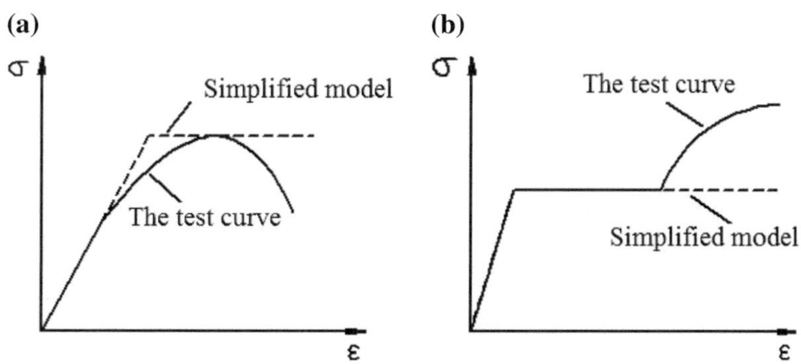

Fig. 4.7 Test curve and simplified model of constitutive relation regarding concrete and steel. **a** Concrete. **b** Mild steel

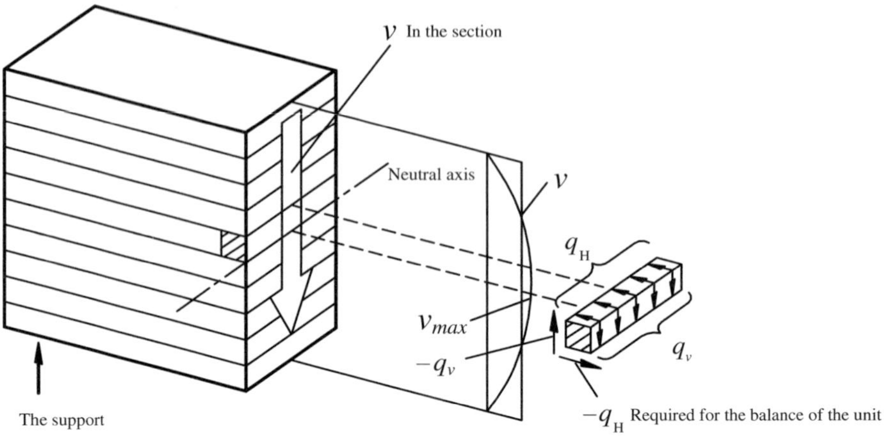

Fig. 4.8 Distribution of shear flow in a beam

$$v = V/(bh) \tag{4.7}$$

$$v_{\max} = 1.5V/(bh) \tag{4.8}$$

where v is the average shear stress; $V_{\max}$ is the maximum shear stress of cross section; V is the shear force; h is the height of cross section; and b is the width of cross section.

As mentioned above, shear force is mainly undertaken by web of cross section. As for components with H-shaped, T-shaped, or I-shaped cross sections, shear force could be assumed to be undertaken by web alone to simplify calculation. Average shear stress and maximum shear stress of H-shaped, T-shaped, or I-shaped cross section, respectively, are as follows:

$$v = V/(th) \tag{4.9}$$

$$v_{\max} = 1.5V/(th) \tag{4.10}$$

where t is the thickness of web of cross section.

After maximum shear stress is calculated by equations displayed above, one has to check whether components meet requirement of shear strength.

$$v_{\max} < [v] \tag{4.11}$$

where $[v]$ is the shear strength of material, which is determined by material employed from concrete or steel. When this condition is not met, more stirrups is usually added to enhance shear-resistance capacity of RC components; when it comes to steel structures, addition to the thickness of web is often used to fortify shear-resistance of components.

5. Truss

When horizontal components are of relatively large span or undertake relatively heavy load, truss could be adopted to serve as transverse bearing component. Truss could be taken as a beam which has gotten rid of inefficiently used web material from a solid beam, retaining upper, lower, and web chords. Since the height of truss is often greater than that of beam, it could undertake larger bending moment and shear force; thanks to the lighter weight truss has compared with solid beam, deadweight of the entire structure could be lessened.

Except for quadrangular open-web truss, truss is usually comprised of upper and lower chords along with inclined web members. Upper and lower chords undertake external bending moment while inclined web members deal with external shear force. Except for a small amount of secondary bending moment given rise by deadweight of members and rigid connection of joints, all components of truss generally undertake axial forces with no bending moment. This is also the reason that truss components are regarded as axially loaded members. As for the joints, they are often simplified as hinged joints. When it comes to the calculation of internal force of components, equilibrating of isolated free-body in structural mechanics could be adopted. Quadrangular open-web truss consists of parallel upper and lower chords as well as vertical web members; upper and lower chords undertake external bending moment while vertical web members, rigidly connected to upper and lower chords, demonstrate rigid frame effect and put up concerted work against external shear force.

Conventional forms of truss include parallel chord truss, triangle truss, open-web truss, and arch truss. Figure 4.9 shows the most common parallel chord truss and the calculation method of equilibrating of isolated free-body.

Fig. 4.9 Parallel chord truss

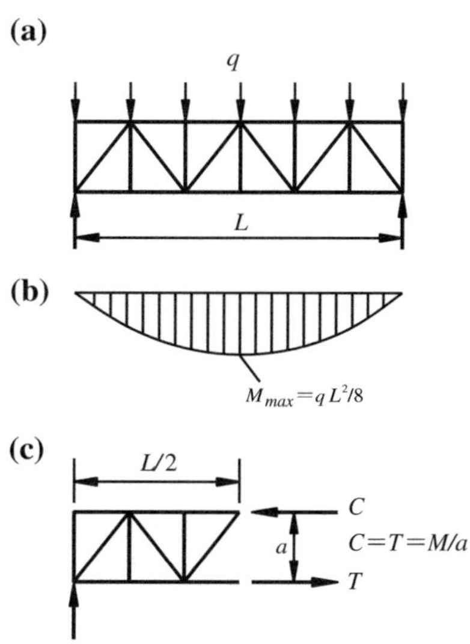

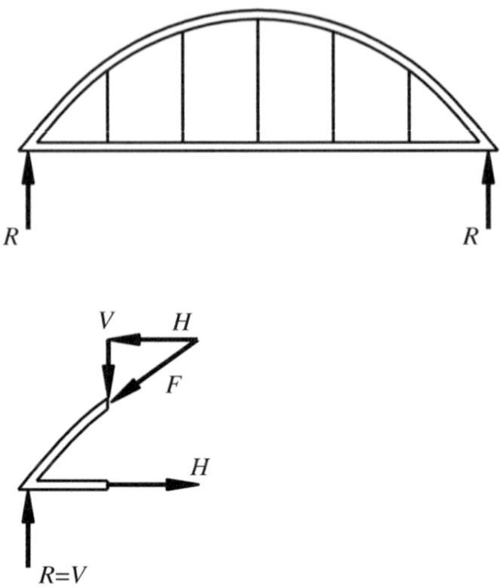

Fig. 4.10 Calculation of arch truss

Arch truss combines the strongpoint of arch and truss to develop a more efficient type of structure in terms of stress state. Arch truss provides a curved upper chord along with an arch effect to resist shear force. If the height of upper chord enables the curve to coincide with pressure line of arch, then the upper chord does not bend and the only force it has to deal with is compression. In practice, due to the influence of live loads, the upper chord, under the action of loads of all sorts, could in no way coincide with pressure line of arch. Nonetheless, height of arch could still be designed to coincide with pressure line of arch under the action of major constant loads, which should considerably reduce bending moment induced by various combinations of stress. Calculation of arch truss is relatively simple and equilibrating of isolated free-body could be employed, as shown in Fig. 4.10.

6. **Deformation of horizontal components**

When beam produces excessive deflection, issues will arise, such as cracking of beam and floor slabs, damage to decoration and installation, uncomfortable feelings of residents, and secondary stress effect. As mentioned before, the fourth step in preliminary design of horizontal components is to check whether the vertical deformation meets the pertinent requirements. When requirements are not met with, measures such as increase in sectional dimensions, modification of sectional configuration, and adoption of other material might be needed. In this case, "deformation control" instead of "strength control" serves as the decisive factor in sectional design.

Deflection deformation limit of beams usually is $L/150$–$L/300$ [21]. Deformation limit varies with operational requirements or levels of significance of structure. To learn more about specific deformation limits, readers could refer to relevant design code of reinforced concrete structure or steel structure [21, 65].

Calculation of deflection deformation of beam is simplified as follows:

$$\delta = \frac{k \cdot q \cdot L^4}{EI} \tag{4.12}$$

where q is the distributed line load on the beam; L is the length of beam; E is the modulus of elasticity of material; I is the moment of inertia; and k is the deformation coefficient, which is determined by end constraint conditions. As for a simply supported beam, $k = 5/384$; when it is a continuous beam, $k = 1/384$; if it is a beam with semirigid constraints, then $k = 3/384$. From the selection of different values of k, we know that under the action of identical load mid-span deflection of a simply supported beam is 5 times that of a continuous beam. Therefore, from the perspective of strength and deformation limit of components, it would be a wise choice to select continuous beam as the form of structure for horizontal components.

To lessen beam deflection, following methods are often adopted:

1. Replace simply supported beam with continuous beam. Deformation could be reduced to as much as 1/5;
2. Decrease beam span. It should reduce deformation to the power of four (refer to Eq. 5.12);
3. Increase beam height. It should significantly increase moment of inertia to the power of three (for that $I = bh^3/12$) and reduce deformation;
4. Widen the beam. It should increase moment of inertia linearly (for that $I = bh^3/12$) and reduce deformation;
5. Replace concrete with steel. Since modulus of elasticity of steel is way greater than that of concrete, refer to Eq. (4.12), deformation could be reduced.

Besides, note that long-term load-induced deformation is usually 2–3 times more than short-term load-induced deformation. Therefore, creep effects under the action of long-term load should be taken into consideration.

4.1.2 Analysis of Vertical Components

1. **Tension members**

Classified by different states of stress, vertical components basically consist of tension members and compression members. In general, tension members are pulled in a uniform fashion within the plane of section. When instability is not a concern, calculation could be quite simple. As for compression members, they are compressed within the plane of section. Strength, as well as buckling instability, should be checked.

Since tension members are uniformly pulled within the section, buckling is not a concern. The strength of material could be fully utilized in tension members, making the most efficient load-bearing components. Steel serves as an ideal tension member, for it, especially stranded high-strength stranded steel able, has remarkable tensile strength. In general, tension members are axially pulled from both ends and the equation of its checking calculation is as follows:

$$\sigma = P/A < [\sigma] \tag{4.13}$$

The equation of checking calculation for deformation of tension members is as follows:

$$\delta = (P/A)(L/E) < [\delta] \tag{4.14}$$

In these equations, P is the tensile force; A is the sectional area of member; σ is the tensile stress within the section; $[\sigma]$ is the tensile strength of material; L is the length of member; E is the modulus of elasticity of material; δ is the axial deformation of member; and $[\delta]$ is the allowable axial deformation of member.

Although high-strength stranded steel cable is the most frequently applied tension member, it has the drawback of insufficient lateral stiffness, especially when the member has to undertake tension, compression, or bending moment induced by eccentric load under the action of different combinations of loads. Due to the lack of bending and compressive strength, steel cable buckles when compressed. In this case, H-shaped, L-shaped, or other fashioned reinforcement is needed to ensure sufficient stiffness and resistance to bending moment induced by eccentric load. Angle steel is another commonly applied tension member. It has a much larger stiffness to resist lateral deformation compared with round steel or steel cable and it facilitates construction.

Since concrete has a low tensile strength (mostly ignored), it is rarely used as tension member alone. Only when concrete components are sufficiently prestressed could they maintain compressed under the action of external tensile load and serve as tension members.

2. **Axially compressed short columns**

When slenderness ratio of compression member is relatively small, failure of compressive strength instead of buckling instability would occur. Compression members of this kind are designated as "short columns." Generally speaking, when slenderness ratio of RC compression member is less than 10 or that of steel compression member is less than 20, it is defined as "short column."

Since short columns are compressed within the cross section without having to worry about buckling instability, the strength of material could be fully utilized, the advantage of which makes them efficient members of load bearing. Steel and concrete both serve as ideal choices for short columns. Wide-flange I-shaped section or latticed composite section is commonly seen in steel compression members. Rectangle or round is more of a choice when section of concrete compression member is concerned. Longitudinal reinforcements and transverse stirrups are to be installed to concrete compression members. There are two types of stirrups: rectangular ones and circular ones. They work to fix longitudinal reinforcements and to resist shear force. Circular stirrups have core concrete constrained, making it compressed in all directions, and thereby considerably enhance compressive strength of core concrete.

Longitudinal reinforcement ratio of RC compression member is usually between 1 and 8 %. Minimum is set to contain cracks and to prevent potential eccentric bending moment. Maximum is set to prevent construction difficulty curing concrete pouring and consolidation. Longitudinal reinforcement ratio of compression members is usually between 4 and 5 %.

When compressed, equation to check strength of short columns is given as:

$$\sigma = P/A < [\sigma] \tag{4.15}$$

Equation to check deformation of compression members is given as:

$$\delta = (P/A)(L/E) < [\delta] \tag{4.16}$$

where P is compressive force; A is sectional area of member; σ is compressive stress within cross section; $[\sigma]$ is compressive strength of material; L is the length of member; E is modulus of elasticity of material; δ is axial deformation of vertically compressed member; and $[\delta]$ is allowable axial deformation of vertically compressed member.

3. **Axially compressed slender columns**

When the slenderness ratio of compression members is relatively large, instability will occur first without failure of compressive strength of material. This phenomenon is referred as buckling of "slender columns." When slenderness ratio of RC compression member is greater than 10 or that of steel compression member is greater than 20, compression members are defined as "slender columns." When the section of slender column is not biaxially symmetrical, buckling instability starts from the direction of weak axis.

To calculate load-bearing capacity for slender columns, Euler's formula could be employed to derive ultimate bearing capacity of compression members.

$$P = \pi^2 EI/L_{\text{eff}}^2 \tag{4.17}$$

where P is the ultimate bearing capacity of compression member; I is the moment of inertia along the weak axis; L_{eff} is the effective length of member; and E is the modulus of elasticity of material.

From the equation, we can see that the ultimate bearing capacity of compressed slender column, when buckling instability, is concerned is largely determined by modulus of elasticity of material, effective length of member, and moment of inertia of the cross section. And it has nothing to do with compressive strength of material.

Effective length L_{eff} is determined according to end constraints of compression members. Figure 4.11 illustrates the way to determine effective length factor K.

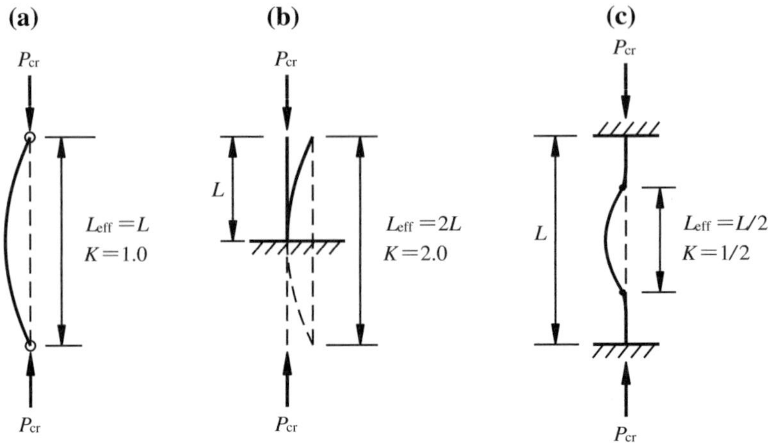

Fig. 4.11 Effective length factor K

$$L_{\text{eff}} = KL \tag{4.18}$$

where L in the equation is the actual length of member.

In construction projects, a majority of concrete columns are cast on-site to ensure the integrity and reliability of connecting joints. In general, steel columns are prefabricated in factory and assembled on-site, which quickens the progress of construction.

To facilitate the calculation of moment of inertia I for members with all fashions of sections, simplified calculation as followed could be employed:

$$I = Ar^2 \tag{4.19}$$

$$W = I/r = Ar \tag{4.20}$$

where I is moment of inertia of the cross section; W is moment of area of the cross section; A is sectional area; and r is sectional radius of gyration.

Next, to meet the requirements in the early stage of design, approximate calculation of ultimate bearing capacity of compression members is introduced.

As for concrete columns, when the slenderness ratio is less than 10 and failure of compressive strength of short columns occurs, the ultimate bearing capacity P could be calculated by Eq. (4.15).

When the slenderness ratio is equal to 20, the ultimate bearing capacity of column is reduced by 30 % of the ultimate bearing capacity of short column in compression.

When the slenderness ratio is between 10 and 20, ultimate bearing capacity of column could be derived from linear interpolation of these two circumstances mentioned above.

When the slenderness ratio is over 20, buckling instability occurs and Euler's formula (4.17) could be employed to calculate the ultimate bearing capacity P of slender columns.

As for steel columns, when the slenderness ratio is less than 20, failure of compressive strength occurs and Eq. (4.15) could be employed to calculate ultimate bearing capacity P of short columns.

When the slenderness ratio is about 80, the ultimate bearing capacity of column is reduced by 30 % of that of short column.

When the slenderness ratio is between 20 and 80, the ultimate bearing capacity of column could be derived from linear interpolation of these two circumstances mentioned above.

When the slenderness ratio is over 80, buckling instability occurs and Euler's formula (4.17) could be employed to calculate the ultimate bearing capacity P of slender columns.

In the application of aforementioned principles of calculation, dimensions and fashion of the section of compression members could be preliminarily determined.

4. **Action of combined loads**

Vertical component is basically pulled or compressed and the aforementioned two approaches of simplified calculation could be employed to estimate the sectional area, shape, and ultimate bearing capacity of vertical components. Nonetheless, since actual construction projects are affected by wind, seismic load, and various combinations of vertical loads, failure of beam or that of column might be induced by different combinations of compression, bending moment, shear force, and torsion. The mechanism of load bearing and failure of components affected by a combination of forces is more complicated than that of the components influenced by a single force. When it comes to the design of horizontal substructure and pier of bridges, special attention has to be paid to the action of combined loads.

In the early stage of design, interaction formula [22–24] could be employed to conduct simplified design of beams or columns under the action of combined loads.

When the component is affected by bending moment and axial force combined, the interaction formula is given as follows:

$$(P/P') + (M/M') = 1 \tag{4.21}$$

where P' is the axial bearing capacity of component when a single axial force is applied; M' is the bearing capacity of bending moment when a single bending moment is applied. P and M are, respectively, the bearing capacity of axial force and that of bending moment when a combination of bending moment and axial force is applied; the two values of which are smaller than those of P' and M'. Figure 4.12 shows an illustrative diagram of the interaction between bending moment and axial force.

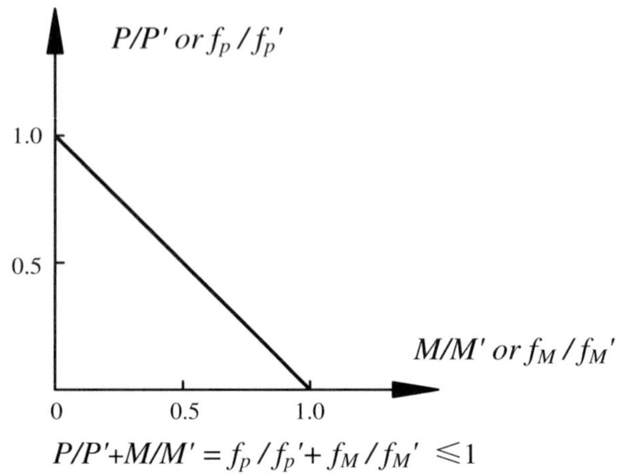

Fig. 4.12 Interaction of bending and axial force

When the component is under the action of bending moment, shear force and torque combined, the interaction formula is given as:

$$(M/M') + (Q/Q')^2 + (T/T')^2 = 1 \tag{4.22}$$

where M' is the bearing capacity of bending moment when the components are under the action of a single bending moment; Q' is the bearing capacity of shear force when the components are under the action of a single shear force; T' is the bearing capacity of torque when the components are under the action of a single torque. M, Q, T are, respectively, the bearing capacity of bending moment, that of shear force and that of torque when the components are acted upon by bending moment, shear force, and torque combined; the values of which are smaller than those of M' and Q' and T'. The equation mentioned above represents a spatial surface of failure. Figure 4.13 shows a diagram of interaction between bending moment, shear force, and torque.

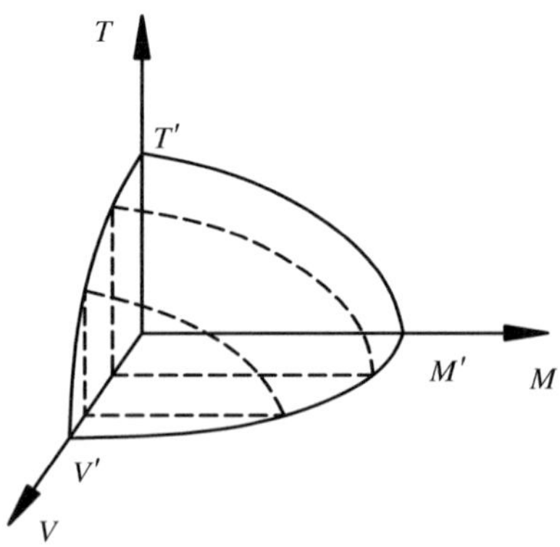

Fig. 4.13 Interaction of bending, shear, and torque

When the component is under the action of axial force, moment, shear force, and torque combined,

$$(P/P') + (M/M') + (Q/Q')^2 + (T/T')^2 = 1 \tag{4.23}$$

where P' is the bearing capacity of axial force when the components are under the action of a single axial force; M' is the bearing capacity of bending moment when the components are under the action of a single bending moment; Q' is the bearing capacity of shear force when the components are under the action of a single shear force; T' is the bearing capacity of torque when the components are under the action of a single torque. P, M, Q, T are, respectively, the bearing capacity of axial force that of bending moment, that of shear force, and that of torque when the components are acted upon by axial force, bending moment, shear force, and torque combined, the values of which are smaller than those of P', M' and Q' and T'. Figure 4.14 shows the diagram of interaction between axial force, bending moment,

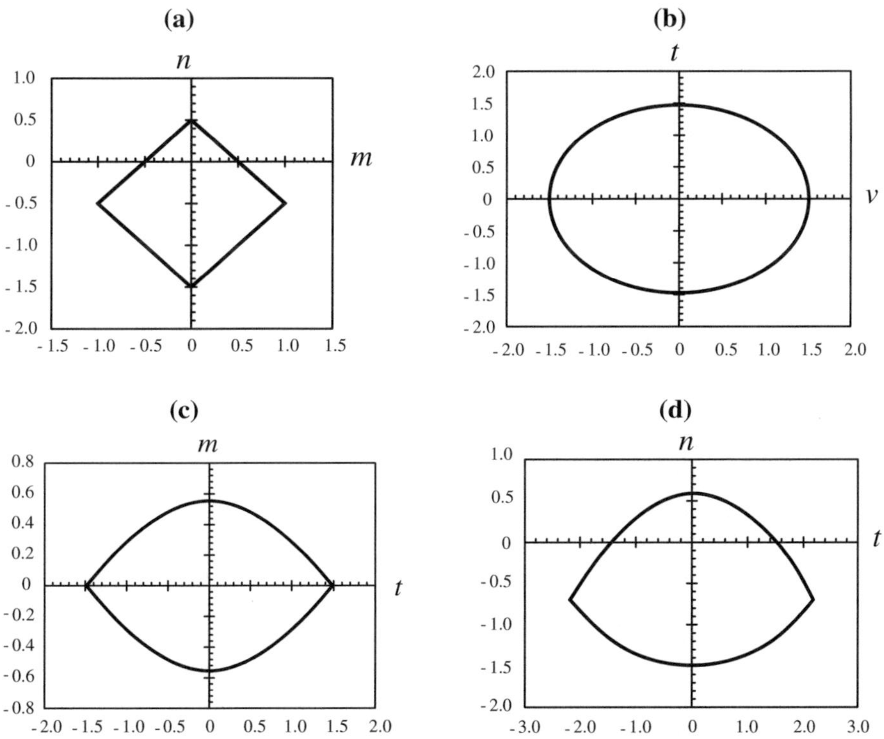

Fig. 4.14 Interaction curve between axial force, bending moment, shear force, and torque. **a** Interaction curve of *m–n*. **b** Interaction curve of *v–t*. **c** Interaction curve of *m–t*. **d** Interaction curve of *n–t*

shear force, and torque. Since there are four variables, respective expressions $m = M/M'$, $n = P/P'$, $v = Q/Q'$, $t = T/T'$ are necessary.

5. **Deformation of vertical components**

Two criteria have to be abided by in structural design: strength criterion and deformation criterion. The former requires that structural components meet the requirement of strength under the action of all unfavorable combinations of loads so that no failure of strength or that of buckling instability should occur. The latter requires that structural components meet the requirement of deformation limitation under the action of all unfavorable combinations of loads so that no oversized deformation of component or that of the structure as a whole should occur. When oversized deformation of component or that of the structure does occur, the structure will not be faced with immediate destruction but operational requirement of the structure is no longer met with, which should lead to uncomfortable atmosphere or damage to decoration or facilities. In structural design, it is often easier to meet the requirement of strength than to meet that of deformation. Therefore, deformation more often than not serves as the control factor in design, especially when ultra-high or ultra-large-span structures are concerned. It is quite easy to meet structural requirement of strength when high-strength material such as steel is applied; to meet that of deformation, sectional area of components as well as material consumption is usually increased or else effective construction measures could be adopted to contain deformation.

In the early stage of structural design, the estimation of deformation of components and the structure as a whole are of critical importance, which should involve the selection of type of structure, consumption of material, and the adoption of effective construction measures to contain deformation. Different operational requirements of structure lead to diverse requirements of deformation limitation. As for specific details, please refer to relevant building codes or standards. In this section, several frequently applied methods of estimation concerning vertical components are introduced, which by the way could also be generalized to the estimation of the overall deformation of structure.

Vertical components generate horizontal and vertical displacement under the action of corresponding loads. As for vertical displacement, it is compressive deformation of material; horizontal displacement serves as the control deformation of vertical components. Under the action of horizontal load, horizontal displacement of vertical components or that of vertical substructure is comprised of two parts: shear deformation and bending deformation (as shown in Fig. 4.15).

To estimate shear deformation of vertical component or that of vertical substructure, when horizontal load is concentrated force V, horizontal shear deformation would be the following:

$$\Delta s = 1.2Vh/(GA) \tag{4.24}$$

where G is the shear modulus of material. $0.4E$ could be granted to the shear modulus of steel or that of concrete and this E represents the modulus of elasticity

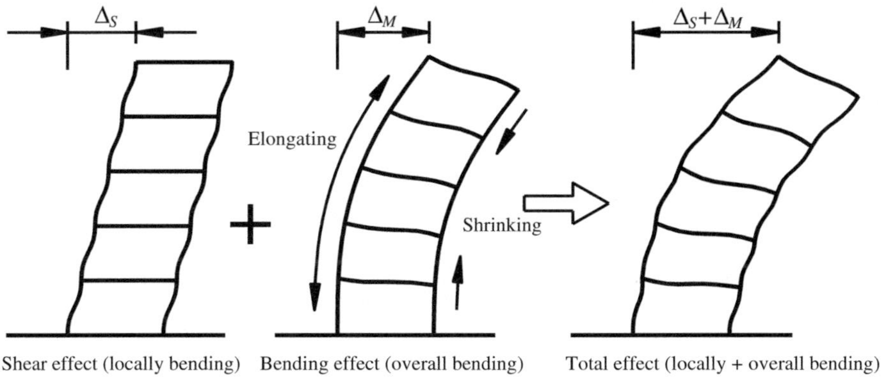

Fig. 4.15 Shear deformation and bending deformation

of steel or concrete. A is the sectional area of vertical component and h is the height of vertical component (as shown in Fig. 4.16).

In the estimation of bending deformation of vertical component or that of vertical substructure, when horizontal load is distributed load p, horizontal bending deformation would be the following:

$$\Delta m = ph^4/(8EI) \tag{4.25}$$

where E is the modulus of elasticity of material; I is the moment of inertia of the cross section; and h is the height of vertical component's height (Fig. 4.17).

In the estimation of bending deformation of vertical component or that of vertical substructure, when horizontal load is concentrated load p applied to the top, horizontal bending deformation would be the following:

$$\Delta m = Ph^3/(3EI) \tag{4.26}$$

where E is the modulus of elasticity of material; I is the moment of inertia of the cross section; h is the height of vertical component or that of vertical substructure.

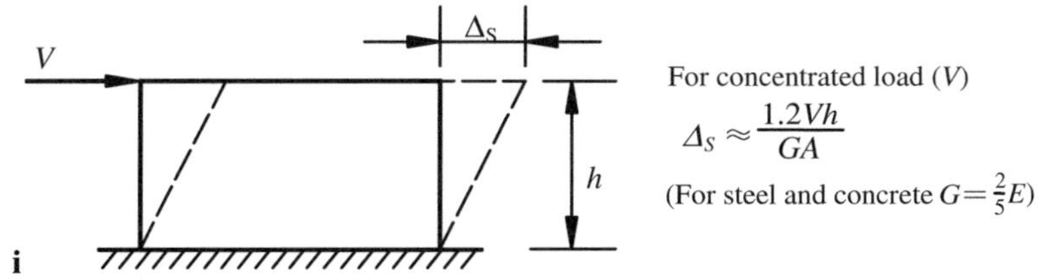

Fig. 4.16 Estimation of shear deformation of vertical component

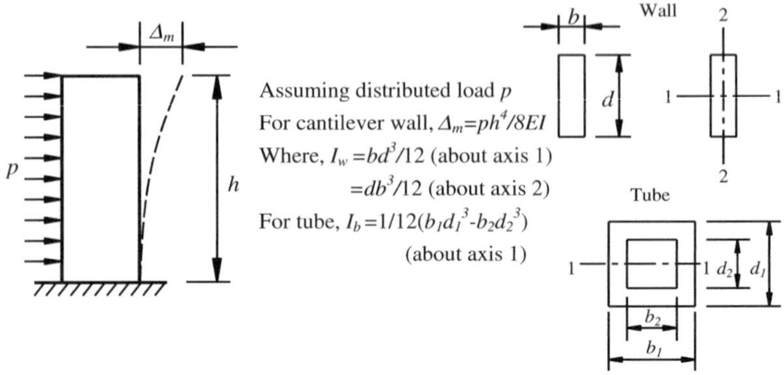

Fig. 4.17 Estimation of bending deformation of vertical component

When the height to width ratio of vertical component or that of vertical substructure is less than 3, horizontal displacement is largely shear deformation Δs, for that bending deformation in this case is very small and could be ignored. When the height to width ratio of vertical component or that of vertical substructure is larger than 5, horizontal displacement is largely bending deformation Δm, for that shear deformation in this case is very small and could be ignored. When the height to width ratio of vertical component or that of vertical substructure is between 3 and 5, horizontal displacement is comprised of bending deformation and shear deformation ($\Delta s + \Delta m$), for that neither of them could be ignored.

As for a regular frame structure, an accurate calculation of deformation needs the help of computer. In the early stage of design, the estimation of deformation could be executed with the help of the assumption of common deformation and the principle of stiffness distribution to see the quality of early stage of design. The assumption of common deformation and the principle of stiffness distribution indicate that when common horizontal deformation occurs to vertical components or vertical substructure within the same building, horizontal load distributed to vertical components is determined by the stiffness of components. In other words, a larger proportion of shear force should be directed to components with larger stiffness and a smaller proportion to components with smaller stiffness. Based on this, the estimation of horizontal displacement of frame structures could be derived as follows (as shown in Fig. 4.18).

Horizontal displacement of external node in the ground floor of frame structure

$$\Delta c = \theta h = V_c h^3/(12EI_c), \quad \Delta g = \alpha h = V_g hL^2/12EI_g \tag{4.27}$$

$$\Delta = (\alpha + \theta)h = V_c h^2(h/I_c + 2L/I_g)/(12E) \tag{4.28}$$

Deformation of frame structure decreases with the increment of layers, for the effects of horizontal loads diminish in that fashion. Thereby, simplified calculation of horizontal displacement at the top of frame structure is given as:

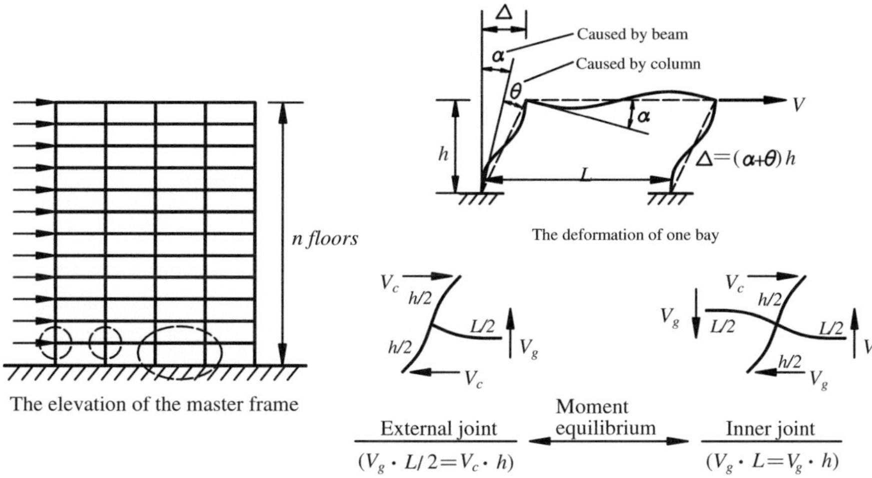

Fig. 4.18 Estimation of deformation of a frame structure

$$\Delta\text{total} = n(\Delta)/2 \tag{4.29}$$

where n is the total number of layers of frame structure; subscripts c and g, respectively, represent column and girder. Vc, Vg refer to the shear force distributed to column and beam; Ic, Ig are the moment of inertia of column and beam; E is the modulus of elasticity of material; and h is the height of vertical component.

The estimation of vertical displacement of beam could employ the calculation method of vertical displacement of continuous beam with semirigid constraints, for that it is of quite some difficulty to make sure that all beams and columns of a frame structure are rigidly fixed and as a matter of fact they are usually in a semi-fixed state.

$$\delta = (3/384)pL^4/EI \tag{4.30}$$

Note that deformation estimation mentioned above has been considerably simplified and could only derive an approximate situation of structural deformation to facilitate comparison and analysis of all proposals in the early stage of design.

4.2 Analysis of Prestressed Structural Elements

1. General principles

In Chap. 3, basic principles of prestressed concrete are introduced. In a perspective, prestress transforms concrete in its application into compressed material of elasticity. Concrete has a weak resistance to tension and a strong resistance to compression. Through the application of prestress by high-strength prestressed reinforcements, concrete, in its working condition, remains in a compressed elastic state and is able to undertake certain amount of external tension. Concrete in this case has transformed from a brittle material into an elastic material. Figure 4.19 illustrates the stress distribution within the section of a concrete beam when a central or eccentric prestress is applied. Prior to the cracking of concrete, in accordance with the plane-section assumption, distribution and magnitude of stress at any point within the section could be calculated.

As for design of prestressed concrete beam or slab, an important concept "balance design" exists, which means the prestress provided by prestressed reinforcements could balance weight-induced bending moment and deformation of beam or slab. Thereby, the value of prestressed befitting "balance design" could be derived as follows:

$$F = w \cdot L^2 / 8 \cdot h' \tag{4.31}$$

where F is the resultant force of prestressed reinforcement within beam or slab; w is the distributed deadweight of beam or slab; L is the span of beam or slab; and h' is the effective thickness of beam or slab.

For instance, a beam installed with parabolic prestressed reinforcement is as shown in Fig. 4.20. Assume that F is the total tension of prestressed reinforcement; L is the span; h is the sag of parabolic prestressed reinforcement. Since we know that the distributed load counterbalanced by the upward bending moment provided by prestress is $w = 8Fh/L^2$, total tension of prestressed reinforcement should be $F = wL^2/(8h)$.

2. **Estimation of prestress loss**

Overall analysis and design of prestressed concrete structure have to take into consideration the effective internal force of prestressed reinforcement in each important loading stage as well as performance of material within the service life span of structure. Important loading steps of prestressed concrete structures include the following:

1. The moment prestress is applied to the section of concrete; tensile force of prestressed reinforcement reaches its maximum while it is unlikely for the concrete to have reached designed strength of 28 day. Checking of the bearing capacity of structure at this instant helps to prevent failure in the process of construction.
2. Under service loads, prestress gradually decreases with the lapse of time and so it is with long-term effective prestress. Thereby, checking of the bearing capacity of structure under the action of long-term loads should prevent failure of strength. Note that the strength of concrete at the moment is already greater than that of designed 28 day strength.
3. Multiple prestress applications might be needed at different stages of some structures. Therefore, multiple checking should be conducted to see the level of

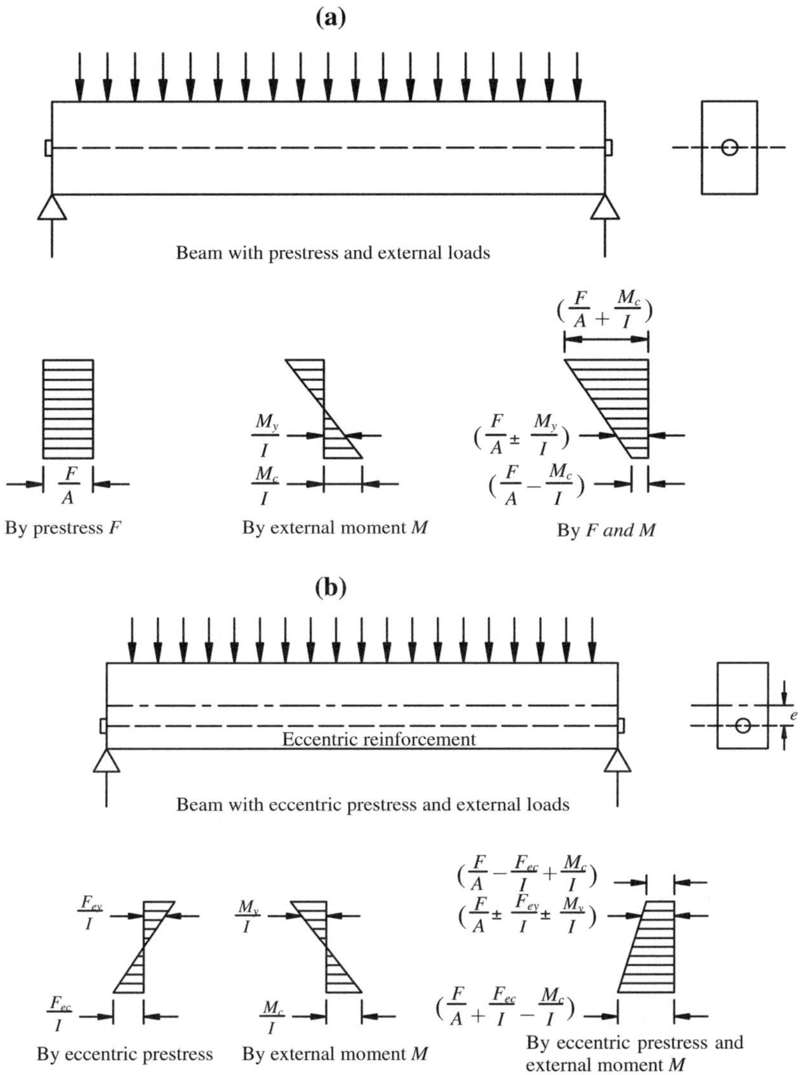

Fig. 4.19 Stress distribution within the section of prestressed concrete beam. **a** When the prestress is applied to the center of section. **b** When the prestress is applied to the section of beam in an eccentric fashion

prestress in each stage. Time is the key factor in deriving an accurate analysis and a correct estimation of prestress loss of concrete is needed in each stage.

Generally speaking, it is quite difficult to precisely calculate the prestress loss of concrete components, for there are so many interactive influencing factors. For instance, on the one hand, the stress within prestressed reinforcement gradually

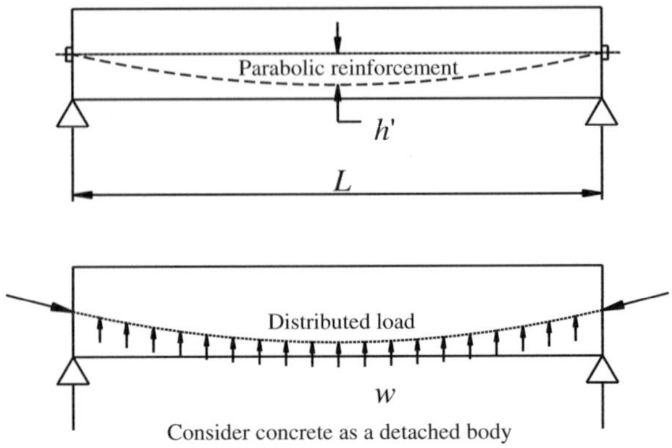

Fig. 4.20 Concept of "balance design" in prestressed concrete beam

decreases with the lapse of time (stress relaxation) and its rate affected by creep-induced level of stress within the section of concrete. On the other hand, the rate of creep in concrete is affected by stress state of the prestressed reinforcement. Therefore, it is quite difficult to distinguish the net prestress loss caused by each factor when uncertain factors such as levels of stress, creep, shrinkage, and loading method exist simultaneously. It is necessary to reasonably simplify the model to determine major influencing factors and estimate the prestress loss of concrete in each step. Although the ultimate strength of prestressed concrete structure is, in general, not affected by prestress loss, a reasonable estimation of prestress loss could ensure that prestressed concrete structure under service loads meets the strength and deformation requirements. Detailed specifications regarding the estimation of the prestress loss are available in relevant building codes and standards. A simplified method of prestress loss estimation is introduced next [24].

Many reasons that will cause prestress lost include elastic compressive deformation of concrete, tensile deformation of steel, creep of concrete, shrinkage of concrete, relaxation of steel, deformation of anchor, and bond slip.

Initial prestress of reinforcement minus the prestress loss is termed as effective prestress or designed prestress. Total prestress loss is affected by the value of initial prestress. The greater the initial prestress is, the greater the proportion of total loss will be. When the initial prestress is in low stress region, the proportion of total prestress loss is relatively small.

As for pretensioning method, to minimize the creep of reinforcement or to balance the loss of friction, excessive tension is initially applied to prestressed reinforcement so that after relaxation designed tensile stress is still achieved. The moment prestress is transferred to concrete, prestress loss due to anchorage set occurs instantaneous. Initial tensile stress minus the loss due to anchorage is the initial prestress prior to relaxation. As for post-tensioning method, when there are other reinforcements needed to be prestressed, loss induced by elastic shortening should gradually occur. Two types exist for elastic shortening of concrete: The elastic shortening induced by direct axial compression and that induced by elastic bending. As for pretensioning method, total loss by elastic shortening occurs the moment when prestress is transferred to concrete. When the prestress prior to elastic shortening is taken as the initial prestress applied to concrete, concrete shrinkage of concrete, creep of concrete, and relaxation of reinforcement are the only prestress loss left. When it comes to points of reinforcement that are contained within the concrete, friction-induced prestress loss has also to be taken into consideration.

The prestress loss could be expressed in four ways as follows:

To express with strain, which in favor of the calculation of the prestress loss induced by creep, shrinkage, and the elastic shortening.
To express with overall deformation, which in favor of the calculation of the loss due to anchorage.
To express with the stress of reinforcement, which converts all kinds of loss into the prestress loss of reinforcement, is a widely used method.
To express with the prestress of concrete, which converts all kinds of loss into the prestress loss of concrete, is also a widely used method.

In the early stage of design, it is quite conventional to express the total prestress loss with the percentage that loss accounts for of initial prestress of concrete. As for steel and concrete widely applied, when they are cured in normal curing environment and applied with medium level of prestress, Table 4.2 could be used to estimate the total loss of prestress.

Table 4.2 is applicable to the estimation of total prestress loss only when the prestress being applied is of medium level. When the prestress being applied is on a relatively high level, the estimated total prestress loss could be up to 30 %. When the prestress being applied is on a relatively low level, the estimated total prestress loss might be reduced to 15 % or so. Engineers are required to inquire and analyze the cause of the loss and make judgment and choice in accordance with specific characteristics of construction project.

3. **Equilibrium prestress design**

(1) **Preliminary estimation of bending section in service state**

Preliminary design of the bending section of prestressed concrete could be easily derived by the equilibrium of force couples comprised of acting tensile force upon the section and the counteractive compressive force. Generally, concrete sections under service loads are required to undertake no tensile force. The arm of the resultant internal force could vary in position within the range of 30–80 % of total height h of

Table 4.2 Estimated value of total prestress loss [24]

Reason for the prestress loss	Pretensioning method (%)	Post-tensioning method (%)
Elastic shortening of concrete	4	1
Creep of concrete	6	5
Shrinkage of concrete	7	6
Relaxation of reinforcement	8	8
Total loss	25	20

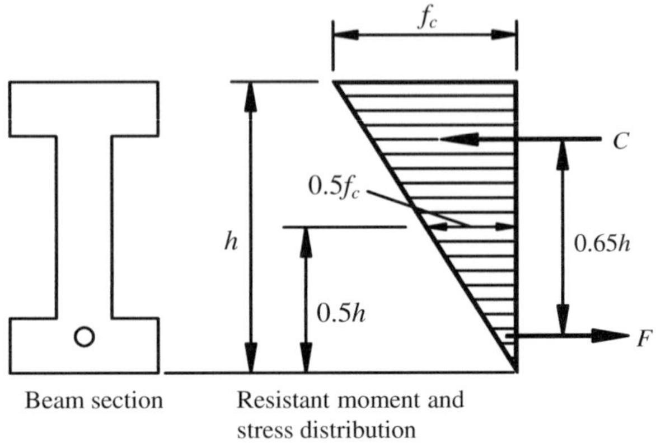

Fig. 4.21 Balance design of prestress in the bending section of concrete

the cross section, which arrives at an average of $0.65h$ or so, as shown in Fig. 4.21. As for the effective prestress F, it could be estimated with the equation below:

$$F = C = M/0.65h \tag{4.32}$$

where M is the bending moment on the section; h is the height of section; F is the resultant force of prestress of reinforcement, and C is the sum of resultant force within the compressed area of concrete.

If the effective prestress of prestressed reinforcement is f_{se}, then the needed area of prestressed reinforcement is given as:

$$A_{ps} = F/f_{se} = M/(0.65h \cdot f_{se}) \tag{4.33}$$

Total prestress F is the resultant force of compression C upon the section and the average compressive stress is given as:

$$C/A_c = T/A_c = A_{ps} \cdot f_{se}/A_c \tag{4.34}$$

As shown in Fig. 4.21, stress at the top fiber of concrete is f_c, which is conventionally required to be smaller than $0.45\,f_c'$. And the average stress is 50 % f_c.

$$C/A = A_{ps} \cdot f_{se}/A_c = 0.5f_c \tag{4.35}$$

$$A_c = A_{ps} \cdot f_{se}/(0.5f_c) \tag{4.36}$$

where f_c' is the designed strength of concrete. The needed area of prestressed reinforcement and the section area of concrete could be estimated by Eqs. (4.33) and (4.36). Note that there are two approximate coefficients 0.65 and 0.5 in the deduction above and they are given by the experience of calculation of prestress regarding components with rectangular section. When it comes to the actual application in construction project, these two coefficients could be regulated in accordance with the section shape to achieve more accurate estimation.

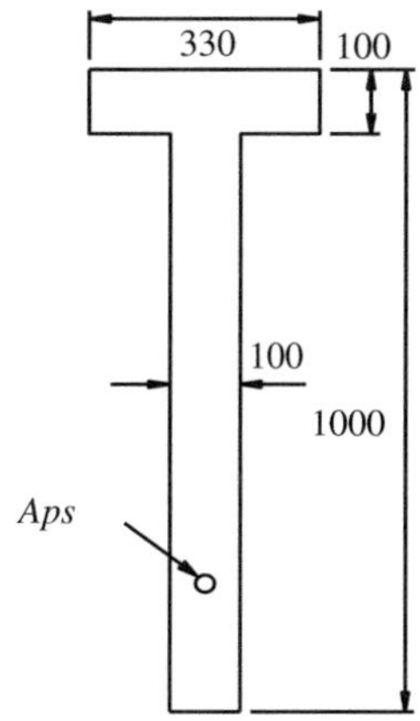

Fig. 4.22 Preliminarily designed T-shaped section

Example 1 Conduct a preliminary design for the section of a prestressed concrete beam using given information. Total bending moment the section has to undertake is 400 kN m; total height of section is 1 m; effective prestress of reinforcement is 850 MPa; allowable stress of concrete under the action of working load is 10 MPa.

Solution: in accordance with Eqs. (4.33) and (4.36)

$$F = T = M/0.65h = (400\,\mathrm{kN\,m})/(0.65 \times 1\,\mathrm{m}) = 615.4\,\mathrm{kN}$$

$$A_{\mathrm{ps}} = F/f_{\mathrm{se}} = (615.4\,\mathrm{kN})/(850\,\mathrm{N/mm^2}) = 724\,\mathrm{mm^2}$$

$$A_c = (615.4\,\mathrm{kN})/(0.5 \times 10\,\mathrm{N/mm^2}) = 123{,}080\,\mathrm{mm^2}$$

Thereby, T-shaped section as shown in Fig. 4.22 is selected as preliminary section of design on the basis of results of calculation.

In the estimation of the height of section of prestressed beam, one general rule is that 70 % of the height of ordinary concrete beam could be a choice. As for the estimation of the thickness of prestressed slab, *L*/55 could be taken as preliminary design of thickness. The thickness of ordinary concrete slab is *L*/35.

(2) **Preliminary estimation of bending section in its ultimate state**

Generally speaking, the application of prestress to the section of component would not be able to increase the ultimate strength capacity; nonetheless, it could enhance the performance under service loads, with reduced deformation and controlled cracking. Preliminary design of the ultimate strength of prestressed concrete component could be carried out via an approximate calculation derived from an empirical equation. The needed area of prestressed reinforcement is given as:

$$A_s = r \cdot M_o/(0.8h \cdot f_{\mathrm{ps}}) \tag{4.37}$$

where r is the safety coefficient or load coefficient; h is the height of section; f_{ps} is the effective prestress of reinforcement; and M_o is the ultimate bending moment bore by the section.

Assume that the neutral axis of the section of concrete is at about 30 % of total height of the section from the top concrete fiber, the needed compressed area of concrete is given as:

$$A_c' = r \cdot M_o / \left(0.8h \cdot 0.85 f_c'\right) \tag{4.38}$$

where $f_c{}'$ is the strength of concrete; r is the safety coefficient or load coefficient; M_o is the ultimate bending moment; h is the height of section.

Example 2 Conduct a preliminary design for the section of prestressed concrete beam in its ultimate state using given information. Total bending moment the section undertakes is 400 kN m; total height of section is 0.9 m; the ultimate strength of reinforcement is 1500 MPa; and the ultimate stress of concrete is 10 MPa.

Solution: in accordance with Eqs. (4.37) and (4.38), 2 is taken as the value of safety coefficient of reinforcement and 2.5 as that of concrete.

$$A_s = 2(400\,\mathrm{kN\,m})/(0.8 \times 1\,\mathrm{m} \times 1500\,\mathrm{N/mm^2}) = 667\,\mathrm{mm^2}$$
$$A_c' = 2.5(400\,\mathrm{kN\,m})/(0.8 \times 1\,\mathrm{m} \times 0.85 \times 30\,\mathrm{N/mm^2}) = 49{,}020\,\mathrm{mm^2}$$

A preliminarily designed section could be sketched as shown in Fig. 4.23.

(3) **Shear design of prestressed concrete component**

Shear design of prestressed concrete component gives the priority of consideration to the shear resistance of concrete and shear reinforcement. Note that dowel effect of prestressed reinforcement is ignored. Three different beams are shown in

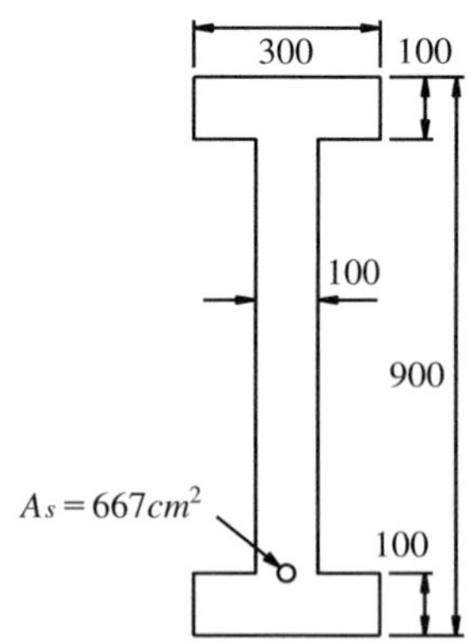

Fig. 4.23 Preliminarily designed section

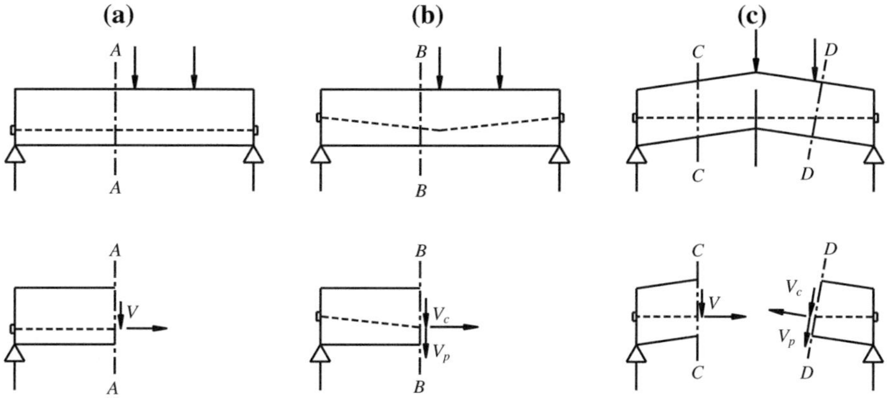

Fig. 4.24 Shear force, respectively, bore by concrete and prestressed reinforcement. **a** Beam with straight tendon. **b** Beam with skewed tendon. **c** Beam with straight tendon

Fig. 4.24. In Fig. 4.24a, prestressed reinforcement is perpendicular to the normal section and the shear resistance of beam is hundred percent provided by that of concrete V_c. In Fig. 4.24b, prestressed reinforcement is skewed toward the normal section and the shear resistance of beam comes from that of concrete V_c and the component of that of skewed prestressed reinforcement V_p, the contribution of which equals to that of bent-up reinforcement to the shear resistance of an ordinary reinforced concrete beam. In Fig. 4.24c, prestressed reinforcement is perpendicular to normal section *C*–*C* and therefore could not undertake the shear force.

The principle of design in Fig. 4.24b is to set prestressed reinforcement in the way that it tilts from the axis of beam so that a portion of shear force acting upon the beam is undertaken by prestressed reinforcement. As shown in Fig. 4.25a, since the prestressed reinforcement has certain degree of sag, under the action of distributed load a portion of the shear force acting upon the beam is undertaken by reinforcement. Figure 4.25b illustrates the situation when the inclination of

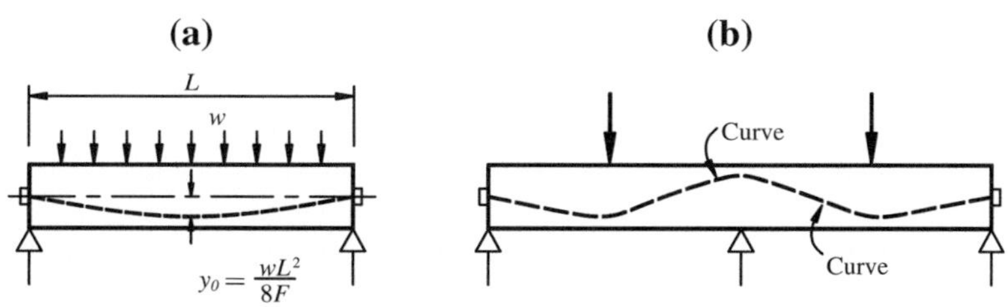

Fig. 4.25 Prestressed reinforcement undertakes part of shear force. **a** Simply supported beam under action of distributed load. **b** Continuous beam under action of concentrated load

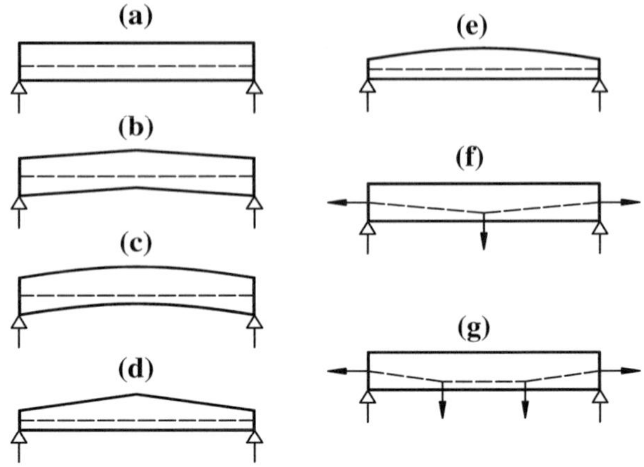

Fig. 4.26 Pretensioning prestressed reinforcement

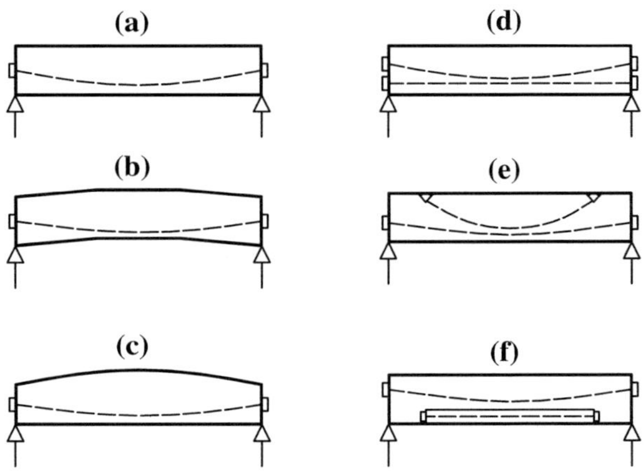

Fig. 4.27 Post-tensioning prestressed reinforcement

prestressed reinforcement is regulated so that it could undertake a portion of shear force acting upon the continuous beam.

(4) **The arrangement of reinforcement in prestressed concrete component**

The arrangement of reinforcement and the process of construction of prestressed concrete component are determined by two critical sections: the section with maximum bending moment and the section at the supports. An eccentric arrangement of reinforcement provides the former with relatively large reverse bending prestress. As for the latter, checking of reliability has to be conducted, including shear resistance of concrete, local bearing capacity, the anchorage of prestressed reinforcement, and construction safety.

Figures 4.26, 4.27 and 4.28 illustrate the arrangement of prestressed reinforcement in simply supported beam and in cantilever beam when frequently used pretensioning method or post-tensioning method is administered.

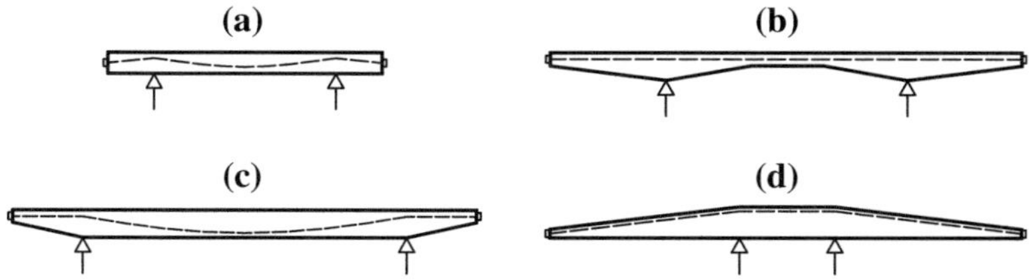

Fig. 4.28 Prestressed reinforcement in cantilever beam with post-tensioning method. **a** Short span. **b** Straight tendon. **c** Cantilever with variable depth of section. **d** Long cantilever

4.3 Structural Joints and Connections

1. Rigidly constrained connection and unconstrained hinged connection

The connection of horizontal components should meet the strength requirement, which demands a same or higher strength of the joints to guarantee that structural damage would not occur at joints. There are two ways of connection of horizontal components in a steel structure: high-strength bolt and welding on-site. Similarly, the connection of horizontal components in a reinforced concrete structure is comprised of casting joints on-site and precast joints. Reinforcement should be reserved within precast component to facilitate on-site casting of joints and to meet the requirement of global stress of structure. Figure 4.29 shows a diagram of joint connection. Pay attention to distinguish between rigidly constrained connection (rigid connection) and unconstrained connection (hinged connection).

The stress state of joints generally is complicated. To reduce the stress at joints, the inflection point where no bending moment exists is the best place to place joints (as shown in Fig. 4.29a). Rigid connection could be adopted as much as needed for frame structures (as shown in Fig. 4.29b, c), which enables the transfer of negative bending moment at joints to reduce the bending moment acting upon the section. However, in actual construction, when the connection is designed to be rigidly constrained, a "real" rigid connection could not be guaranteed due to the action of

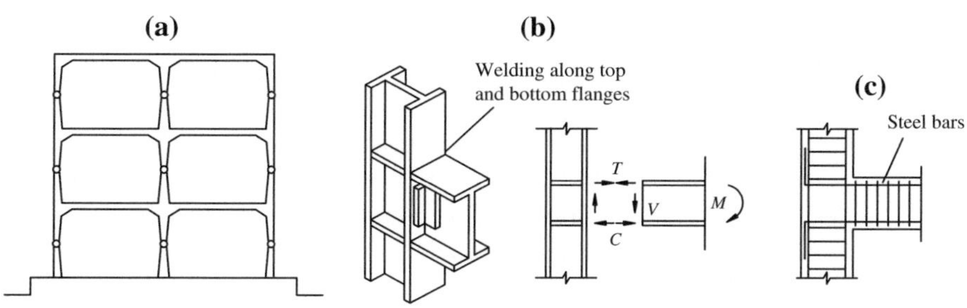

Fig. 4.29 Connection between components

various factors. As a matter of fact, connections are usually found to be semirigidly constrained. Therefore, when we check deformation, rigidly constrained connection is usually taken as semirigid constraint, which resembles in a more realistic way the state of stress and deformation.

2. **The toughness of structure and its application**

The toughness of structure refers to a potential capacity to prevent failure by absorbing energy under the action of external factors (usually comes as the input of a huge amount of energy within a short span of time [26]). Practical experience proved that many a structure failed due to the wreckage of joints. Therefore, a sound design of connection is of significant importance to prevent sudden failure of structure (please refer to Chap. 2). Only when joints function in a reliable way as expected could the toughness of structure be utilized. Thus, prior to the discussion of the toughness of structure, we have to understand the importance of an ensured working state of joints.

To utilize the toughness of structure, plastic deformation and plastic hinge produced in the cross section due to the development of plasticity is allowed [26–28]. After the formation of plastic hinges, apart from the guarantee that the ultimate bending moment could still be transferred in that section, rotation resistance is lost at plastic hinge, which enables a free rotation of structure in that section. In this way, plastic deformation of structure is increased, which absorbs much more energy and turns it into internal energy of plastic deformation. Note that in the process of plastic deformation, a certain amount of bearing capacity still remains within the structure until excessive plastic deformation gives rise to destruction. Structure capable of plastic deformation of such kind is designated as ductile structure.

When it comes to the design of structure in seismic region, a ductile design of structure is often employed so that the structure could be up to the criteria of seismic fortification, specifically "to be structurally intact in minor earthquakes; to be repairable in medium-level earthquakes; to remain standing in major earthquakes." Only ductile structures could accomplish "cracking without collapse" when they are struck by major earthquakes and guarantee the safety of people and property [29–31]. The ductility of structure is usually measured by the magnitude of lateral displacement at the top of structure. Ductility ratio could be defined as: $\mu = \Delta u/\Delta y$, where Δu refers to the lateral displacement at the top when the structure is in its ultimate state and Δy refers to the lateral displacement at the top of the structure when one section of it generates plastic yielding (plastic hinge appears) [32]. It is more convenient to repair when the plastic hinge is generated in the beam as shown in Fig. 4.30a and it is more difficult when the plastic hinge is generated in the beam as shown in Fig. 4.30b, which should be avoided in all circumstances. Plastic hinge could raise ductility ratio of structure and thereby enhance the seismic performance of structure. In general, the ductility ratio μ of lateral displacement at the top of anti-seismic reinforced concrete structure could be up to 3 or 4; when it comes to seismic design for steel structure, μ could be as high as 7 or 8 or even more.

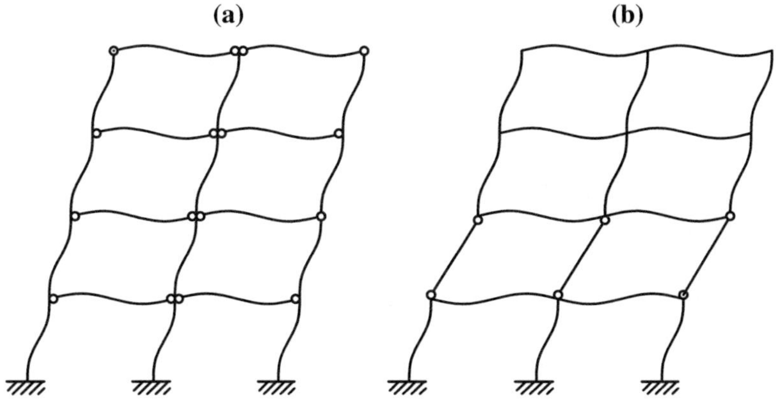

Fig. 4.30 Plastic hinges and their positions in a rigid frame

The appearance of plastic hinge in a structure indicates that the structure has entered the stage of plastic deformation. When this occurs, deformation multiplies rapidly with the addition of load. However, since columns are the major load-bearing components, plastic hinges are expected to appear in beams rather than columns. In this case, columns could be working in an elastic state or with relatively minor plastic deformation. When major deformation manifests in columns, it does not only raise more difficulty to repair but also induces additional bending moment or even the collapse of the entire structure.

4.4 Decomposition of Structural Systems

In regard of a complicated structural design project, designers should not be puzzled by the complexity of the structure. In the early stage of plan selection, designers are recommended to take design requirements into consideration in an objective and all-around manner. Based on the previous experience, sketching of major components could be conceived.

Take the suspension bridge in Fig. 4.31 for example, the preconditions are as follows:

1. Demand: wide span (typically more than 500 m);
2. Foundation: bedrock, favorable condition.
3. Terrain: no islands in river, relatively even terrain;
4. Construction: equipment and technology for large-scale marine construction are available;
5. Material: supply and production of pipes, cables, piles, and corresponding building materials are available.

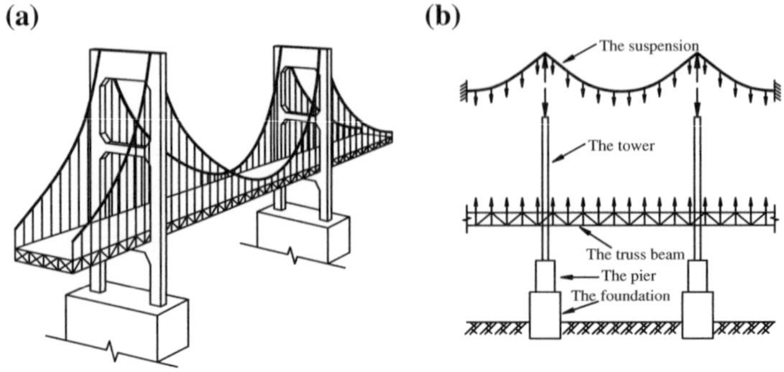

Fig. 4.31 Structural design of this suspension bridge. **a** Suspension bridge. **b** Decomposition of its major components

Conception process is as follows: (1) from previous experience, suspension bridge in Fig. 4.31a would serve as a perfect candidate to meet all the requirements of this case; (2) Fig. 4.31b shows the force decomposition of the main components of this suspension bridge; (3) loads are transferred to truss beams which are lifted through nodes by vertical tension rods of the main suspension cable; (4) the main suspension cable is supported by towers and anchored to the foundation at both sides; and (5) towers are constructed upon piers and the foundation.

Therefore, the overall structure of this suspension bridge can be broke down to substructural horizontal load transferring system and vertical one, as shown in Fig. 4.32. Horizontal system, including bridge truss beams, corresponding suspenders, and the main suspension cable, is to transfer the deck load to towers, piers, and then to the foundation. Vertical system, including towers, piers, and the foundation, is to transfer the load from horizontal substructural system to the foundation. The force transfer between these two systems is quite clear.

From the analysis in Figs. 4.31 and 4.32, during the preliminary design state, designers should process general ideas about the major substructures, their compositions, and interactions. Then, an overall structural design could be formed.

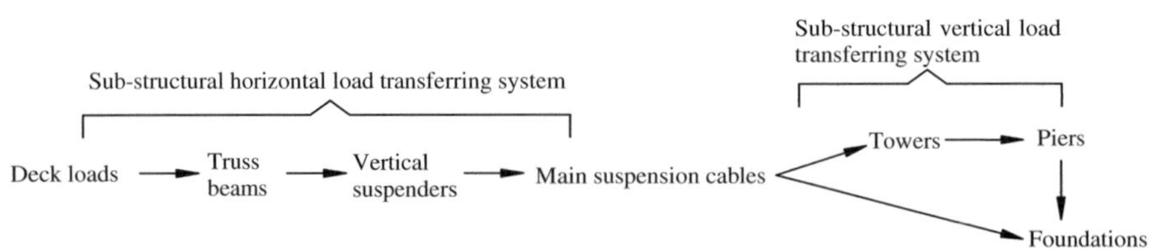

Fig. 4.32 Overall structure of suspension bridge decomposited into substructural systems

Having a clear understanding of the space and load transfer path, attention will be given to major components of vertical or horizontal substructures along the load transfer path.

Although the detailed design of substructures or major load transfer components is not the mission in the stage of conceptual design, it will be helpful to make an estimation of major vertical or horizontal substructures to perform feasibility assessment and preliminary cost estimation.

When it comes to a more specific stage of conceptual design, more attention should be given to the specific structural members, analyzing and calculating the each major component in details.

In sum, as the structural design proceeds, the subject of analysis will change from the overall structure to substructures and then to components. More and more focus should be given to details, which take from space conception, to plan components, and then to specific structural components (from linear to nonlinear). In conceptual design, the emphasis is the transition from the overall structure to substructures, or in other word, from three-dimensional space to two-dimensional components. Stress analysis will only be performed to specific major components when necessary.

Focus during different design stage could be summarized as follows:

1. Conceptual design: Focus is on overall space and structure conception, including load-bearing capacity of the entire structural system, and interactions between the overall structure and its substructures.
2. Preliminary design: Focus shifts to horizontal and vertical substructures so as to establish the interactions between major components.
3. Detailed design: Focus is now on the design of linear or nonlinear structural components and connections, during installation and construction.

It is crucial for designers to understand the characteristics of the overall structural entity and the interactions between the entity and its substructures during every design stage.

In the stage of conceptual design, designers' concern should be to establish a structural entity of three dimensions to achieve the purpose of being expressive of the space formation of the structure and being in harmony with the surroundings. In Fig. 4.31, horizontal substructural system transfers the load to vertical substructural system and then to the foundation, which could be seen as a continuous beam supported by four pedestals. In this case, the overall structural system could be taken as a three-dimensional entity bearing loads from vertical and horizontal directions.

In the stage of preliminary design, force analysis of horizontal and vertical substructures together with mechanical characteristics of major components is to be dealt with. Towers and piers in Fig. 4.31 are the major components of the vertical substructure while deck truss beams, suspenders, and cables function as the major components of the horizontal substructure. Estimations of major components sections should be given in this stage.

In the stage of detailed design, attention is to be paid to details such as structural components and connections. The completion of construction layout is expected. Further refinement of sections and connections is required so as to everything works properly. As shown in Fig. 4.31, material selection and the determination of the sizes of towers, suspenders, cables, and connecting nodes are all on the agenda of this stage.

These three stages of design and the derived results are interdependent upon and complementary to each other. Layouts of the structural entity and substructures are proposed in the stage of conceptual design; substructural systems are determined and sections of the major components are estimated in the stage of preliminary design; sections of the structural members and details of the connections are mapped out in the stage of detailed design. The latter does not only refine the former but also provide feedbacks and revisions, which mean to modify the previous design or to conceive new ones in view of the issue emerged from the latter stage. For instance, in the detailed design stage of high-rise structures, when the story height of the original design is found insufficient, modifications might be suggested to be made to the horizontal substructures. For instance, more efficient horizontal substructure system could be adopted to increase usable story height. When necessary, the entire structural system could be changed (e.g., replace the integral frame system with tube-in-tube system).

4.5 Integrated Structures and Substructures

Structure, no matter how complicated it might look like, can always be broke down to substructural systems. Quality of substructural systems design directly affects that of the overall structure. This chapter discusses the principle of the substructural design of horizontal substructures and vertical substructures, emphasizing the approach of "starting from the overall, focusing on the detail."

In the early stage of structural design, the architect usually thinks from a global perspective and endeavors to make an overall project that will meet the needs of structural function as well as esthetic expression. Substructure is the first item on the agenda when it comes to project feasibility and economy. The overall conception will be improved through the refinement of substructural design.

A vast majority of the buildings are expected to meet various seemingly contradictory application requirements. For instance, structures are not only expected to have large space and wide span but also small story height and low cost. Day-lighting and ventilation are well anticipated together with the requirements for heat preservation and insulation. Engineers have to balance these demands, offering varied conceptions of structures as well as their substructures.

After laid down the groundwork for conceptions of the overall structure and its substructures, it is necessary to determine the dimensions of major structural components so as to have a comparatively accurate estimation of the costs.

The process of architectural design starts from the overall concept, then proceeds to the conception of its substructures, and then to the details of the structural components. It is an analytical decomposition process of three dimensions to two dimensions and then to one. The overall functional requirements have to be met throughout the entire process so that the resulting integrated structure will suffice.

The relation between the design of the overall structure and local substructures is that the former controls the latter and the latter perfects the former. The roof design shown in Fig. 4.33 serves as a substructural design example. Being able to meet all the functional requirements, these three conceptions and their varied costs represent different overall designs. Figure 4.33a shows a flat roof system. Rigid connection between the roof and column has to be considered in order to create a frame structural system. Flat roof system needs to be able to resist a larger bending moment. Figure 4.33b, c shows, respectively, the folding-board roof system and arched roof system. In these two systems, horizontal pushing force from the roof to the column has to be taken into consideration. Compared with the flat roof, the bending moment of the folding-board roof and the arched roof is smaller, so the roof board can be made relatively thinner. The appearance of arched roof is more appealing, but the construction is more demanding. Therefore, the determination of a local conception is indispensable to that of an overall design; they are complementary to each other.

A rational overall conceptual design could be achieved through efficient load-bearing and transferring systems, especially for major components. As shown in Fig. 4.34a, compared with the flat roof, if reasonable force-bearing arched roof is chosen, a better load-bearing efficiency is anticipated. As shown in Fig. 4.34b, compared with linear shear wall, the horizontal force bearing of vertical substructures of arc shear wall is more rationalized and the façade of the overall design will be more diversified and the horizontal stiffness outside the shear wall plane will also be improved. As shown in Fig. 4.34c, compared with the frame structure, the force bearing of vertical substructures of core tube is more efficient and the vertical and

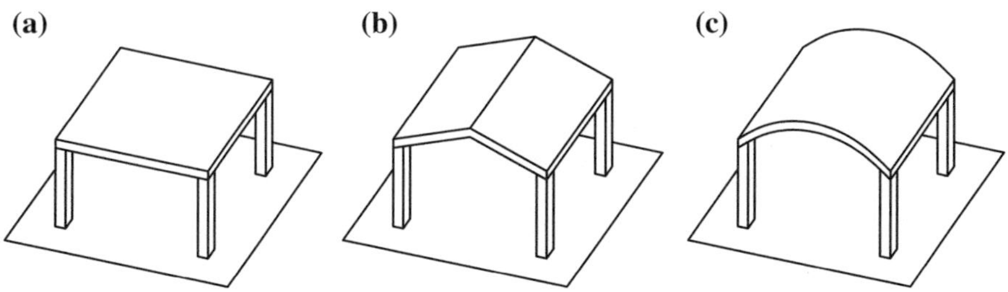

Fig. 4.33 Designs of the roof system. **a** Flat roof. **b** Folding-board roof. **c** Arched roof

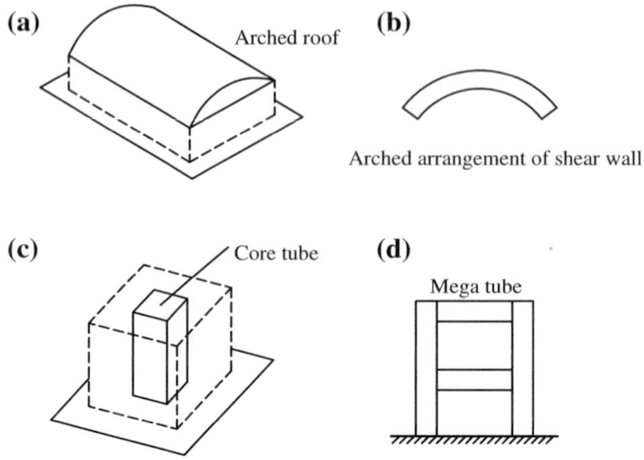

Fig. 4.34 The relationship between the overall analysis and local analysis

horizontal load resistance of the overall structure will also be improved. As shown in Fig. 4.34d, compared with the independent tube system, horizontal giant tubes can better connect independent vertical tubes and constitute a giant framework and the horizontal load resistance of the whole design is to be greatly enhanced.

The purpose of all structures is to provide a safe and comfortable space for different operational functions. Thus, apart from meeting the requirements for strength and deformation, structures have to provide available internal space. And this is closely related to selected load transferring systems. The investigation of spatial load transfer of the overall structure must be integrated with the internal

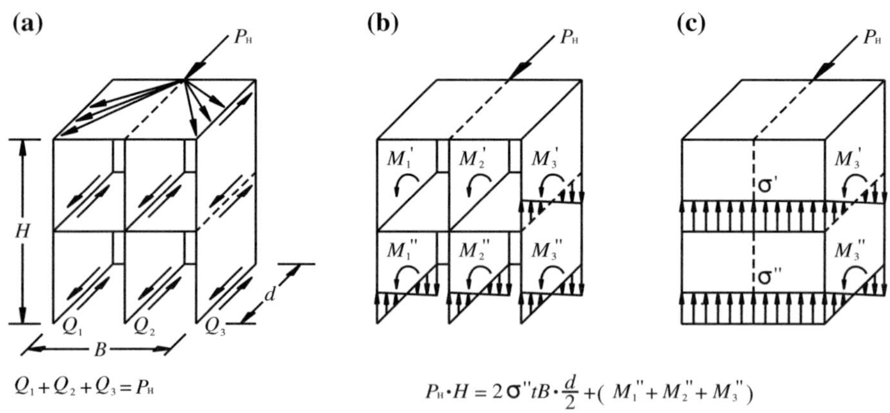

Fig. 4.35 Spatial load transfer systems of the structures

spatial decoration of the substructures, namely the plane layout and the load transfer of the substructural system.

Figure 4.35 shows a two-span and two-storied building surrounded by a shear wall. The thickness of the shear wall is *t*, the height of the building is *H*, the width is *B*, and the depth is *d*. We will analyze the process of the internal load transfer of the building when the horizontal load P_H is acted upon the roof as well as the interaction between the horizontal and the vertical substructural systems in their interaction to resist bending moment and shear force.

Figure 4.35a shows the distribution and transition of the horizontal load P_H along three pieces of longitudinal shear walls. As shown in the graph, the horizontal load P_H is borne by the shear wall in the same direction; P_H is transferred from the wall to the foundation when Q_1, Q_2, and Q_3 counteract and balance with the reaction from the foundation. Figure 4.35b shows the moment aroused by the horizontal force P_H, M_1', M_2', and M_3' of the second floor and M_1'', M_2'', and M_3'' of the foundation. Figure 4.5c shows the normal stress σ', σ'' of the shear wall derived from the moment caused by the horizontal force P_H.

When the shear wall is thick enough, local buckling will not happen and vertical loads will be evenly distributed down the shear wall. However, when the horizontal load is imposed upon the top, due to the insufficient plane stiffness of thin-walled shear wall, buckling at the application point and deformation of the tube are prone to happen. At time like this, transverse reinforcement of the top or vertical bracing should be executed so that the horizontal load on the top will be able to transfer to the foundation through the wall, as shown in Fig. 4.35a.

Compared with open structure, enclosed box-like structure is more effective in strength and in buckling resistance. Even though the horizontal and the vertical systems are comprised of thin-walled planes, horizontal and vertical loads still can be easily transferred from the top to the foundation. Thus, this kind of space structure is very effective in force bearing.

As shown in Fig. 4.35, if the horizontal load is evenly distributed along the height of the structure, shear force of the foundation will be proportional to the height of structure with the bending moment proportional to the square of the height. Therefore, the lower half of vertical substructures of a high-rise need to be enhanced. This is also true for the foundation, for it bears the largest bending moment and shear force. Proper attention should be given during design process.

It should be pointed out that when the horizontal force P_H produces a bending moment, vertical planes of the shear wall will bear compressive stress, which may give rise to a local buckling. However, due to the clapping effect of the surrounding shear wall, local instability usually will not happen. The thin-walled structure is like a closed box. For the two-way resistance of bending moment and shear force, the structure shows good load-bearing capacity. Mechanical characteristics of box-like thin-walled structure under the action of horizontal loads can be summarized as follows:

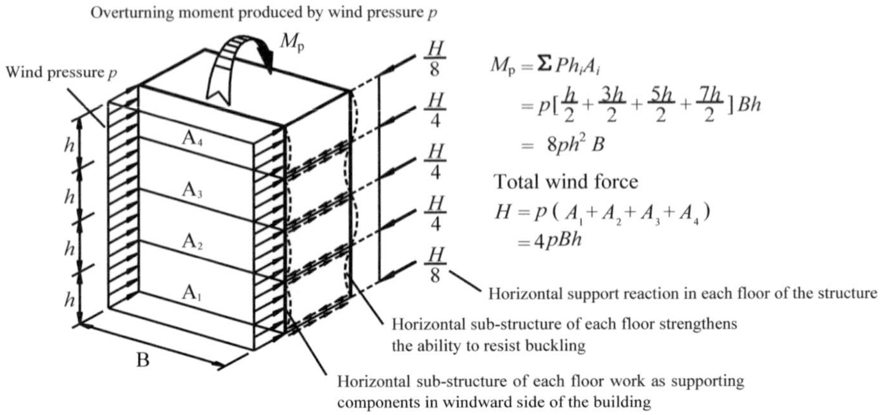

Fig. 4.36 Interaction between horizontal substructures and vertical substructures

1. Horizontal substructures should be able to bear and transfer the horizontal loads to vertical substructures while maintaining the geometry of the outer surface;
2. Vertical substructures should be able to transfer the vertical loads to the foundation and resist the shear forces derived from the horizontal loads;
3. Vertical substructures should be rigidly connected to the horizontal substructures so as to improve the overall resistance to bending moment and the shear force;
4. As for the connecting parts between horizontal and vertical substructures, measures should be taken to ensure no unreasonable stress concentration;
5. Components such as shear wall and columns need to be properly designed to resist overturning moment produced by horizontal loads.

As shown in Fig. 4.36, horizontal substructures (the floor) will not only enhance the local strength and the stability of vertical substructures but also build a load-resisting system by rigid connections. Besides, vertical substructures should be able to transfer the wind load to horizontal substructures. The shear force undertaken by the shear wall could transfer the shear force to the foundation.

The core of the space composition is to materialize structures with diversified appearances and to transfer loads to the foundation, by fully utilizing the combination of horizontal and vertical substructures. In fact, the same external appearance can be designed with different internal structures. Figure 4.37 illustrates different internal structures with the same outside. Figure 4.37a shows the tube-in-tube structure with one internal and one external tube. Figure 4.37b shows the tube-in-tube structure with four internal tubes and one external tube. Figure 4.37c shows the bundled-tube structure using shear walls.

Mechanical characteristics, classification, and simplified calculation of horizontal and vertical substructural systems will be introduced next.

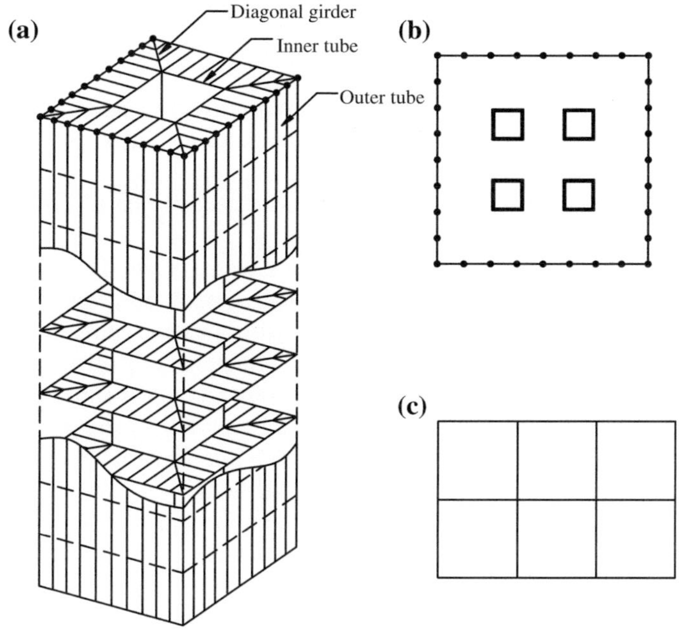

Fig. 4.37 Different internal structures. **a** The tube-in-tube structure with internal and external tube. **b** Tube-in-tube structure with four-fold internal tubes and one-fold external tube. **c** Bundled-tube structure

4.6 Characteristics and Simplified Analysis of Horizontal Substructure Systems

4.6.1 Characteristics of Horizontal Substructure Systems

The mechanical characteristics of horizontal substructures are to resist the bending moment and shear force so that the requirements of strength and stiffness are to be met.

The fundamental simplified approach of structural design is to take the structure as a simplified entity. The same approach could be applied to the conceptual design of substructural systems. In this section, we will take horizontal substructural system as a two-dimensional simplified entity and analyze its resistance of bending moment and shear forces.

We will take the horizontal substructure as a two-dimensional board system to analyze. Horizontal substructure is usually comprised of slabs, beams, or trusses. It bears the bending moment and shear forces and bears and transfers the vertical loads from the floor or the roof. Meanwhile, horizontal substructure is still required to bear and transfer the horizontal wind load or seismic load from vertical substructural system. An effective and efficient load-bearing structure should integrate horizontal and vertical substructures. Therefore, the design of horizontal substructure is closely related to the layout of vertical substructural systems, such as columns, frames, and the shear walls.

From perspective of economy, denser column grid will lead to effective beam–slab horizontal substructure. On the other hand, from perspective of structural functions, the greater the size of the column grid is, the easier the arrangement and alteration of the interior space will be. A balance needs to be found between cost and function.

Common horizontal substructures are plane system, beam–slab system, primary and secondary beam–slab system, two-way slab system, and space truss system. Table 4.3 shows the minimum thickness of cast-in-place reinforced concrete slab specified in the code for the design of concrete structures GB50010-2002 [21]. We can see the difference in material usage of various horizontal substructures due to their varied force-bearing mechanical characteristics. For example, the minimum thickness of cast-in-place reinforced concrete slab in beamless floor (plane system) is 150 mm, while that of the multi-rib slab is 40 mm.

In this section, we will discuss some common characteristics of horizontal substructures. And in the next, we will see about mechanical characteristics and simplified calculation methods of different horizontal substructures, such as plane system, beam–slab system, primary and secondary beam–slab system, two-way slab system, and space truss system.

1. The resistance analysis of the bending moment of horizontal substructure

The overall analysis of horizontal substructure mainly concentrates on the influence of vertical supports on the resistance of bending moment and shear forces of horizontal substructure. Generally, horizontal substructure is assumed to be an ideal slab. The force bearing of an ideal slab is directly related to the location and the arrangement of supports. The force bearing of the same slab is different when it is in different support systems. Assuming the maximum bending moment of the slab, supported on four sides bearing a uniformly distributed load, is m, as shown in

Table 4.3 Minimum thickness of cast-in-place reinforced concrete board [21]

The category of boards		Minimum thickness (mm)
One-way slab	Roof sheathing	60
	Slab in civil buildings	60
	Slab in industrial buildings	70
	Slab under the carriageway	80
	Two-way slab	80
Multi-ribbed slab	Rib spacing ≤700 mm	40
	Rib spacing >700 mm	50
Cantilever slab	The length of cantilever slab ≤500 mm	60
	The length of cantilever slab >500 mm	80
Slab without beams		150

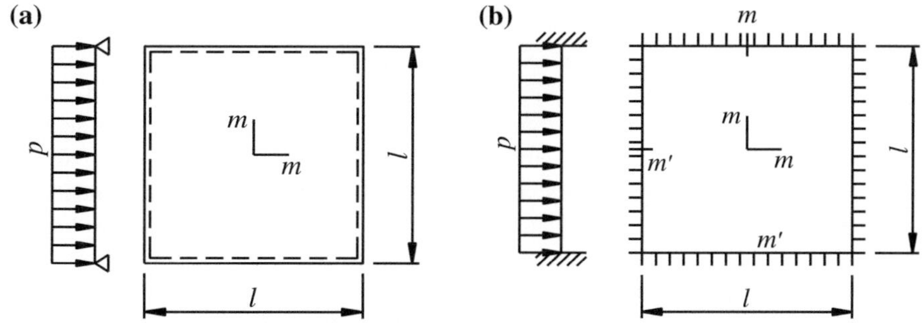

Fig. 4.38 Differently supported slabs under stress. **a** Four-edge simply supported slab under stress. **b** Four-edge fixed slab under stress

Fig. 4.38a. The mid-span maximum positive bending moment of the slab, fixed on four sides bearing the same uniformly distributed load, as shown in Fig. 4.38b, is about (1/3)m and the maximum native bending moment of this slab is about (2/3)m.

The force bearing of an ideal slab is directly related to the load transfer which in turn relates with its supports. Figure 4.39 illustrates the different forms of load transfer of one-way and two-way slabs. The force flow of one-way support slab is one-dimensional, and that of two-way support slab force is two-dimensional. Therefore, the bearing capacity of a two-way support slab is about two times that of one-way support slab of the same size and the same material. The load transfer of a two-way support slab is related to the length/width ratio. When the length/width ratio of a two-way support slab is greater than 2, this two-way support slab will transfer more than 90 % of the total bending moment to the support of the short side and only less than 10 % to that of the long side. In this case, the load transfer of this two-way support slab is more like that of one-way support slab and can be simplified as a one-way slab for the sake of calculation.

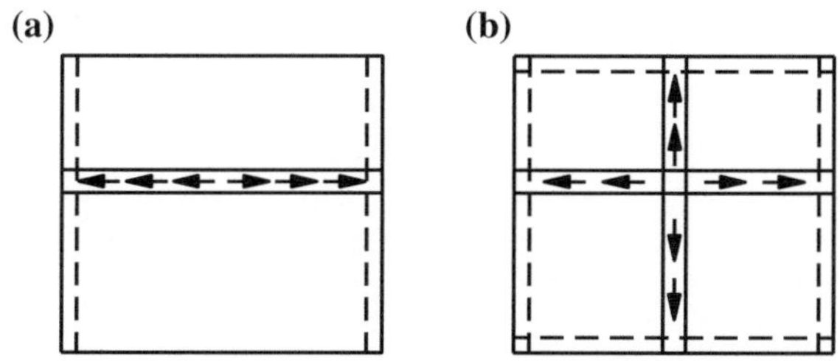

Fig. 4.39 One-way- and two-way-supported slabs. **a** One-way-supported slab. **b** Two-way-supported slab

2. The resistance analysis of shear forces of horizontal substructure

Since local stress concentration will be derived from the support of slabs, the resistance of punch and shear forces of the support or that of the slab around the top of the column has to be checked. Usually we will check the resistance of punch and shear forces of the concrete and reinforcements on the 45° punching-shear surface formed in slab due to supports. As shown in Fig. 4.40, the 45° punching-shear area is 1, the intersecting perimeter of the punching-shear surface with the mid-plane of the slab is 2 and we multiply the perimeter 2 with the thickness of the board to obtain the results of the punching-shear area of the concrete. The shear resistance of the concrete on this surface should be greater than the punching-shear force of the board.

The following measures of anti-punching-shear forces will enhance the supports or that of the slab around the top of the column:

1. Allocate shear-resisting reinforcements in the slab around the top of the column (Fig. 4.40b);

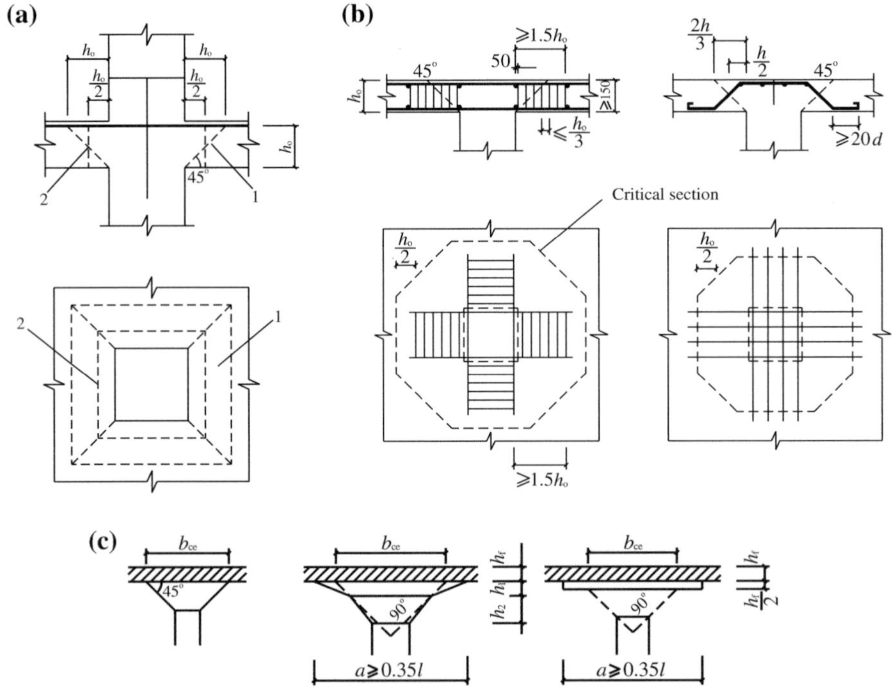

Fig. 4.40 Punching-shear calculation and enhancement measures of the slab. **a** Punching-shear critical section. **b** Allocate shear-resisting reinforcements. **c** Add column caps or pallets to the *top* of the column

2. Add column caps to the top of the column to increase the shear-resisting capacity of concrete slab (Fig. 4.40c);
3. Add cross-beams to the top of the column to minimize the effect of punching-shear forces;
4. Bend prestressed reinforcements at the top of the column to enhance the punching-shear resistance of the concrete slabs.

4.6.2 Simplified Analysis of Horizontal Substructure Systems

1. **Flat-slab system**

General features

Flat-slab system is a commonplace horizontal substructure in buildings. It features a flat slab directly supported by the column. The contradiction often met in flat-slab design is to decide the size of the column grid. From the architectural perspective, the greater the size of the column grid is, the easier the arrangement and alternation of the interior space will be. However, it will increase the thickness of the slab, which in turn will increase the cost. Functional and economical requirements should therefore be attended to in the design in a balanced manner.

Figure 4.41 shows a typical classification and simplified analysis of the flat-slab system. Figure 4.41a shows the flat-slab system without column caps; Fig. 4.41b shows the flat-slab system with column caps; Fig. 4.41c shows the reinforced flat-slab system with tributary areas; Fig. 4.41d shows the reinforcement arrangement in flat-slab system without column caps. As shown in Fig. 4.41c, due to the uneven distribution of the bending moment within the slab, reinforcement calculation should be conducted to check the area on the top of the column and that of the mid-span. More reinforcements are needed in the area on the top of the column since the bending moment borne by the slab is comparatively greater. The reinforcements can be reduced in the mid-span area for the bending moment borne by the board is comparatively smaller. Prestressed reinforcements once adopted in the flat-slab system will greatly increase its span. General thickness of the flat-slab system is 15–24 cm and the span is 4–8 m. When prestressed reinforcements are adopted, the span will be extended to 8–10 m.

Key points of calculation

When the flat-slab system directly supported by the top of the column is dealt with, calculation of the bending moment can be simplified as shown in Fig. 4.42 in accordance with the ideal elastic–plastic theory. Slab will be divided into column area and mid-span area. The former will bear 75 % of the total bending moment and the latter bears the rest 25 %. In addition, the total bending moment of the flat area actually is the total bending moment of the simply supported board ($wL^2/8$). Positive and negative bending moments will be distributed to the continuously

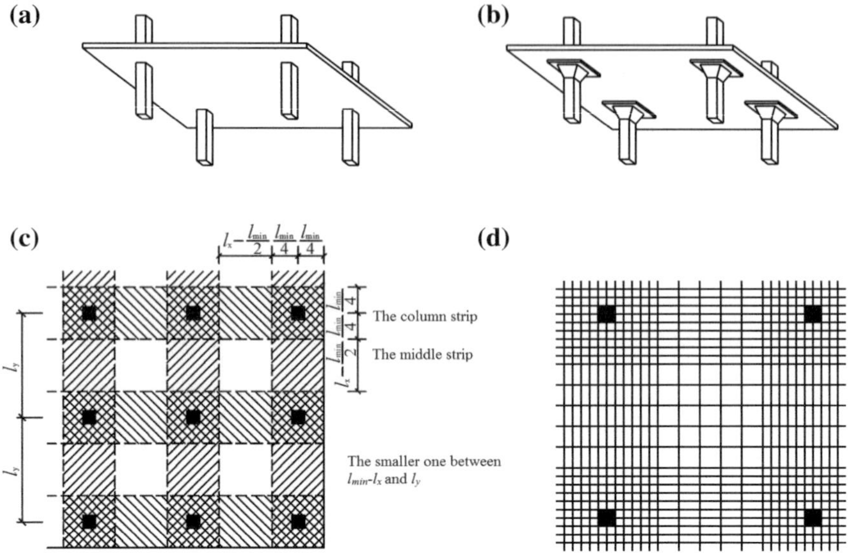

Fig. 4.41 Typical classification and reinforcement design of the flat-slab system. **a** Flat-slab system without column cap. **b** Flat-slab system with column caps. **c** Area division of reinforced floor. **d** Reinforcement sketch of the flat-slab system

supported slab. Negative moment for the board on the top of the column is (−2/3) (wL^2/8) and positive bending moment for the mid-span is (1/3)(wL^2/8). This simplified calculation of the bending moment is able to meet the requirements for preliminary design and analysis of the structure. Specific calculation formulas are given as follows:

$$\text{Column strip}\, M_{\max} = 75\,\% \times (1/3) \times (wL^2/8) = 0.25(wL^2/8) \tag{4.39}$$

$$M_{\min} = -75\,\% \times (2/3) \times (wL^2/8) = -0.5(wL^2/8) \tag{4.40}$$

$$\text{Middle strip}\, M_{\max} = 25\,\% \times (1/3) \times (wL^2/8) = 0.08(wL^2/8) \tag{4.41}$$

$$M_{\min} = -25\,\% \times (2/3) \times (wL^2/8) = -0.16(wL^2/8) \tag{4.42}$$

Taken the influence of the side span on the bending moment of the board into consideration, Formula (4.39)–(4.42) are modified as follows (4.43)–(4.46):

$$\text{Column strip}\, M_{\max} = 0.5(wL^2/8) \tag{4.43}$$

$$M_{\min} = -0.7(wL^2/8) \tag{4.44}$$

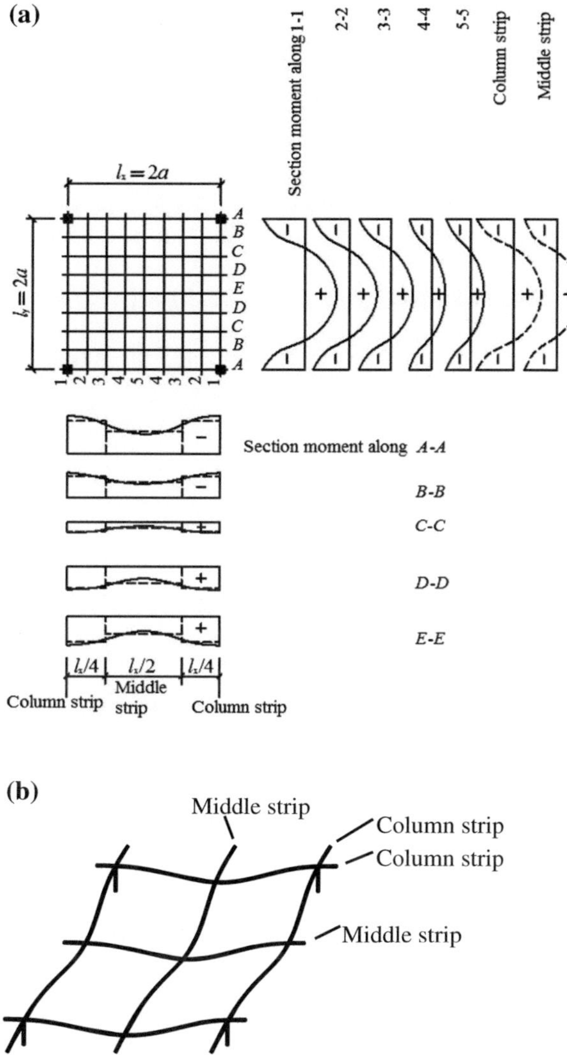

Fig. 4.42 Calculating methods of strip moments (Strip method). **a** Slab moments change along span and with sections. **b** Supports and deformation of the strips

$$\text{Middle strip}\, M_{\max} = 0.1(wL^2/8) \tag{4.45}$$

$$M_{\min} = -0.2(wL^2/8) \tag{4.46}$$

The influence of the column caps

Concrete column caps are usually added to the support on the top of the column so as to meet the anti-punching-shear requirement of the reinforced concrete slab. Column caps will not only greatly enhance the anti-punching-shear capacity of the board but also improve the bending capacity of the slab. As shown in Fig. 4.43, column caps and the board work together to achieve an arch effect so as to redistribute the stress and reduce 15–25 % of the maximum bending moment. The bending moment calculation of the slab can be simplified as follows:

$$\text{Column strip}\, M_{\max} = 75\,\% \times (1/3) \times (wL^2/8) \times 0.85 = 0.21(wL^2/8) \quad (4.47)$$

$$M_{\min} = -75\,\% \times (2/3) \times (wL^2/8) \times 0.85 = -0.42(wL^2/8) \quad (4.48)$$

$$\text{Middle strip}\, M_{\max} = 25\,\% \times (1/3) \times (wL^2/8) \times 0.85 = 0.07(wL^2/8) \quad (4.49)$$

$$M_{\min} = -25\,\% \times (2/3) \times (wL^2/8) \times 0.85 = -0.14(wL^2/8) \quad (4.50)$$

Example The dimensions of a floor are as shown in Fig. 4.44a. Vertical uniformly distributed load is designed to be 10 kN/m^2. Estimate the maximum bending moment of the control section.

Solution:
Divide the board as mid-span stripe (stripe 1) and side-span stripe (stripe 2). Then subdivide them into column area and mid-span area.

1. Mid-span stripe (stripe 1), linear uniformly distributed load is w_1 = 7 m × 10 kN/m^2 = 70 kN/m, L = 7 m.

Total bending moment is M_o = 1/8(w_1 L^2) = 1/8(70 kN/m × 7 m × 7 m) = 428.75 kN m.

Taken into consideration the arch effect, the bending moment of the mid-span control sections is calculated in accordance with the Formula (4.39)–(4.50) and the results are as shown in Table 4.4.

2. Side-span stripe (stripe 2), linear uniformly distributed load is w_1 = 3.5 m × 10 kN/m^2 = 35 kN/m, L = 7 m.

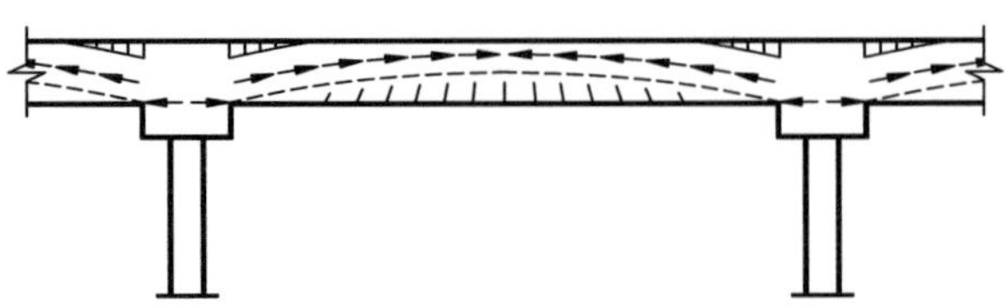

Fig. 4.43 Arch action by column hats and slabs

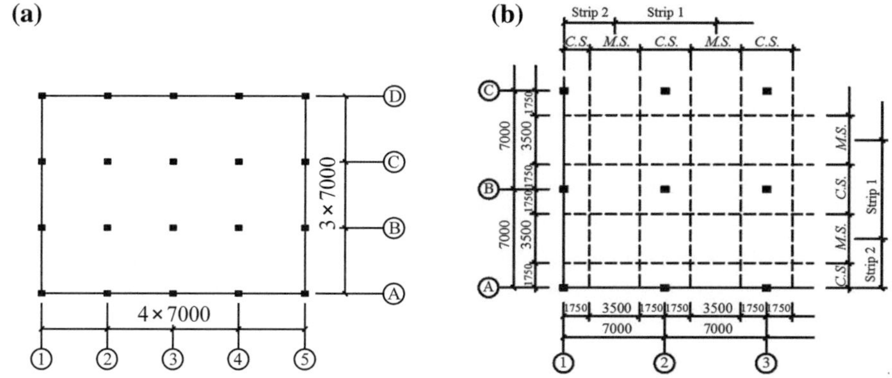

Fig. 4.44 Simplified calculation of reinforced concrete board. **a** Column grid layout of a slab. **b** Division of strips

Total bending moment is $M_o = 1/8(w_1\ L^2) = 1/8(35\ \text{kN/m} \times 7\ \text{m} \times 7\ \text{m}) = 214.38\ \text{kN m}$.

Taken into consideration the arch effect, the bending moment of the side-span control sections is calculated in accordance with the Formula (4.39)–(4.50) and the results are as shown in Table 4.4.

2. **Beam–slab system**

General features

In a beam–slab system, loads are transferred from slabs to beams and then to columns and to the foundation. In light of different forms of load transfer from slabs to beams, the system is divided into one-way beam–slab system and two-way beam–slab system. When the length/width ratio of the board supported by the beam is greater than 2, the load borne by the board will largely be transferred to the supporting beam over the width (shorter side) and rarely to that over the length (longer side). The system is therefore designated as one-way beam–slab system. When the length/width ratio of the board supported by the beam is less than 2, the load borne

Table 4.4 Bending moment of the control sections in mid-span area and side-span area

Strips	M_o (kN m)	Section bending moment (kN m)				
		Outer support in end span $-0.3M_o$	Mid-span in end span $0.4M_o$	Interior support in end span $-0.6M_o$	Middle span in mid-span $0.21M_o$	Interior support in mid-span $-0.42M_o$
Mid-span slab (Strip 1)	428.75	−128.63	171.5	−257.25	90.04	180.08
Side-span slab (Strip 2)	214.38	−64.32	85.8	−128.63	45.02	90.04

by the board will be transferred to the supporting beams on both sides and the system is therefore designated as two-way beam–slab system. Beam–slab systems, distinguished by the material in application, consist of wood beam–slab system, steel beam–slab system as well as the mixed steel-and-concrete beam–slab system.

Key points of calculation

When the length/width ratio of the slab is equal to or greater than 2, the force bearing of the slab is to be calculated as one-way beam–slab system. Assuming the load of the one-way slab will only be transferred to the supporting beam over the width, ignoring the bending moment and shear forces of the supporting beam over the length. The calculation would be as follows:

$$M_{\max} = (1/3) \times \left(wL^2/8\right) = wL^2/24 \tag{4.51}$$

$$M_{\min} = -(2/3) \times \left(wL^2/8\right) = -wL^2/12 \tag{4.52}$$

$$M_{\max} + (-M_{\min}) = wL^2/8 \tag{4.53}$$

The calculation of the two-way beam–slab system (Fig. 4.45a) will take two steps: (1) the bending moment of a two-way slab; (2) the bending moment and shear forces of the supporting beams on both sides (calculate as a continuous beam under the load shown in Fig. 4.45b). It is very difficult to accurately calculate the bending moment of a slab in a two-way beam–slab system. In the early stages of design, simplified calculation as shown in Eqs. (4.54)–(4.57) could be adopted. That is, in every major direction of force bearing, two-thirds of the board bears 30 % of the total bending moment and a third of the board together with the beam bears 70 % of the total bending moment. Calculation formulas are as follows:

$$\text{Strips over the beams } M_{\max} = 70\,\% \times (1/3) \times \left(wL^2/8\right) \tag{4.54}$$

$$M_{\min} = -70\,\% \times (2/3) \times \left(wL^2/8\right) \tag{4.55}$$

$$\text{Middle strips } M_{\max} = 30\,\% \times (1/3) \times \left(wL^2/8\right) \tag{4.56}$$

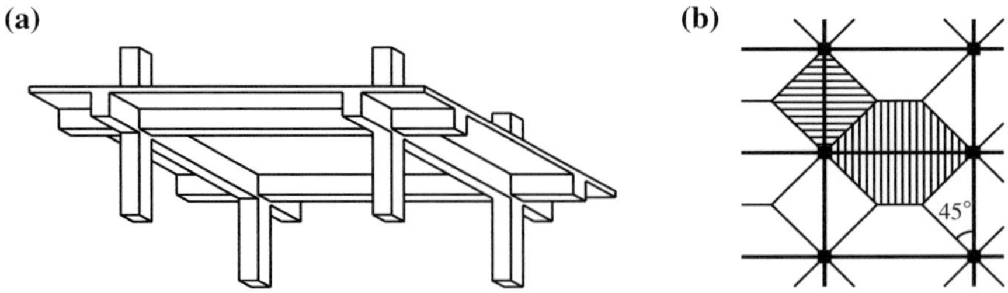

Fig. 4.45 Two-way beam–slab system. **a** Two-way beams and slabs. **b** Calculating area of beams

$$M_{\min} = -30\,\% \times (2/3) \times \left(wL^2/8\right) \tag{4.57}$$

3. **Primary and secondary beam–slab system**

General features

In a primary and secondary beam–slab system, the load transfers from the slab to the secondary beam and then to the primary beam and then to the column and the foundation. Since the space between secondary beams can be set comparatively small, the thickness of the board could be greatly reduced. Compared with the abovementioned flat-slab system and beam–slab system, the thickness of the slabs in primary and secondary beam–slab system is smaller. The thickness of the boards in this system is up to 6–10 cm, the span of secondary beams is up to 10–20 m, and the space between them is 1–3 m. Secondary beams are supported by primary beams, which will greatly reduce the number of columns, and the spacing between columns could be increased to 9–12 m.

Key points of calculation

Figure 4.46 shows a diagram of primary and secondary beam–slab system. The board in primary and secondary beam–slab system conducts one-way load transfer and the bending moment of the mid-span and that of the supports could be calculated as the bending moment of a one-way slab. The bending moment transferred from the slab to the secondary beam could be calculated as that of a continuous beam or that of a simply supported beam. The bending moment transferred from the secondary beam to the primary beam could be calculated as that of a continuous beam or that of a simply supported beam.

Steel primary and secondary beam–slab system

Figure 4.47 shows the primary and secondary beam–slab system commonly seen in a steel structural design. Calculation is the same as the aforementioned. Pressed

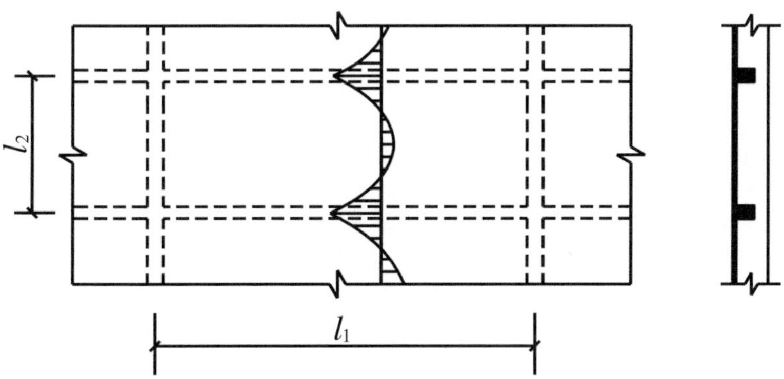

Fig. 4.46 Primary and secondary beam–slab system

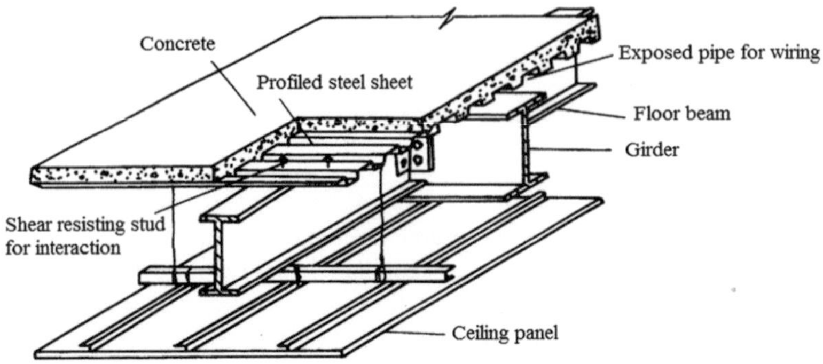

Fig. 4.47 Primary and secondary steel beam–slab system

steel plates are allocated on steel secondary beams and could serve as templates in the construction of cast-on-site concrete floor. At the completion of the hardening process, the concrete floor will bear the forces with pressed steel plates and further form a primary and secondary beam–slab system, reducing the quantity of concrete and reinforcements applied to the floor. On the other hand, structural pipelines could be preburied in pressed steel plates in this system. These pressed steel plates are to be welded to steel secondary beams with shear-resisting studs in construction so as to ensure the concerted work of the floor and secondary beams.

The beneficial influence of pressed steel plates on the force bearing of the floor is often neglected in the design and calculation of the floor system. The reason is that pressed steel plates when taken as structural components will need fire-refractory coating and the extra expenses will exceed the cost reduction from the replacement of reinforcements with pressed steel plates. Therefore, the beneficial influence of pressed steel plates on the force bearing of the floor usually does not count and the plates are only taken as one-time templates for the floor in order to expedite the construction.

4. **Two-way ribbed slab system**

General features

The feature of two-way ribbed slab system is the small spacing between supporting beams. The depth of a two-way board can be reduced to 5–6 cm and the arch effect could be utilized to diminish the bending moment of the mid-span. Two-way ribbed slab system is the optimized form of concrete flat-slab system, for it removes redundant concrete from the system (Fig. 4.48). Two-way ribbed slab system is therefore widely used. On the other hand, construction techniques of this system have been well-established. Tools such as plastic mold facilitate construction, as shown in Fig. 4.48c.

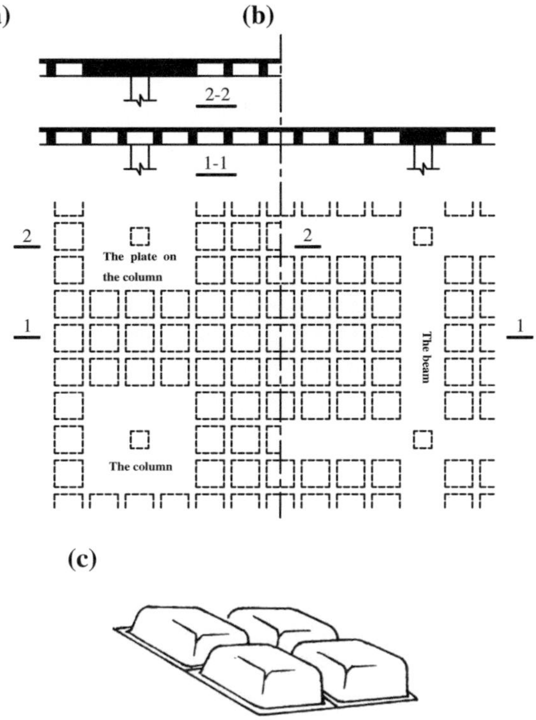

Fig. 4.48 Two-way ribbed slab system. **a** Column strips. **b** Additional secondary beams. **c** Injection molding plastic shuttering

Key points of calculation

The calculation of two-way ribbed slab system is the same as that of flat-slab system and that of two-way beam–slab system. Reference could be made to Figs. 4.42 and 4.45 in calculation. As shown in Fig. 4.48a and b, the anti-punching-shear requirement in this system is often realized through enhanced support at column, or additional secondary beams.

5. **Space truss system**

General features

It is usually more economical and more feasible to adopt steel space truss in a horizontal substructural system for large span (more than 30 m). The abovementioned flat-slab system, beam–slab system, and two-way ribbed slab system have a relatively heavy deadweight and therefore have difficulty in maintaining a wider span and the economy of the structure at the same time. According to the difference in load transfer, steel space truss system is divided into one-way space truss system and two-way space truss system (Figs. 4.49 and 4.50). One-way space truss system unilaterally transfers the load from the floor to the secondary truss beam and then to the primary truss beam, while two-way space truss system transfers the load to the truss beam in two directions.

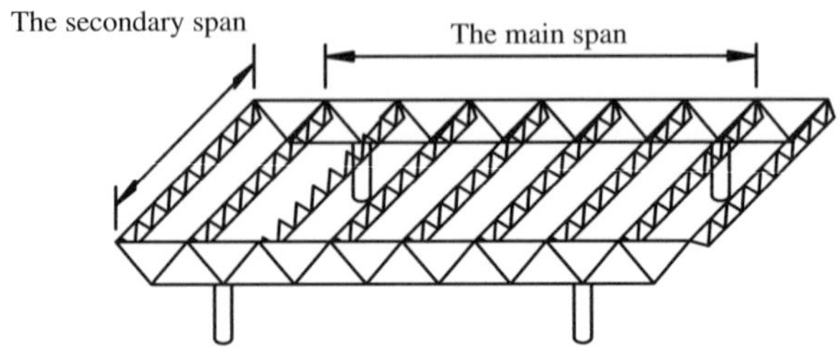

Fig. 4.49 One-way space truss system

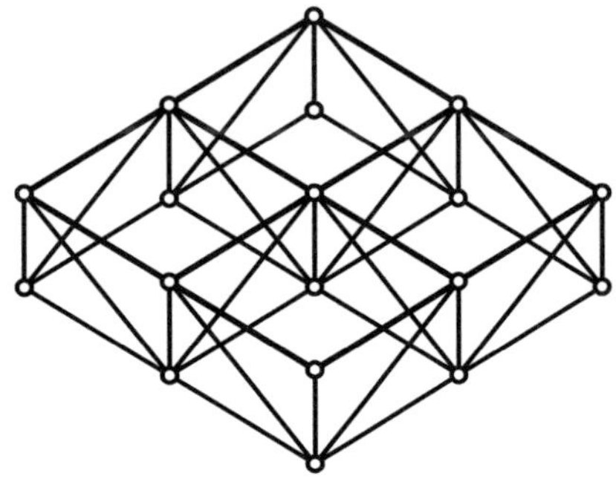

Fig. 4.50 Two-way space truss system

Three-dimensional load transfer system is constructed through reasonable arrangement of the upper and the lower chords of the space truss system. This system will transform the bending of horizontal substructures to the axial tension or compression of the truss chord, which will take a better advantage of the material and bring forth the characteristics of the space truss system, such as excellent force-bearing capacity, small deformation, light deadweight, and large span.

Through intensified trusses, as shown in Fig. 4.51, the space truss system can better control the deformation. To partially intensify trusses, rods with lager dimensions could be applied. Also, the increment of the height of the truss intensifies the truss and it serves as a local enhanced truss beam. Figure 4.51a shows a four-sided intensified space truss; Fig. 4.51b shows a four-sided and axial intensified space truss; Fig. 4.51c shows the intensified space truss for Pakistan Islamabad gymnasium, in which trusses were fortified along axial directions.

Key points of calculation

In light of the multiple indetermination of the space truss system, it is not easy to conduct precision calculation. Computerized finite element method is applied to solve the problem. In the early stages of design, space truss system could be simplified as a flat-slab system or as a primary and secondary beam–slab system to calculate the bending moment and shear force of the control sections. Sections of the upper and the lower chords have to be estimated so as to roughly derive the costs of the space truss system.

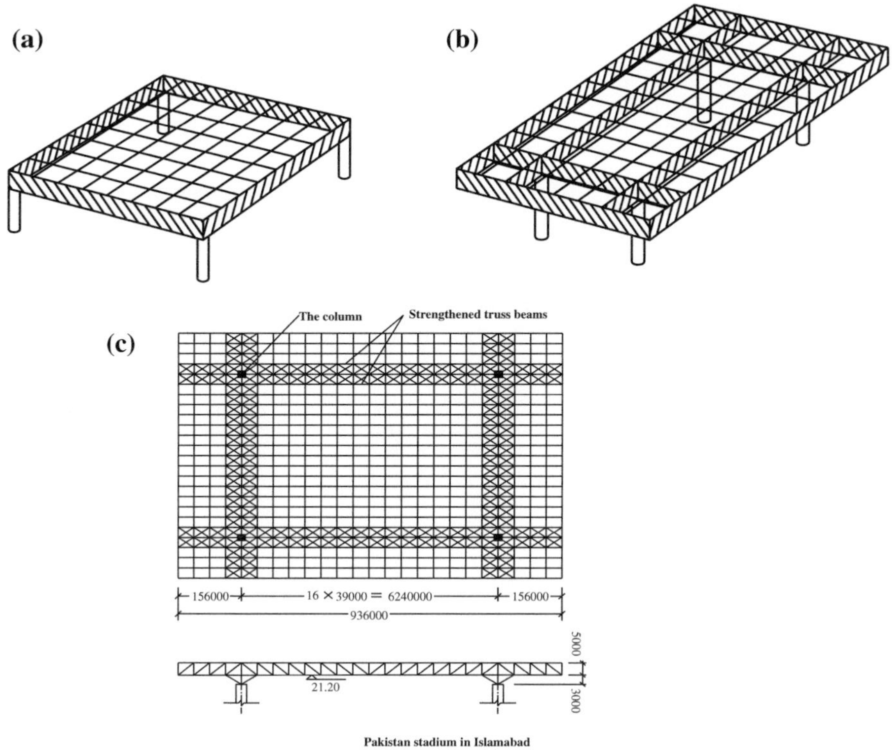

Fig. 4.51 Intensified arrangements of space truss systems. **a** Four-sided intensified space truss. **b** Four-sided and axial intensified space truss. **c** Example for intensified space truss

4.7 Characteristics and Simplified Analysis of Vertical Substructure Systems

4.7.1 Characteristics of Vertical Substructure Systems

Vertical substructures need to meet the requirements of strength, stiffness, local or global stability, bearing vertical load, horizontal bending moment, and the shear force.

In general, vertical substructures consist of many slender components. Therefore, integration with horizontal substructures is needed to achieve reliable stability.

Vertical substructures usually classified as follows:

1. Shear wall system;
2. Tube system;
3. Frame system; and
4. Column system.

To fully utilize all features of different systems, a combination is usually adopted, such as the frame- and core-tube structure, the frame-and-shear wall structure, and the mega-frame structure. The major function of the core tube or the shear wall in a frame- and core-tube structure or a frame-and-shear wall structure is to provide strong resistance to horizontal loads. Another example is mega-frame which utilize the high stiffness of connecting beams between slender tubes. It combines the merits of the tube structure as well as that of the frame structure. Mechanical characteristics and simplified calculation of these vertical substructures are to be discussed next.

4.7.2 Classification and Simplified Calculation of Vertical Substructure Systems

1. **Shear wall system**

General features

The major role of the outer walls is to enclose the structure and create usable space in the building, while the inner wall is to divide the available space. A shear wall system is achieved when the outer and inner walls are well connected, then this shear wall system, together with roof system, functions to resist vertical and horizontal loads.

The shear wall has a strong resistance of horizontal shear forces and bending moment in-plane while the resistance is greatly diminished out of plane. For instance, the length and the thickness of a solid shear wall are 10 m and 30 cm. The moment of inertia outside its own plane is $Ix = Lt^3/12 = 0.0225\ \text{m}^4$ and that in its own plane is $Iy = tL^3/12 = 25\ \text{m}^4$. $Iy/Ix = 1111$, meaning these two moments of inertia have a difference of 1111-fold, and that in turn indicates their resistance of bending moment has a difference of 1111-fold. Therefore, in-plane resistance of the shear wall to horizontal shear forces and bending moment usually is the only concern in the analysis of vertical substructures while neglecting the resistance out of plane.

Shear Wall Classification

The solid wall and the truss wall in Fig. 4.52 are two conventional types of the shear wall. The solid wall is often seen in reinforced concrete structures while the truss wall is commonly applied in steel structures. The difference lies in their different ways of load transfer: Under the action of vertical loads, the solid wall will be

Fig. 4.52 Two types of the shear walls

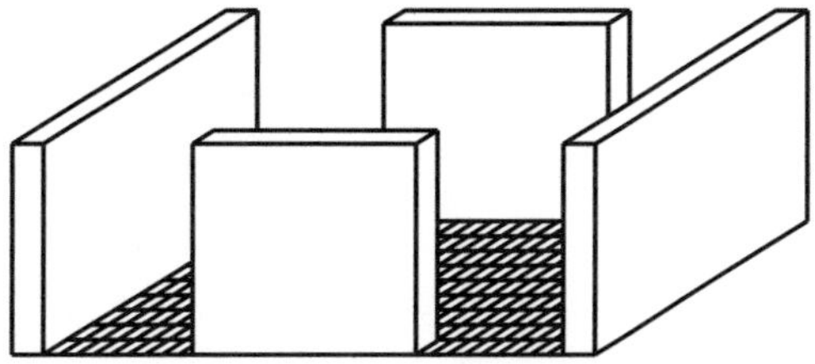

Solid shear wall

Trussed shear wall

uniformly compressed while the truss wall transfers the loads through the axial force of the components. Both of them are to fortify the horizontal strength and stiffness.

Simplified analysis of the shear wall

Under the action of vertical loads, walls are uniformly compressed. Under the action of horizontal loads, the walls are to be simplified as cantilever beams, simply supported beams, or continuous beams in calculation according to different support assumptions. Figure 4.53 is an example of calculation in which the truss wall is simplified as a cantilever beam. To simplify a truss wall, axial tension or stress is assumed to be the only force and free body diagram is applied to calculate the internal forces. In the early stage of design, locations with largest internal forces are the major concern. In the example shown in Fig. 4.53, the largest internal force locates at the bottom. Therefore, cross section could be selected and costs could be estimated.

Figure 4.54 shows the influence of the stiffness of horizontal diaphragm on the force bearing of the shear wall. For flexible diaphragm, vertical or horizontal forces borne by the shear wall in the middle double that borne by shear walls on the two sides. While for rigid diaphragm, vertical or horizontal forces borne by the shear wall in the middle is the same as that borne by shear walls on both sides.

The arrangement of the shear wall system

The "footprint" analysis introduced in Sect. 3.4 is very important in the early stage of structural design. The arrangement of the shear wall in the "footprint" has great influence on the force bearing of the whole structure and is therefore worth careful analysis. The arrangement of the shear wall directly relates to the torsion design of the overall structure. Since the structure will almost inevitably bear asymmetric

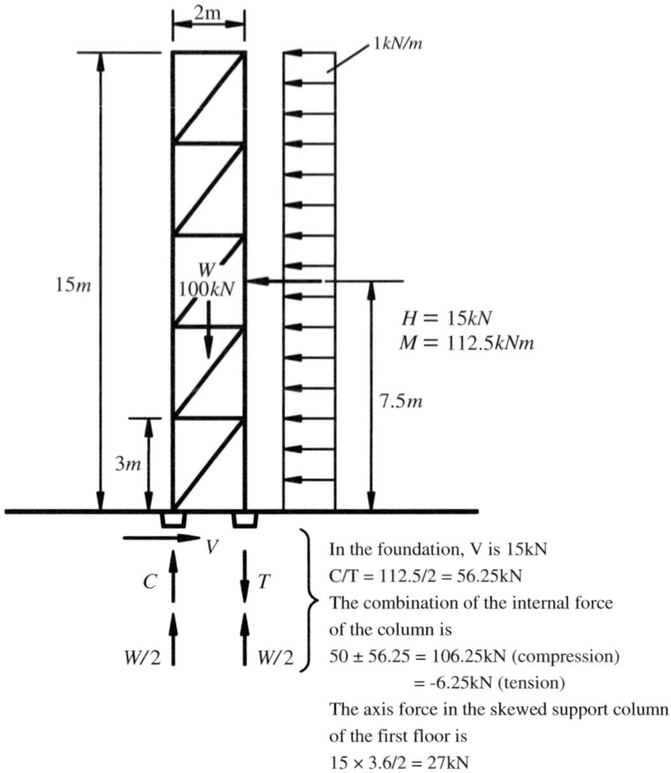

Fig. 4.53 Calculating method of truss shear wall

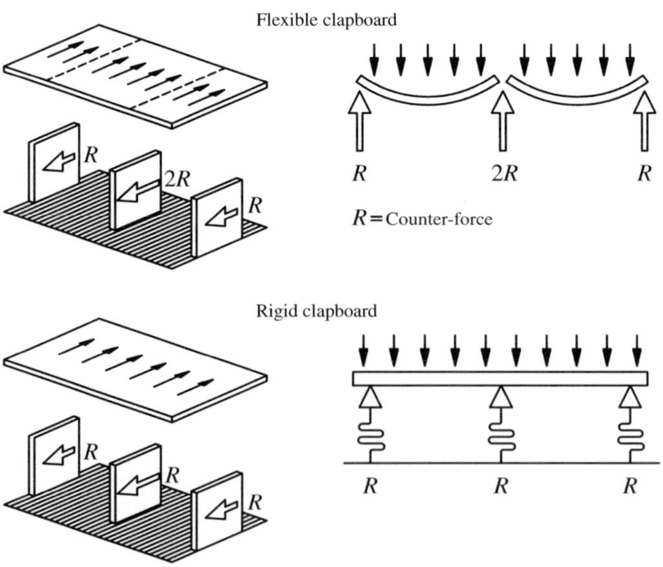

Fig. 4.54 Influence of the stiffness of horizontal diaphragm on the force bearing of shear wall

loads, additional torque is easily generated. To meet the torsion requirement, shear walls are to be located in a symmetrical, uniform, paired, and two-way pattern to facilitate a better formation of shear resistance and an overall torsion resistance. Figure 4.55 illustrates several ways to arrange the shear wall (the solid parts in the figure represent the load-bearing shear walls). Some of them are reasonably allocated (such as a and f shown in Fig. 4.55) while the others are not and would not help with the resistance of torsion (such as b, c, d, e shown in Fig. 4.55). Additional torque will be generated: (1) when the centroid and the center of rotation do not collide; (2) shear walls are not allocated in pairs. Occurrences like this should be avoided in structural design.

2. **Tube system and tube effect**

Tube effect

Tube structure is the optimal among vertical substructures in respect of force bearing and resistance to deformation, due to the "tube effect."

When it comes to the load transfer, tube functions as a space structure with high stiffness and force-bearing capacity in every direction. It not only effectively transfers the vertical loads but also transfers the horizontal loads from all directions. Figure 4.56 displays the characteristics of load transfer of the tube structure. Two tube walls, which are in parallel with the horizontal loads, mainly transfer horizontal shear forces to the foundation and would effectively resist about 90 % of them. Another two tube walls, which are perpendicular to horizontal loads, would largely resist the horizontal bending moment. Through the formation of the couple of forces, about 75 % of the total bending moment would be resisted by these two vertical pieces of the tube wall. This is the so-called tube effect. In other words, these two pairs of the tube wall would, respectively, resist the shear forces and the bending moment; meanwhile, since these four pieces of the tube wall are supported and restrained by the same horizontal substructure, they are able to work together. Therefore, the tube structure has huge force-bearing capacity minimum buckling risk.

References [6, 9] compared and analyzed the stiffness and force-bearing capacity of wide column frame structure and that of tube structure with the same material and the same sectional area. The stiffness and force-bearing capacity of the tube structure have always been greater than that of the wide column frame structure, which indicates that the tube structure would perform better under the load.

If the aspect ratio of the tube structure is less than 3, the tube structure, under the action of horizontal loads, would largely manifest shear deformation, in which case the shear forces are in control. If the aspect ratio of the tube structure is greater than 5, bending moment is in control. When the aspect ratio lies between three and five, both shear deformation and bending deformation need to be checked. When the aspect ratio of the tube structure is larger than 7, the tube structure under the action of horizontal loads would display considerable bending deformation. At times like this, rigid connecting beams are constructed to form a mega-frame structure. Shanghai Overseas Chinese Building, as illustrated in Fig. 4.57, connected several slender tubes with huge steel trusses at the top of building to constitute a mega-frame system which would greatly reduce the horizontal deformation and

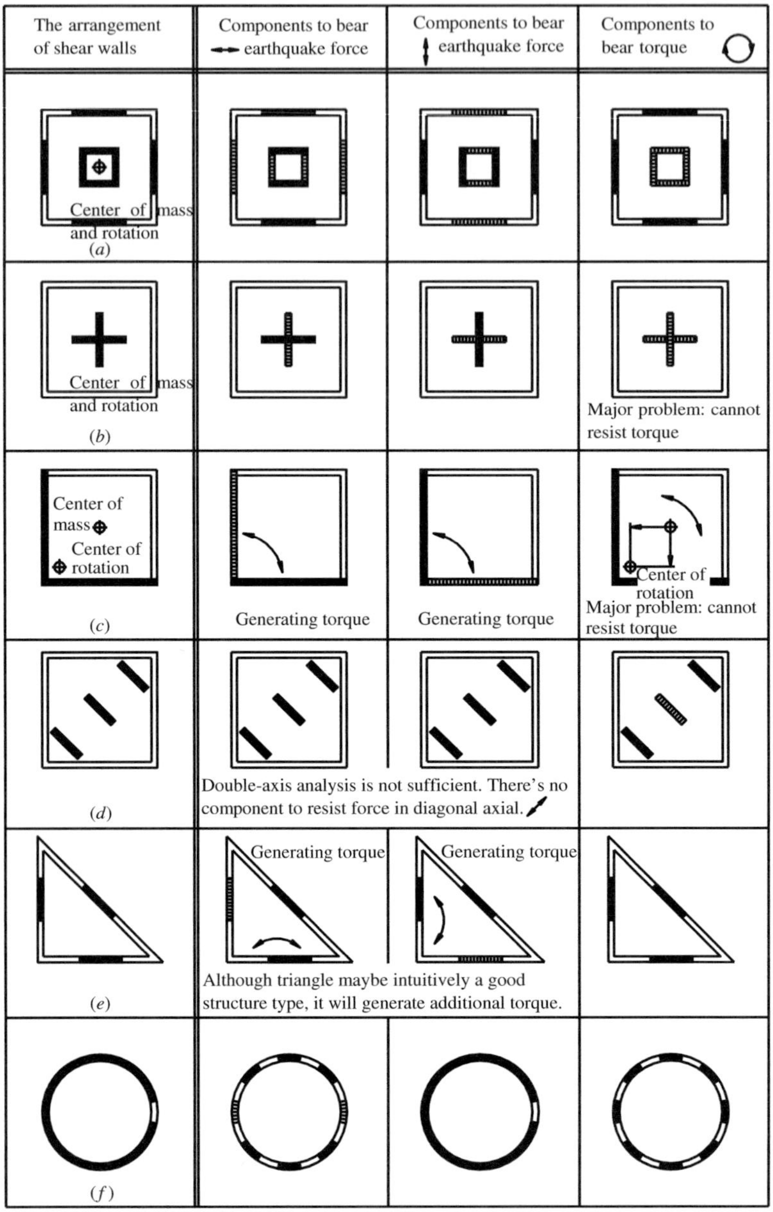

Fig. 4.55 Several layouts of the shear wall (*Solid parts* represent the load-bearing shear walls)

improve the bending stiffness. Figure 4.58 explains how the mega-frame structure diminishes the horizontal deformation through the rigid connection between the cap truss (or waist truss) and the slender tubes, in which case inflection points will take place and the horizontal deformation will be constrained.

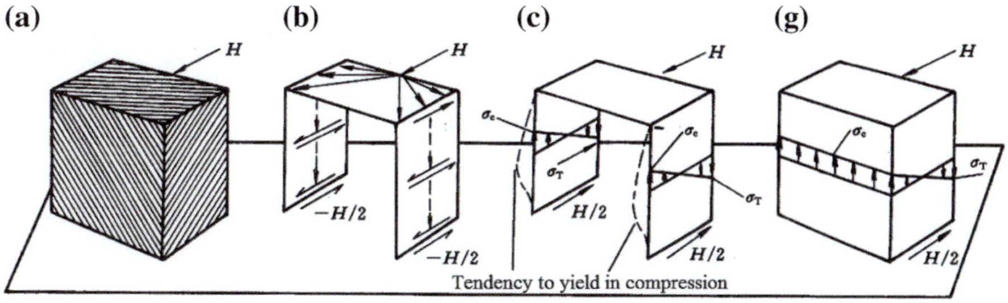

Fig. 4.56 Characteristics of load transfer of tube structure and tube effect [6]

Fig. 4.57 Shanghai Overseas Chinese building

Calculation of the tube structure

Exact computation of the internal forces of the tube structure is quite complex and the finite element method is generally adopted in calculation. In the early stage of design, "tube effect" is to help simply the calculation of the internal forces.

As shown in Fig. 4.59, the tube structure is divided into two pairs of the tube walls. One pair of them would resist the shear forces and the other would resist the bending moment. Since their functions are similar to that of the web and the wing plate of I-section, they are also called the web and the wing of the tube structure. How to calculate their resistance of the bending moment? When a square tube is under the function of a bending moment, the tensile stress distribution of the web

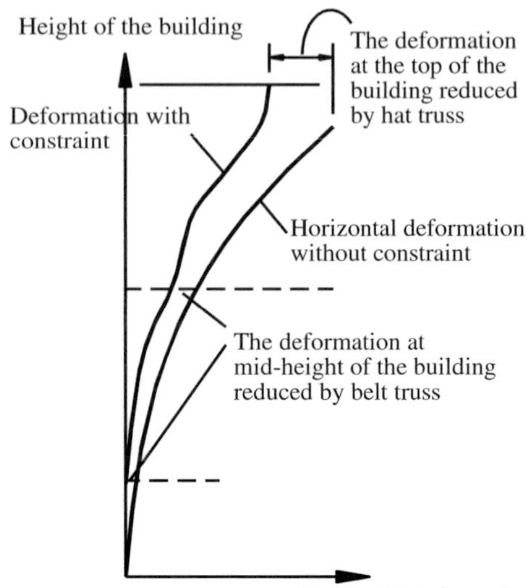

Fig. 4.58 Mega-frame diminishes the horizontal deformation of structure

would be a triangle (the stress close to the neutral axis is zero) while the tension or the stress is evenly distributed in the wing (as shown in Fig. 4.59c). The bending resistance of the wing is three times that of the web. In other words, when material of the same strength and the same size is adopted, within the range of elasticity, the stress distribution of the web under the function of the bending moment would be a triangle while the tension or the stress of the wing would be evenly distributed. When the tube is in its critical elastic state, the maximum stress in the triangle distribution would reach a critical value with a majority of the web in an elastic state, whereas the evenly distributed stress of the wing is at its maximum. Therefore, the bending moment borne by the web only constitutes as one-third of that borne by the wing with the same area, meaning the web would bear one-fourth of the total bending moment and the wing would bear the rest. That is why we say the efficacy ratio of the bending resistance of the web to the wing is 1:3 (Fig. 4.59b).

When the vertical axial load W is the only load, uniform compressive stress will appear in the section of the tube. When the square tube bears both the bending moment M and the shear force V, calculation of the bending moment could be simplified as shown in Fig. 4.59c. Specifically, quoting the 1:3 efficacy ratio of the web to the wing and the simplified calculation in which the wing bears 3/4 of the total bending moment, the total compressive or tensile force borne by the wing of the tube can be simply calculated as follows:

$$C \text{ or } T = \frac{3M}{4d} \tag{4.58}$$

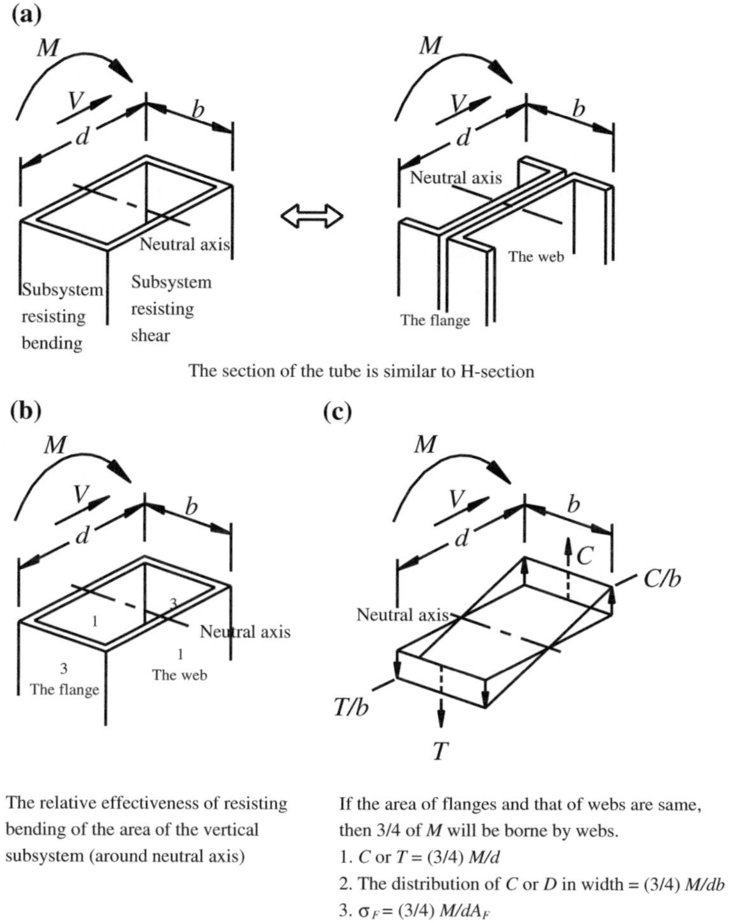

Fig. 4.59 Simplified calculation of tube internal force

$$\sigma_{\max} = \frac{3M}{4dA_f} \tag{4.59}$$

The calculation of the shear force is given as follows:

$$\tau_{\max} = \frac{1.5V}{2A_w} \tag{4.60}$$

where M and V are the bending moment and shear force borne by the foundation of the tube structure; d is the width of the square tube; A_f is the sectional area of a wing (the sectional area perpendicular to the direction of the stress); A_w is the sectional area of a web (the sectional area in parallel with the direction of the stress). The total addition of the maximum bending stress in Formula (4.58) and vertical uniformly distributed stress needs to be smaller than the compressive strength of the concrete. The maximum shear stress in Formula (4.60) is to be smaller than the shear strength of the concrete. The maximum shear stress will appear in the middle of the web and the assumption is that it is 1.5 times the average shear stress.

Classification and arrangement of the tube system

Same with the classification of the shear wall structure, tube structure can be divided into the solid tube structure and the truss tube structure. Square, rectangular, or circular section has different ways of arrangement in the "footprint" (Fig. 4.60). Figure 4.60a shows a triangle core tube, *b* is a circular core tube, *c* is a triangle core tube and an outer frame tube, *d* is a circular tube-in-tube, *e* is an internal multi-tube, *f* is a three-layer tube-in-tube, *g* is a frame core tube, *h* is a hexagonal tube-in-tube, *i* is a frame core tube, *j* is a frame tube, *k* is a mega-frame comprised of double tubes and horizontal constraints, and *l* is a typical square tube-in-tube structure.

When the building has only one tube structure, the tube is usually allocated in the center of the structure. When there is more than one tube, they can be dispersed ideally symmetrically. Since the tubes usually serve as the channels of vertical transportation or service facilities, holes (doors and windows) are often needed. In the conceptual design, when the open area of the tube is less than 30 %, it could be approximated as one without holes. However, the calculation of the strength and the stiffness will have to be conducted with a reduction of the thickness of the wall. When the open area of the tube is more than 50 %, mechanical characteristics of the tube are more like that of the frame structure. Since the overall strength and stiffness have been significantly reduced, the tube should be calculated as a frame structure. When the open area of the tube lies between 30 and 50 %, the reduction of the strength and the stiffness could not be neglected and it should be calculated as a frame-tube structure. And the shear-lag effect (see Fig. 6.40) should be taken into consideration.

3. **Frame system and rigid frame effect**

Rigid frame effect

The simplest way to create a living space for people is to support a horizontal substructure such as roof with four independent columns. These columns must have the strength to bear the vertical loads and resist buckling under the vertical loads. Under the horizontal loads, columns will be like cantilever beams embedded in the earth and will produce horizontal displacements, which could be considerable due to relatively low horizontal rigidity of the cantilever beams. If the horizontal substructure is hinged to the top of the supporting columns, the overall horizontal displacement of the structure would not be prevented even when the stiffness of the horizontal substructure is large enough. This is because that the column would still produce as much displacement like cantilever beam, as shown in Fig. 4.61a. However, if the horizontal substructure is rigidly connected to the top of the columns, the situation will change that rotation constraint produced by the top of the columns will greatly improve the overall horizontal stiffness of the structure ($\Delta_2 < \Delta_1$). And this is the so-called rigid frame effect, as shown in Fig. 4.61b. In this case, the columns and the beams constitute a frame structural system.

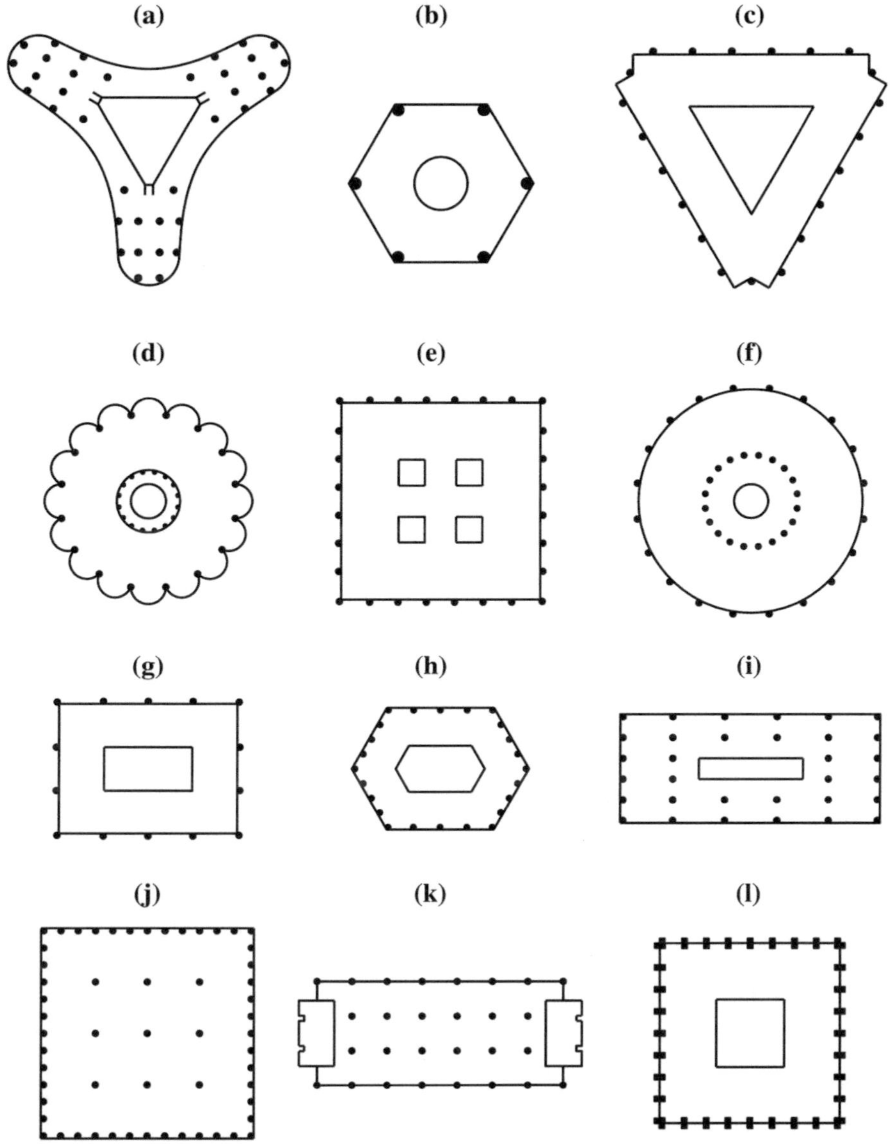

Fig. 4.60 Different shapes and arrangement of tubes

"Rigid frame effect" comes from the inflection of vertical columns under the influence of the inflection of horizontal beams. As shown in Fig. 4.61b, the inflection of horizontal and vertical components are reflected as points of inflection (P.I.) in their deformation curve. Due to the existence of P.I., the effective length of column (length without restraints) is shortened. As a result, the flexural strength and

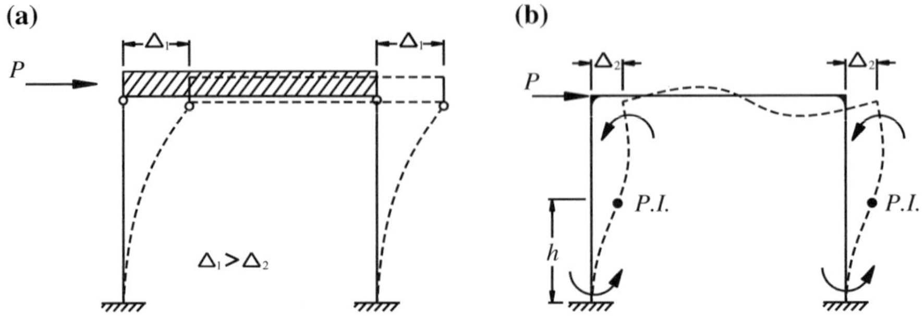

Fig. 4.61 Frame action constituted by columns and beams

the stiffness of the columns are enhanced and the horizontal displacements are restrained. The mutual constrain of beams and columns gives rise to mutual inflection constraints and the overall flexural strength and stiffness are therefore improved. This is the hallmark of the "rigid frame effect" in a frame structure. The higher the constraint degree is, the P.I. in column is closer to its midpoint, as shown in Fig. 4.61b. Under perfect rigid condition (assuming the line stiffness of the beam is 4 times greater than that of the column), the P.I. will be at the midpoint of the column and the displacement of the top will be $\Delta_2 = \Delta_1/4$. And it is then called "infinite rigid frame effect." All these frame structures could be of two or three dimensions.

Rigid frame structural system

Three methods could be applied in design to increase the stiffness of the frame and to restrain the horizontal displacement under horizontal loads:

1. Add column in between original columns (Fig. 4.62a). In this case, the beam on top of the columns will be divided into several short beams. The line stiffness of the beam (the line stiffness is adversely proportional to the length of the beam) is to be increased, due to rigid frame effect. At the same time, the unrestraint length of the columns is reduced, and the flexural stiffness is increased. In this way, the stiffness of the overall structure is increased and the horizontal displacement is decreased. Moreover, since the horizontal loads are borne by multiple columns, the shear distributed to each column will be reduced, which will in turn decrease the transverse shear displacement of the structure.
2. Increase the dimension of the column in the bending plane (Fig. 4.62b). Since the flexural stiffness of the column is proportional to the moment of inertia of its section and the latter is proportional to the cubic of the height of its section, the increment of the height of the column will significantly increase (to the order of 3) the flexural stiffness of the column and thus improve the overall stiffness and restrain the horizontal displacement of the frame.
3. Add rigid beams or other horizontal connecting members along the column height (Fig. 4.62c). Columns are divided along the height into several short columns, which will increase the flexural stiffness of each part as well as the

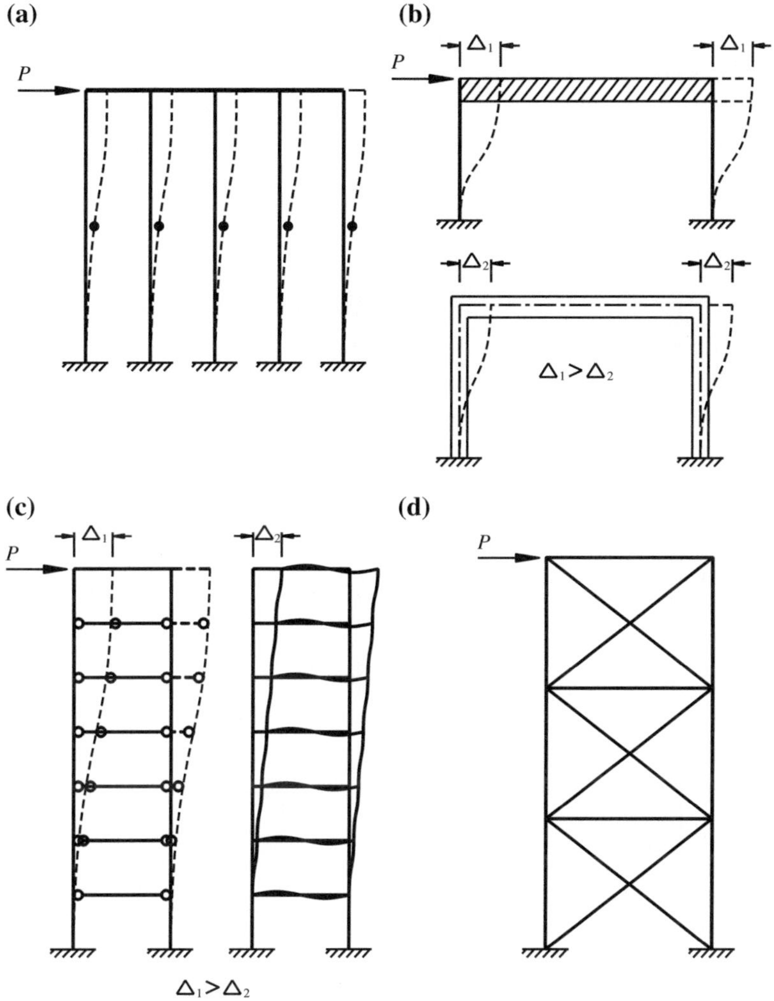

Fig. 4.62 Methods to increase the stiffness of frames

stiffness of the nodes connecting the beam and the column. Meanwhile, it will cause the "rigid frame effect" for this multi-story frame, increasing the overall stiffness and restrain horizontal displacements. It is noted that only through a rigid connection between these reinforced beams and the column could a frame system be realized.

4. Add diagonal bracing to the frame (Fig. 4.62d). The addition of diagonal bracings to the frame will form a high stiffness truss–shear wall system with enhanced resistance to horizontal deformation. This approach has been widely applied in modern steel high-rises design.

Frame structure and its simplified calculation

Frame structure is formed through a rigid connection between beams and columns. This is quite common structural system with the abovementioned "rigid frame effect." Figure 4.63 shows a diagram of the internal forces of the frame structure under different kinds of load. Figure 4.63a shows the plane arrangement; Fig. 4.63b

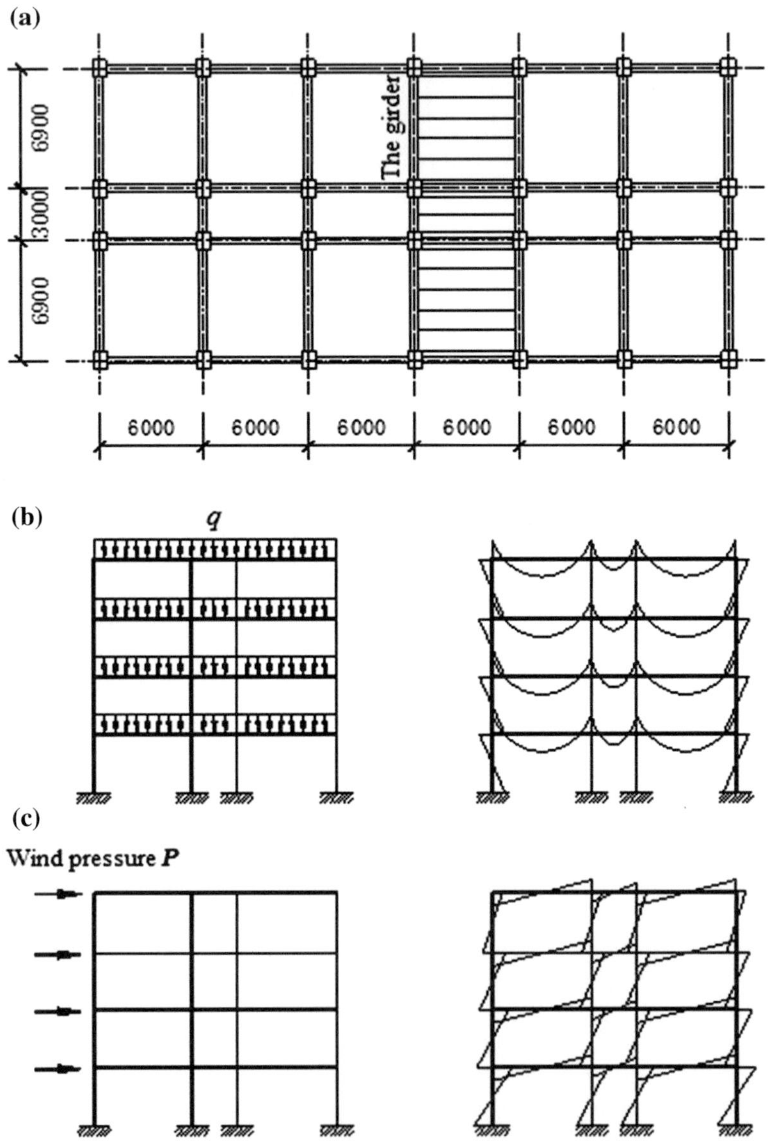

Fig. 4.63 Diagrams of internal forces of frame structure under *vertical* and *horizontal* loads

shows a diagram of the distribution of the bending moment under vertical loads on the floor, in which we can see that the bending moment is evenly distributed and does not decrease as the height increases; Fig. 4.63c shows a diagram of the distribution of the bending moment under horizontal loads, in which the bending moment decreases as the height of the structure increases.

Due to multiple indeterminations of frame structure, precise calculation of the internal forces of the frame structure is quite complicated. In this section, we will introduce a simplified calculation of the frame structure under horizontal and vertical loads. This will not only meet the calculation requirements in the early design but also enable a preliminary estimation of the dimensions of the structural members as well as the construction costs. Appendices A and B have applied this simplified calculation to the conceptual design of an actual single- and multilayer steel factory building. How to make a preliminary decision on the dimensions of the structural members and quantity of the steel needed are also introduced, followed by a feasibility analysis.

After the confirmation of the conception, further precise computation will need the help of computer programs. Preliminary sectional dimensions will be input to the computer as the initial sectional data to calculate the internal forces of the frame structure and derive an accurate design of the structure.

Simplified calculation of the frame structure under vertical loads

In this section, we will discuss simplified calculation of the frame structure under the action of vertical loads. It is based on the following calculation assumptions:

1. Components of same segment have same cross section, moment of inertia, and symmetrical stiffness;
2. The forces borne and the deformations are symmetrical; deformation outside the plane is ignored;
3. The forces are mainly borne by the component directly connected to it;
4. The connections and components are elastic and the deformation of nodes is also within the elastic range.

Figure 4.64 shows the simplified principle of transmission and distribution of the bending moment. The bending moment at the end of the components is related to the constraint of the other end. When the other end is completely fixed, the transmission coefficient of the bending moment is 1/2. When the remote end is semi-fixed, the transmission coefficient is 1/3. When the other end is free of constraints, the transmission coefficient is zero. Table 4.5 is a diagram of the bending moment in the components under different conditions of loads and supports. Table 4.6 gives the transmission coefficients of the other end under different supporting conditions and the fixation coefficients of the other end.

How to determine the fixation degree of the other end of the components in a frame structure? In conceptual design, simplified assumption is applied: (1) When the total line stiffness of the other component connected with the beam is equal to or greater than four times the line stiffness of that beam, it could be regarded as a complete fixation at the other end; (2) when the total line stiffness of the other

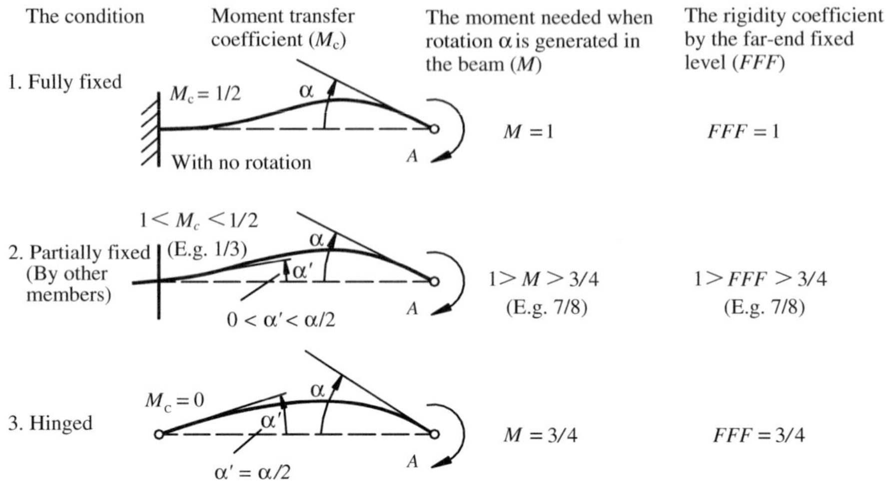

Fig. 4.64 Simplified principle of transmission and distribution of bending moment

components is less than that of the beam, it would be regarded as constraint-free at the other end; and (3) when the total line stiffness of the other components is between 1 and 4 times the line stiffness of the beam, it is regarded as semi-fixed at the other end.

In the calculation of the stress of components in a frame structure, the assumption of two-end fixation has to be made and the bending moment of the end of the component would be $M_o = wL^2/12$ under a uniformly distributed load. Next, in accordance with the actual end restraints, redistribution of the internal forces would take place and release a larger bending moment at the end with the help of fixed constraints. The rotation degree of the restraints at the nodes is proportional to the stiffness ratio of the node constraints (the ratio of the stiffness of the restraining components at the node to the total stiffness of the node). Specific calculation is that the line stiffness of component i (K_i), the total line stiffness of the components at the node (ΣK), and the redistribution of the bending moment at the end of the components (M_o) is $M_{oi} = M_o\,(\Sigma K - K_i)/\Sigma K$. The distribution of the bending moment of the other j components connected with the same node is conducted according to the line stiffness of each component $M_{oj} = M_{oi}\,K_j/(\Sigma K - K_i)$. The transmission of the bending moment (M_c) to the other end of the other components is to be executed as shown in Fig. 4.64. Here, $K = EI/L$ represents the relative line stiffness of the component. Note that if the line stiffness in each span of the beams in this frame structure is the same and each span bears the same evenly distributed load, the nodes at the end of each span will not rotate. In this case, these nodes serve as complete fixation restraints and the bending moment at the end of the component is $M_o = wL^2/12$. There is no need to reduce the bending moment at the end of the component or to transfer the bending moment.

By applying there basis principles, the simplified calculation of frame structure under vertical load could be fairly quickly done, then distribution of the bending

Table 4.5 Bending moment diagrams of members under different kinds of loads and supporting conditions

	No.	Schematic diagram	Fixed-end moment (set counterclockwise as positive)	Fixed-end shear force
Both ends fixed	1	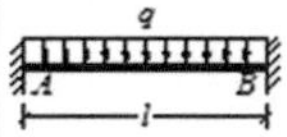	$m_{AZ} = -\frac{ql^2}{12}$ $m_{ZA} = +\frac{ql^2}{12}$	$Q_{AZ} = +\frac{ql}{2}$ $Q_{ZA} = -\frac{ql}{2}$
	2	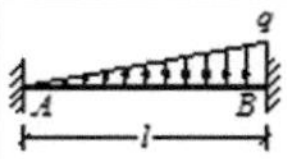	$m_{AZ} = -\frac{ql^2}{30}$ $m_{ZA} = +\frac{ql^2}{20}$	$Q_{AZ} = +\frac{3ql}{20}$ $Q_{ZA} = -\frac{7ql}{20}$
	3	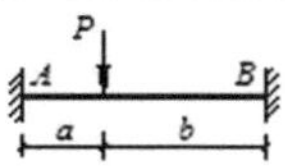	$m_{AZ} = -\frac{Pab^2}{l^2}$ $m_{ZA} = +\frac{Pa^2b}{l^2}$	$Q_{AZ} = +\frac{Pb^2}{l^2}\left(1+\frac{2a}{l}\right)$ $Q_{ZA} = -\frac{Pa^2}{l^2}\left(1+\frac{2b}{l}\right)$
	4	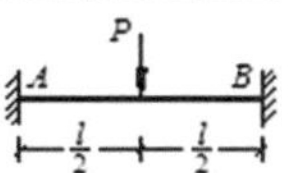	$m_{AZ} = -\frac{Pl}{8}$ $m_{ZA} = +\frac{Pl}{8}$	$Q_{AZ} = +\frac{P}{2}$ $Q_{ZA} = -\frac{P}{2}$
	5	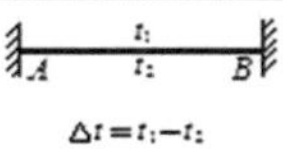	$m_{AZ} = \frac{EI\alpha\Delta t}{h}$ $m_{ZA} = -\frac{EI\alpha\Delta t}{h}$	$Q_{AZ} = 0$ $Q_{ZA} = 0$
One end fixed and the other hinged	6	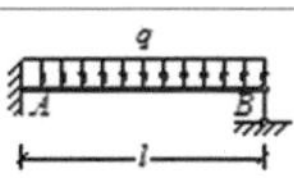	$m_{AZ} = -\frac{ql^2}{8}$	$Q_{AZ} = +\frac{5}{8}ql$ $Q_{ZA} = -\frac{3}{8}ql$
	7	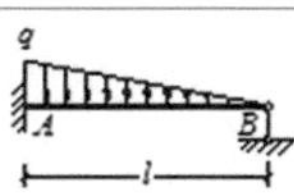	$m_{AZ} = -\frac{ql^2}{15}$	$Q_{AZ} = +\frac{2}{5}ql$ $Q_{ZA} = -\frac{1}{10}ql$
	8	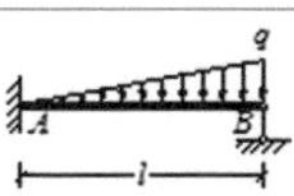	$m_{AZ} = -\frac{7ql^2}{120}$	$Q_{AZ} = +\frac{9}{40}ql$ $Q_{ZA} = -\frac{11}{40}ql$

moment diagram could be drawn. Examples will explain this simplified calculation in detail.

Example Figure 4.65a shows a diagram of distribution of the bending moment under a vertical uniformly distributed load in a simplified calculation. The bending moment at the end of the beam with complete fixation restraints is $M_o = wL^2/12 = 100$ kN m. Conducting a redistribution of M_o according to the ratio of the line stiffness of the beam to the total line stiffness of the other components, the bending moment at the end of the beam with actual constraints is $M_{oi} = M_o\,(\Sigma K - K_i)/\Sigma K = 100 \times (10 - 3)/10 = 70$ kN m. The distribution of the bending moment of the

Table 4.6 Moment transfer coefficients and fixed far-end strength coefficients under different supporting conditions

No.	Schematic diagram	Moment transfer coefficient K	Fixed far-end strength coefficient F
1	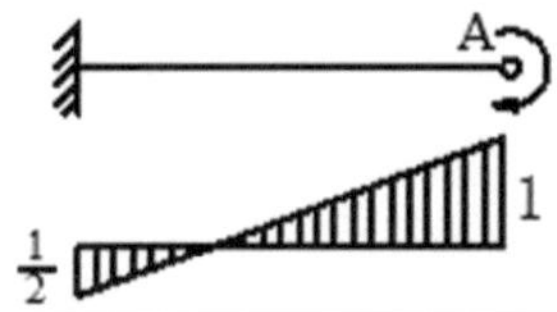	$\frac{1}{2}$	1
2	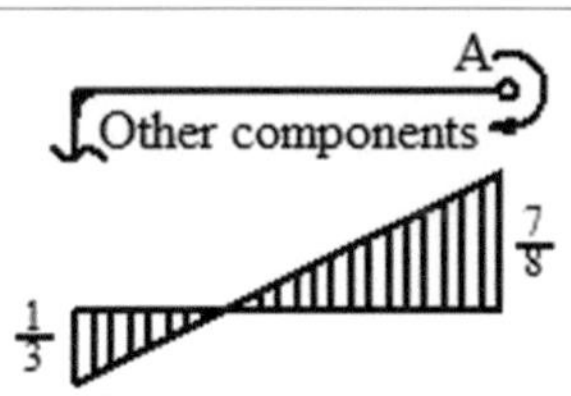	$0<M_c<\frac{1}{2}$ $M_c\approx\frac{1}{3}$	$1>F>\frac{3}{4}$ $F\approx\frac{7}{8}$
3	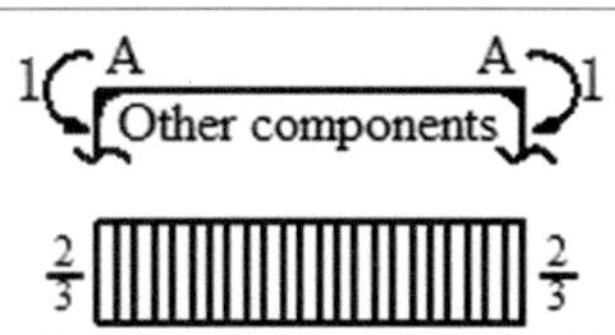	$\frac{2}{3}$	$\frac{2}{3}$
4	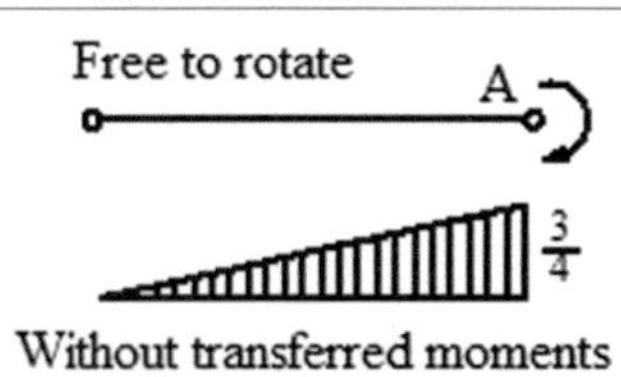	0	$\frac{3}{4}$
5	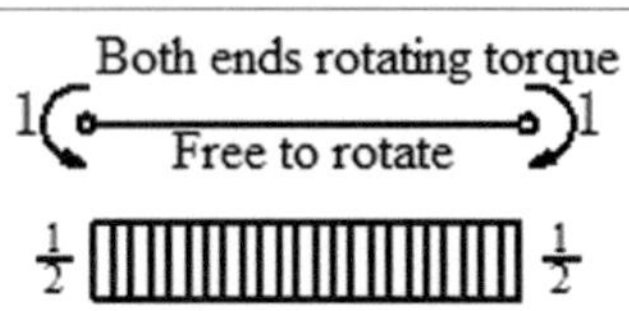	$\frac{1}{2}$	$\frac{1}{2}$

other components at the same node is made in accordance with the stiffness of each component. For example, the bending moment distributed to component BD at point B is $M_{oj}=M_{oi}\,K_j/(\Sigma K-K_i)=70\times 3/(10-3)=30$ kN m. As for the other components, the transfer of the bending moment to the other end will be processed according to the principles indicated in Fig. 4.64. For example, the transitive coefficient of the bending moment of component BD to point D is $M_c=1/3$, so the bending moment transferred to node D is $(1/3)\times 30=10$ kN m. Note that since the bending moment at the end of component AB is deducted 30 % in view of its surrounding constraints, the bending moment at the midpoint of component AB will correspondingly be added 30 % of the bending moment at the end of the component (that is $0.5M_o+30\ \%M_o=80$ kN m) to meet the equilibrium conditions. A diagram of the distribution of the bending moment in simplified calculation is as shown in Fig. 4.65b.

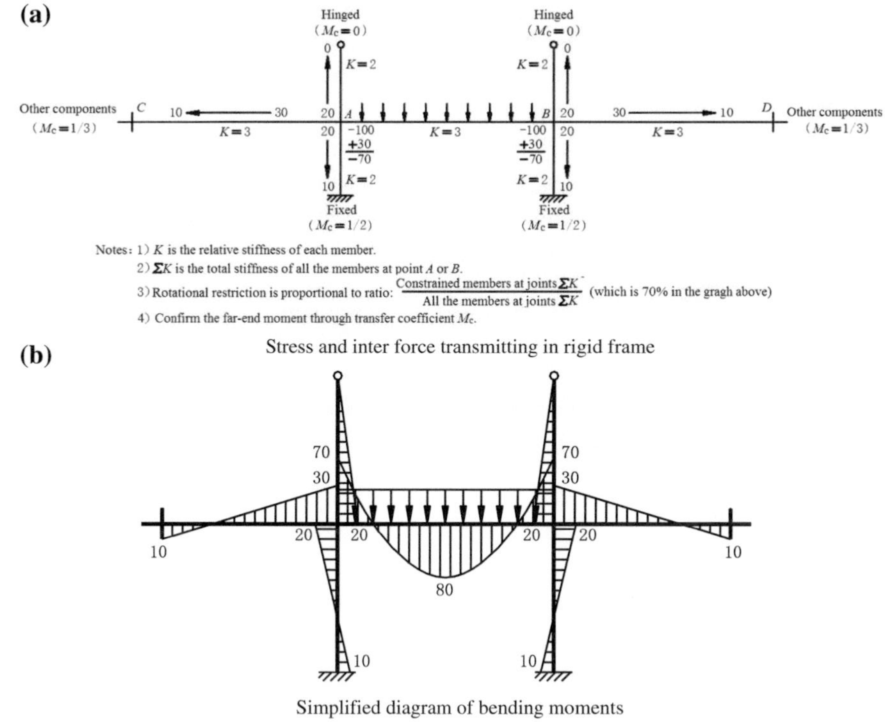

Fig. 4.65 Simplified calculation of the bending moment under a vertical uniformly distributed load. **a** Stress and internal force transmitting in rigid frame. **b** Simplified diagram of bending moments

Simplified calculation of the frame structure under horizontal loads

As for frame structures under horizontal loads, two simplified calculation methods exist: portal-rigid frame method and cantilever beam method. Portal-rigid frame method is suitable for calculation of bending moment and shear forces of the beams and columns under horizontal loads. Cantilever beam method is suitable for calculation of axial force distribution of columns under the action of the bending moment caused by horizontal loads, for the axial force of columns calculated in this method is conservative. These two simplified calculation methods could be applied in conceptual design. Note that generally they would only be applied to regular frame structures with traditional proportions.

Portal-rigid frame method

Points of inflection will appear in beams and columns when the frame structure is under the action of horizontal loads. To simplify the calculation, about the location of the point of inflection and the stiffness of the components, assumptions are as the following:

1. Each span of the frame structure could be regarded as an independent portal-rigid frame (consisting of two columns and a beam);
2. All the points of inflection of beams and columns appear at midpoints;
3. The shear forces of all interior columns in a multi-span frame structure are the same, which is assumed to be half of the exterior column shear force.

Based on the abovementioned assumptions, Figs. 4.66 and 4.67 give the simplified calculation of the bending moment and the shear forces of single-layer or multilayer multi-span structure, using portal-rigid frame method. This could also be applied to the other types of structure. The calculation includes: (1) the shear forces of all columns under horizontal loads are to be determined; (2) since all the points of inflection of columns are assumed to be at the midpoint, the bending moment at both ends are to be calculated by the shear force; (3) the bending moment at the end of the beam could be derived from that at the end of the column according to the equilibrium principle of the bending moment at the node; (4) the shear force in the beam

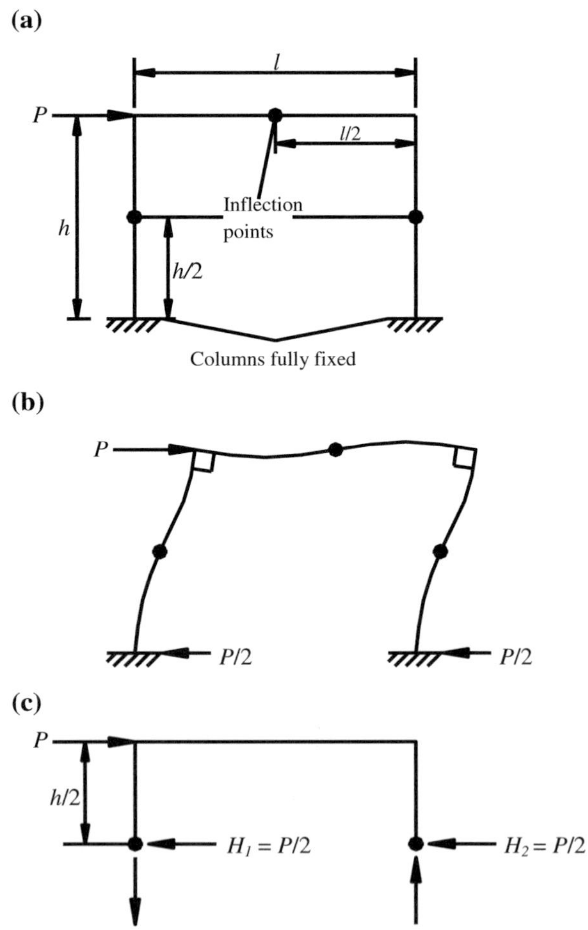

Fig. 4.66 Distribution diagrams of bending moment and shear force of single-layer rigid frame

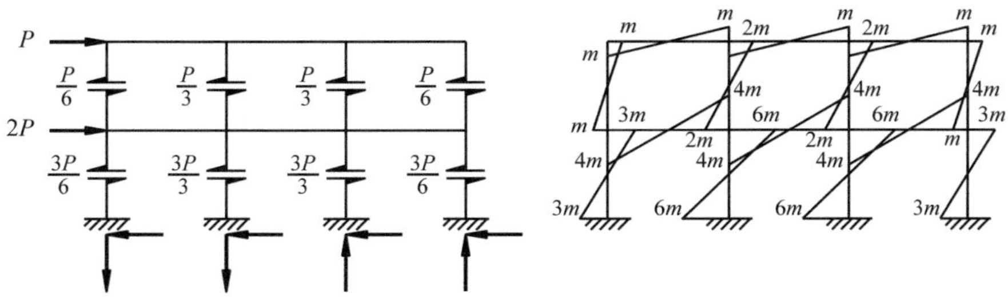

Fig. 4.67 Multilayer multi-span rigid frame and its bending moment diagram

could be calculated in accordance with the equilibrium condition of an isolated body; (5) at the very last, diagrams of the bending moment and the shear forces could be drawn. To mark the bending moment in a simplified way, $m = Ph/12$ is adopted in Fig. 4.67 to represent unit bending moment.

<u>**Cantilever beam method**</u>

Under horizontal bending moment, the calculated results of the axial forces of columns using the cantilever beam method are more conservative. As the name suggests, the cantilever beam method conducts the analysis of forces regarding the whole frame structure as a cantilever beam standing on its foundation. Several assumptions are: (1) the bending moments at both ends of the columns on each floor are neglected; (2) the horizontal bending moment is only counterbalanced by the moment produced by the axial forces of the columns; (3) the axial force of each column on each floor is proportional to the distance between the column and the neutral axis of the whole structure, displaying a triangular distribution.

Fig. 4.68 Calculation of axis force in columns using cantilever beam law

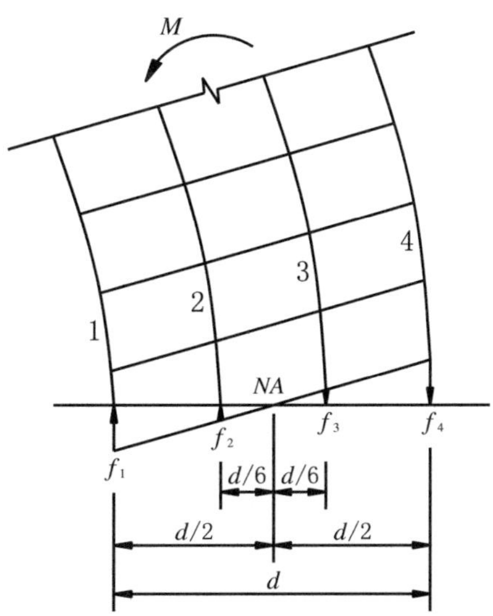

As shown in Fig. 4.68, when the whole frame structure bends to the left, column 4 has the highest tension, while column 1 has the highest compression. The axial stress in column 2 and column 3 could be calculated using linear interpolation, whereas, since the calculated bending moments at the end of the columns using portal-rigid frame method tend to be greater, the axial forces of the columns tend to be smaller and therefore less conservative. This is the reason that the portal-rigid frame methods should be used to estimate bending moment and shear force, while cantilever beam method should be used to estimate axial forces. Readers can compare portal-rigid frame method with cantilever beam method in the calculation of the axial forces of the columns in a single frame. The difference between the calculated results is in the order of 2.

For multilayer multi-span structure, one layer of the structure could be regarded as an isolated body of a cantilever beam to conduct the analysis and derive the distribution of the axial forces of all layers of this frame structure. As shown in Fig. 4.69, we will first analyze the load borne by the isolated body of this floor. Vertical load from the upper structure is 1000, 200 kN of which will be assigned to the side columns and 300 kN to the central columns according to the area (the sectional area of a central column is 1.5 times that of a side column). Total horizontal force borne by the upper structure is 336 kN; according to the assumption of portal-rigid frame method (the shear force borne by the central columns is two times that borne by the side columns), 56 kN will be assigned to the side columns and 112 kN to central columns. Since the horizontal shear force of this floor has a 12 kN increment, the shear forces passed on to the side and central columns of the floor below are calculated as 58 and 116 kN through the horizontal shear force equilibrium of the isolated body of this floor. Through the balance of the bending moment of the isolated body on this floor, the axial forces of the side columns F_1 and F_4 caused by the bending moment are $\pm(336 \times 4 + 12 \times 2)/(22.8 + 7.6/3) = \pm 54$ kN, and the axial forces of the central columns F_2 and F_3 are $\pm 54/3 = \pm 18$ kN. Balanced with the axial forces aroused by the vertical loads from the above, the axial force of the left side column is $(200 + 54) = 254$ kN, that of the left central column is $(300 + 54/3) = 318$ kN, that of the right central column is

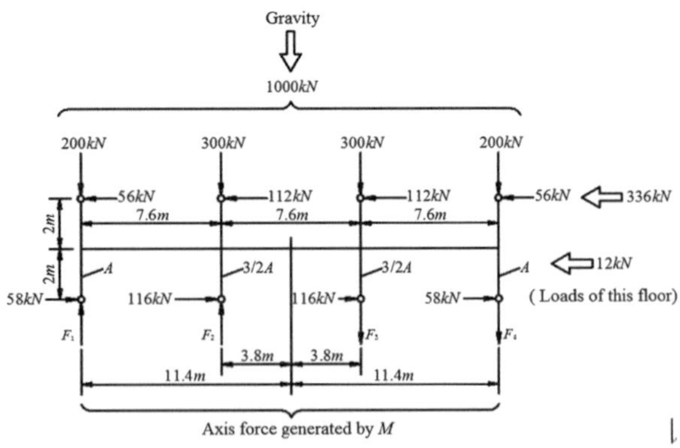

Fig. 4.69 Calculation of axis forces of multilayer multi-span rigid frame

(300 − 54/3) = 282 kN, and that of the right side column is (200 – 54) = 146 kN. Note that all the axial forces of the columns in this example are compressive forces without any uplifting tension. Balance requirements of this structural design are met. The largest axial compressive force lies on the left central column.

In the next section, readers will have more examples of frame structure calculation and some exercises in Appendix A to practice hand computation, sketch of diagrams of the bending moment, determination of the critical section, and estimation of sectional area.

Examples of simplified calculation of the frame structure

Example 1 Figure 4.70 shows the state of symmetrical cross-rigid frame when a concentrated moment M_A is applied at the central node A. Due to the action of M_A, rigid node A produces four identical angular displacements $\theta_{\mathrm{AB}} = \theta_{\mathrm{AC}} = \theta_{\mathrm{AD}} = \theta_{\mathrm{AE}}$; Since the far ends of all components are rigidly connected, the bending curve can be obtained directly, as shown in Fig. 4.64a. The bending moment diagram is drawn in light of the abovementioned principles, as shown in Fig. 4.70b. Since their far ends are rigidly connected, half of the bending moment is transferred. Considering $M_{\mathrm{AB}} = M_{\mathrm{AC}} = M_{\mathrm{AD}} = M_{\mathrm{AE}}$, we have

$$\begin{aligned} M_{\mathrm{BA}} &= 0.5M_{\mathrm{AB}} = 0.125M_A \\ M_{\mathrm{CA}} &= 0.5M_{\mathrm{AC}} = 0.125M_A \\ M_{\mathrm{DA}} &= 0.5M_{\mathrm{AD}} = 0.125M_A \\ M_{\mathrm{EA}} &= 0.5M_{\mathrm{AE}} = 0.125M_A \end{aligned} \quad (4.61)$$

Equation (4.61) utilizes the symmetry of the structure to simplify calculation.

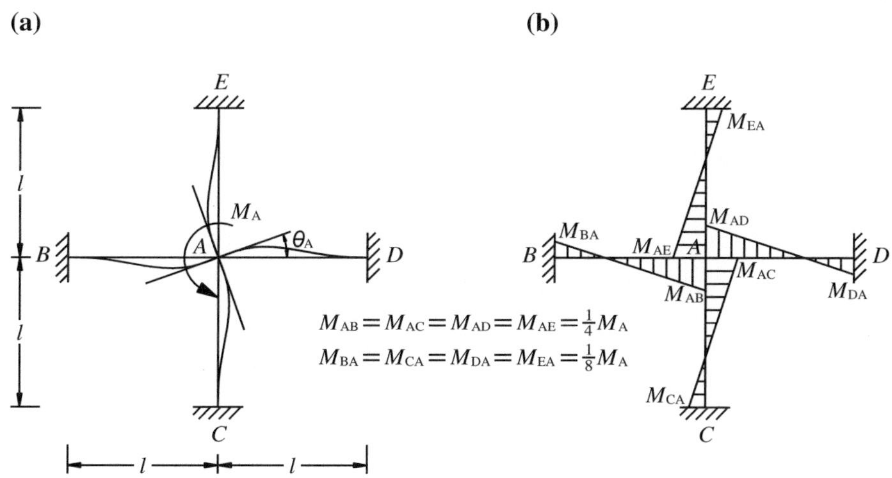

Fig. 4.70 Approximate joint analysis of rigid frame

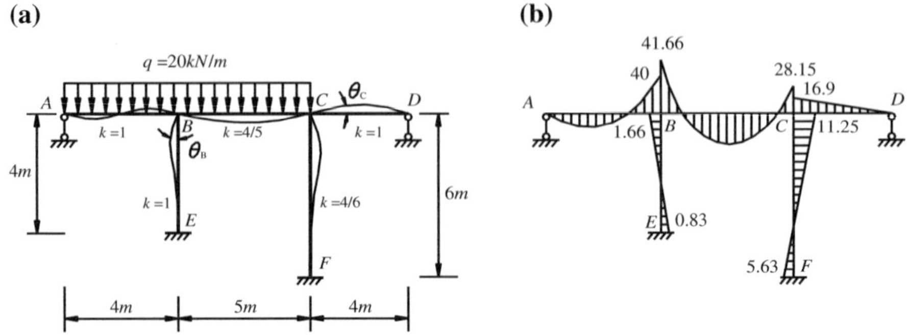

Fig. 4.71 Approximate force analysis of plane frame

Example 2 The geometric dimension of a frame structure is as shown in Fig. 4.71. It is under a vertical uniform load q = 20 kN/m. Try and draw the diagram of the bending moment of this structure.

When we look at the two-dimensional frame structure as shown in Fig. 4.71, we will first draw its deformation curve under the action of this vertical uniform load q pursuant to the characteristics of this structure. Since the span BC is greater than span BA and CD, the rotor angle of node B (θ_B) should be clockwise. Node A is a hinged support. The bending curve of component BA should show point of inflection under this vertical uniform load q. E is rigidly fixed. The upper half of the bending curve of component BE should convex to the left and the lower to the right. Under the action of vertical load q, the bending curve of span BC should convex downward and the rotation angle of node C (θ_C) should be anticlockwise. Therefore, curve CF is similar to curve BE only with an opposite rotating direction. Since node D is a hinged support, curve CD convex upward. The bending deformation curve of this structure derived from qualitative analysis is shown in Fig. 4.71a.

Before the quantitative analysis of this bending moment diagram, we need to speculate in a qualitative manner the distribution of the bending moment with the help of the bending deformation diagram as shown in Fig. 4.71a. The determination of the bending moment of each component can be described as follows:

$$M_{\mathrm{BA}} = qL^2/8 = 20 \times 4^2/8 = 40\,\mathrm{kN\,m}$$

Component AB: Under the action of vertical load q within the span, node B almost does not rotate and thus approximately considered as rigidly fixed. With reference to the stress state of Number 6 in Table 4.5,

$$M_{\mathrm{BA}} = qL^2/8 = 20 \times 4^2/8 = 40\,\mathrm{kN\,m}$$

Component BC: Same with component BA, node B is considered as rigidly fixed. With reference to the stress state of Number 1 in Table 4.5,

$$M_{\mathrm{BC}} = qL^2/12 = 20 \times 5^2/12 = 41.66\,\mathrm{kN\,m}$$

Since there is no vertical load on span CD, there will be a moment release of M_{CB} at node C. Assuming that the relative line stiffness of components AB, BE, and CD is 1, the relative line stiffness of components BC and CF will be 4/5 and 4/6. And $M_{\mathrm{CB}} = M_{\mathrm{BC}}(\Sigma K - K_{\mathrm{i}})/\Sigma K = 41.66 \times (1 + 4/6)/(1 + 4/5 + 4/6) =$ 28.15 kN m.

$$M_{\mathrm{CD}} = 28.15 \times (1/1 + 4/6) = 16.9\,\mathrm{kN\,m}.$$

Component CD: The distribution of the bending moment of the other components connected at node C will be conducted according to the line stiffness of each component. For example, the bending moment distributed to component CD at point C is $M_{\mathrm{CD}} = 28.15 \times (1/(1 + 4/6)) = 16.9$ kN m. The transitive coefficient of the bending moment to point D is $M_c = 0$, so the moment transferred to node D is 0.

Component CF: The bending moment distributed to node C is $M_{\mathrm{CF}} = 28.15 \times (4/6)/(1 + 4/6) = 11.25$ kN m. The transitive coefficient of the bending moment to node F is $M_c = 1/2$, so the moment transferred to node F is $(1/2) \times 11.25 =$ 5.63 kN m.

Component BE: Node B is balanced and $M_{\mathrm{BE}} = 41.66 - 40.00 = 1.66\,\mathrm{kN\,m}$

With reference to the stress state of Number 1 in Table 4.6, at the far end of component BE $M_{\mathrm{EB}} = 0.5$ and $M_{\mathrm{BE}} = 0.83$ kN m.

The simplified distribution graph of the bending moment is as shown in Fig. 4.71b. Comparing with the accurate solution shown in Fig. 4.72, the outlines of these two diagrams match satisfactorily, with only a slight difference in the

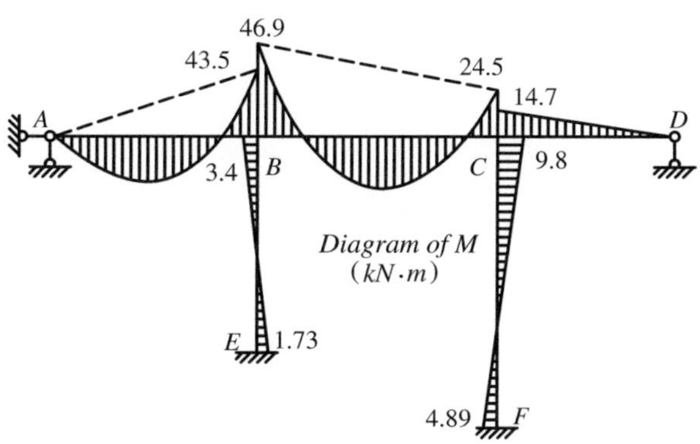

Fig. 4.72 Bending moment diagram of plane frame

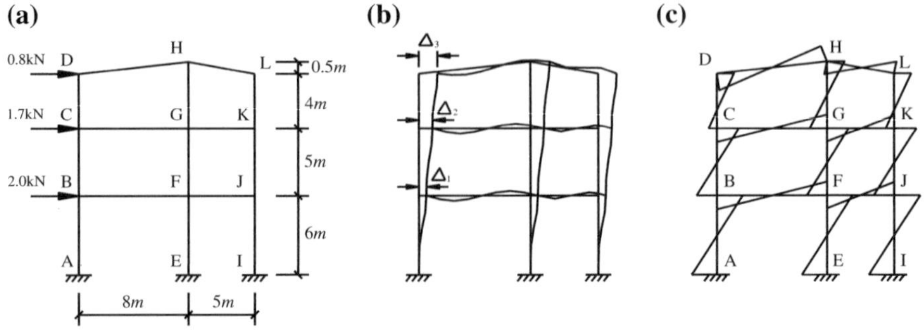

Fig. 4.73 Approximate analysis of multilayer multi-span plane frame

maximum value of the bending moment of node B, being 10 % or less. In the conceptual design of a complicated structure, the only purpose of approximate structure analysis is that the designer will have an explicit idea of the mechanical properties of the designed structure so as to determine the feasibility of the whole conception. Thus, 10 % of error is usually acceptable.

Example 3 Figure 4.73a illustrates the stress state of a multilayer multi-span rigid frame under the action of horizontal loads. Draw an approximate diagram of the bending moment of this structure.

Qualitative deformation curve of the bending moment is as shown in Fig. 4.73b. In light of the horizontal concentrated loads the structure bore, the lower parts of the bending curves of all columns convex to the left while the upper parts of all columns manifest inflection due to the rigid frame effect of beams and the curves convex to the right. For the same reason, the left ends of all bending curves of the beams convex downward while the right ends convex upward. From this qualitative bending deformation curve, we will be able to determine an approximate diagram of the bending moment of this frame, in which the bending moment is linearly distributed as shown in Fig. 4.73c.

As for the columns, we see that the left of the bottom and the right of the top are being pulled. As for the beams, the bottom to the left and the top to the right are being pulled. Without uniformly distributed loads, the bending moments are linearly distributed.

Ignoring the elasticity of beams and assuming that the horizontal displacements of node B, F, and J are the same, the shear forces at the end of the columns are distributed as the central columns undertake twice the shear forces that bore by the side columns.

As for the bottom columns:

$$Q_{\mathrm{AB}} = Q_{\mathrm{IJ}} = (0.8 + 1.7 + 2.0) \times (1/4) = 1.125\,\mathrm{kN}$$
$$Q_{\mathrm{EF}} = (0.8 + 1.7 + 2.0) \times (2/4) = 2.25\,\mathrm{kN}$$

Under the assumption that the points of inflection of columns and beams are at their midpoints, the bending moments bore by the bottom columns should be derived from their corresponding shear forces with identical value at both ends:

$$M_{AB} = M_{BA} = M_{IJ} = M_{JI} = 1.13 \times 6/2 = 3.39\,\text{kN m}$$
$$M_{EF} = M_{FE} = 2.25 \times 6/2 = 6.75\,\text{kN m}$$

As for the columns on the second floor:

$$Q_{BC} = Q_{JX}(0.8 + 1.7) \times (1/4){=}0.63\,\text{kN}$$
$$Q_{FG} = (0.8 + 1.7) \times 2/4 = 1.25\,\text{kN}$$

The bending moments bore by these columns should be derived from their corresponding shear forces with identical value at both ends:

$$M_{BC} = M_{CB} = M_{JK} = M_{KJ} = 0.63 \times 5/2 = 1.58\,\text{kN m}$$
$$M_{FG} = M_{GF} = 1.25 \times 5/2 = 3.15\,\text{kN m}$$

As for the columns on the third floor:

$$Q_{CD} = Q_{KL} = 0.8 \times 1/4 = 0.2\,\text{kN}$$
$$Q_{GH} = 0.8 \times 2/4 = 0.4\,\text{kN}$$

The bending moments bore by these columns should be derived from their corresponding shear forces with identical value at both ends:

$$M_{CD} = M_{DC} = M_{KL} = M_{LK} = 0.2 \times 4/2 = 0.4\,\text{kN m}$$
$$M_{GH} = M_{HG} = 0.4 \times 4/2 = 0.8\,\text{kN m}$$

In light of the equilibrium of the bending moment at node *B*, *C*, *D* and node *J*, *K*, *L*, the bending moments at the ends of beams can be determined.

$$M_{BF} = M_{FB} = M_{FJ} = M_{JF} = M_{BA} + M_{BC} = 3.39 + 1.58 = 4.97\,\text{kN m}$$
$$M_{CG} = M_{GC} = M_{GK} = M_{KG} = M_{CB} + M_{CD} = 1.58 + 0.4 = 1.98\,\text{kN m}$$
$$M_{DC} = M_{DH} = M_{HD} = M_{HL} = M_{LH} = 0.4\,\text{kN m}$$

The approximate diagram of the bending moment is as shown in Fig. 4.74a. The accurate diagram of the bending moment is as shown in Fig. 4.74b for comparison. From these two diagrams, we know that the former is quite close to the latter. The maximum bending moment of the beams in the former is 4.97 while that in the latter is 4.98 with a slight difference of 0.2 %. Therefore, the simplified calculation can be used in early conceptual design.

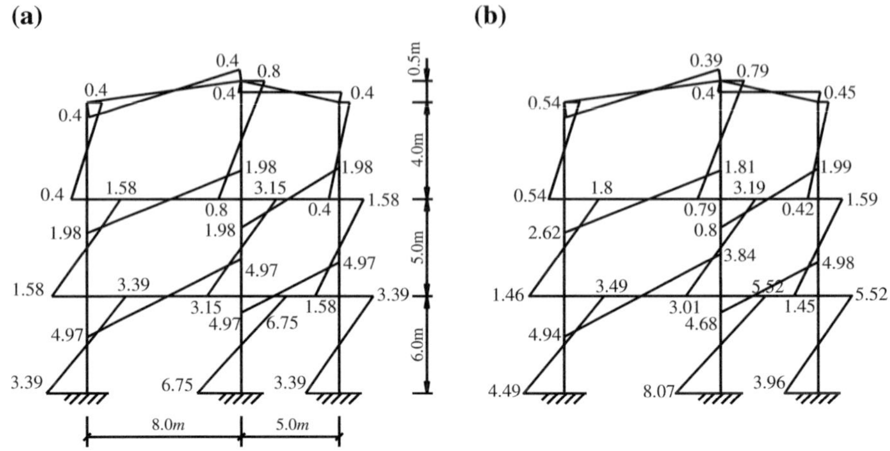

Fig. 4.74 Comparison between approximate and accurate calculated diagram of bending moment

4.8 Conclusions

In this chapter, we discussed the decomposition process of a structural system together with mechanical characteristics of varied types of horizontal and vertical substructures in detail. The relation between the global and the local was analyzed. And a design concept of "starting from the overall, focusing on the detail" was emphasized.

Mechanical characteristics of horizontal substructures, such as flat-slab system, slab beam system, primary and secondary beam system, two-way ribbed slab system, and space truss system, are to resist the bending moment and shear forces so as to meet vertical displacement requirements. Mechanical characteristics of vertical substructures, such as shear wall system, tube system, and frame system, are to provide strength and stiffness to resist horizontal loads so as to meet horizontal displacement requirements and transfer the loads to the foundation. Horizontal substructures must be integrated with vertical substructures to ensure an overall performance of the structure. In this chapter, we also introduced mechanical performances and simplified calculations of vertical substructures with effects such as tube effect and rigid frame effect. Learning these fundamental mechanical properties of horizontal and vertical substructures will greatly benefit a comprehensive understanding of mechanical properties of the structure as a whole.

In this chapter, we also presented simplified calculations of the bending moment, shear forces, and axial forces of a frame structure under the action of vertical and horizontal loads. Examples were used to demonstrate how to simplify the assumption as well as the analysis and the calculation with reference to mechanical properties of structural components. The distribution graph of the bending moment could be completed through manual calculation without the help of computer. Critical sections could be determined; sections of structural components and the costs could be preliminarily estimated; mechanical rationality and the economy of varied designs could be compared in the early stage of design.

After reading this chapter, readers should be able learn more about mechanical properties of different substructures as well as how they affect the features of the entire structure.

Chapter 5
Comparative Design and Analyses of High-Rise Buildings

5.1 Characteristics of High-Rise Structures

5.1.1 General Characteristics

Generally speaking, structures can be classified as low-rise buildings with less than six stories, mid-high-rise buildings with stories between six and fifteen, and high-rise buildings with more than fifteen stories. As the number of stories increase, horizontal substructures are no longer decisive in overall structure design. Meanwhile, the vertical substructure design becomes significantly important. Reasons are listed as follows:

1. High-rise buildings are required to resist large vertical loads, and therefore, high-strength columns and thick walls are required. Moreover, shaft-tube structures like the elevator shafts have to be arranged to bear vertical loads, too.
2. Due to horizontal loads, the maximum overturning moment and the horizontal shear force will appear at the foundation of the structure, which will in turn cause serious horizontal displacement at the top of the building. Roughly estimated, the overturning moment at the foundation is proportional to the square of the height of the building, while the horizontal displacement at the top of the building is proportional to the fourth power of the height.

In terms of the material consumed, unit area of a high-rise building is more than that of a low-rise building. This is because vertical components such as columns, walls, and shafts in a high-rise building are required to bear all the vertical loads. Moreover, the material used to resist horizontal loads will greatly increase with the increment of the stories. In accordance with related statistics and analysis [1], the amount of material applied to resist the vertical loads generally increases in a linear way with the increment of the stories, while that applied to resist the horizontal loads increases in a nonlinear way with the increment of the stories. However, those

S. Lin and Z. Huang, *Comparative Design of Structures*,
DOI 10.1007/978-3-662-48044-1_5

designs are usually conservative. If effectively optimized, material applied to resist horizontal loads will have a huge deduction.

Material in reinforced concrete structures mounts with the increment of the stories. So it is with the deadweight of the structure. And it has to be pointed out that the addition of material to resist vertical loads in a reinforced concrete structure is greater than that in a steel structure. This is due to the larger deadweight of the former. However, compared with the steel structure, the material applied to resist horizontal loads in a reinforced concrete structure does not increase much with the increment of the stories. This is because the deadweight of the reinforced concrete structure counterbalances some of the overturning effect derived from horizontal loads. Nonetheless, in seismic design, the enormity of the deadweight of a reinforced concrete structure might give rise to serious problems, tremendously increasing the seismic response of the structure.

In the seismic design of a high-rise building, restricted horizontal displacement is of the utmost importance. Appropriate adoption from the following methods will enhance horizontal resistance of the structure.

1. Add enclosing giant truss. Apart from a common structural system, an enclosing giant truss comprised of enormous horizontal substructures (load-bearing diaphragm and deep beams) and vertical substructures (elevator shafts and giant columns) will effectively restrict horizontal displacement of the structure (Fig. 5.1).
2. Design components which are able to transfer forces in three dimensions. The horizontal stiffness of vertical linear members like the columns is very small. Although vertical plane members such as shear walls and frames have great stiffness within their own plane, the stiffness of these members outside the plane is quite small. If outside-the-plane support can be realized for columns or shear walls, so will be the force transfer in space and a stable structural anti-overturning system. The framed tube system illustrated in Fig. 5.2 had its outer tube enclosed with a series of plane frames with shear walls. The addition of high-strength rigid connection to the frames in its horizontal direction will ensure the integrity of the structure for the rigid connection will largely reduce the horizontal displacement of the plane frames.
3. Utilize bracing system to enhance horizontal stiffness. As for structural layers for shear, the best way to avoid excessive horizontal displacement is to add diagonal braces to bear the shear forces. It is not economical to use vertical flexural members to resist shear forces.
4. Optimize arrangement in "footprint" of the foundation. Since horizontal overturning moment is resisted by the moment of force produced by the force of reaction from the columns at the foundation, the space between the columns arranged for the foundation will directly affect the anti-overturning capacity of the structure. Sections of the components in the "footprint" are to be marginalized, which means to augment the distance between the centroid of the foundation and the columns, so as to significantly increase the anti-overturning capacity of the foundation. Typical design is to keep the bottom wide and the

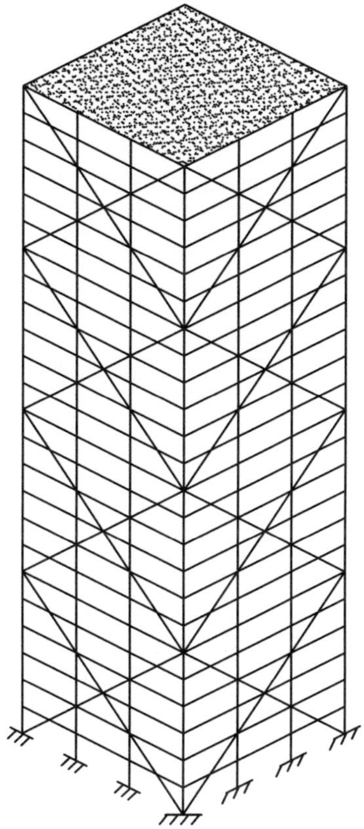

Fig. 5.1 External giant truss

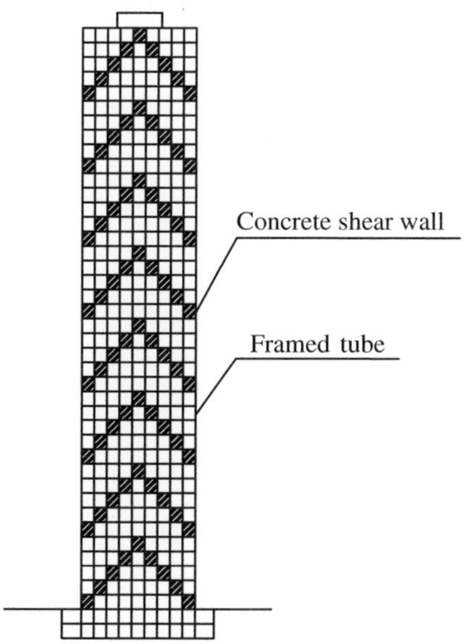

Fig. 5.2 Outside tube frame with the shear wall boards

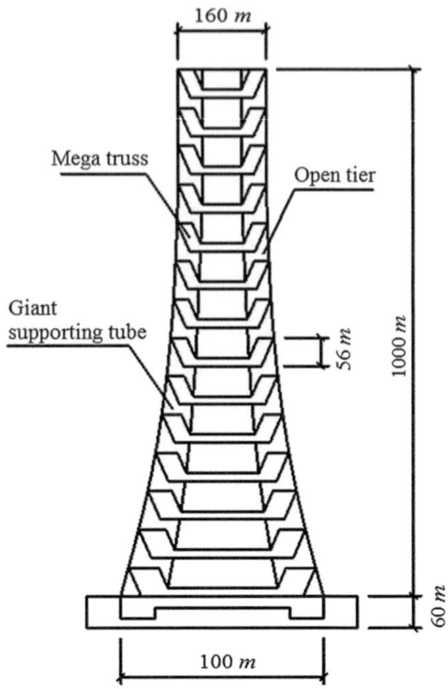

Fig. 5.3 Conception of Japan "Building of City in the Air-1000"

top narrow to effectively reduce horizontal displacement due to the long distance between columns and the centroid of the "footprint." The gradual cone-shaped tube structure (Fig. 5.3) and the high rise with base annex (Fig. 5.4) are two examples of this kind. Figure 5.3 is a diagram of the conception of a Japanese architect. It illustrates the "Building of City in the Air-1000" with enormous weight and ultra-many stories. The plane layout is circular, the elevation facade

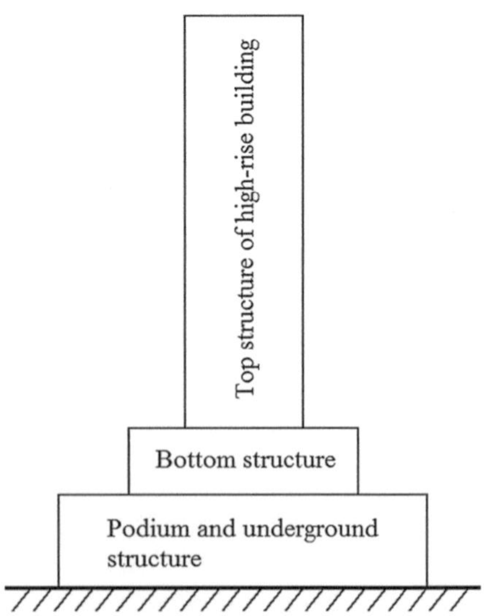

Fig. 5.4 Conventional façade arrangements in high-rise buildings

is a hyperbolic frustum, the depth of foundation is −60 m, the height above ground is 1000 m, the diameter of the bottom of the tube cone is 400 m, and the diameter of the top is 160 m. A horizontally penetrating open tier of four stories (16 m) is to be installed every fourteen stories (56 m). Major load-bearing system is a mega-frame system with giant supporting tubes as the columns and giant space trusses as beams. The high-rise building in Fig. 5.4 can attend to various architectural requirements. It utilizes the annex structure at the base as well as the stability of the foundation to restrain the horizontal displacement. The connection of vertical components with different sections will need some extra attention.

5. Strength critical connections. Rigid connection is better ensured at the nodes of beams and columns. For instance, widening or thickening of the undersurface of the beams or the addition of reinforcements to the nodes of beams and columns in a steel structure will enhance the stiffness of the connecting nodes and directly reduce horizontal displacement derived from horizontal loads.
6. Increase prestressed effect of columns. The direct transmission of vertical loads to the columns which bear the overturning flexural moment of the whole building will prestress the columns, reducing the pulling effect of the overturning moment and make the structure more stable.

No matter which style high-rise structures take, the mechanics and the simplified calculation model are a cantilever beam embedded in the foundation. To restrain the horizontal displacement under the horizontal loads, the only way is to improve the horizontal flexural stiffness of this cantilever beam. All the above-mentioned structures and structural measures are to achieve this purpose. And practice has proved that such designs can deliver satisfactory results in all aspects.

The design of high-rise buildings will also have to meet functional requirements. Functions and space are directly related and are in connection with the structural system adopted. The conceptual design has to meet functional requirements of the structure. Structural systems and the characteristics of the internal space commonly seen in high-rise buildings are listed in Table 5.1 for reference.

The height of a high-rise building will be affected by materials applied and the structural system adopted. The steel structures and the reinforced concrete structures are adaptable to different stories and different heights. In addition, different structural systems will also exert different influence. In regard to this, "Technical Specification for Concrete Structures of Tall Building" JGJ3-2010 [33] has made specific requirements (Tables 5.2, 5.3, 5.4 and 5.5). Table 5.2 lists the maximum height applicable to varied reinforced concrete structural systems. Table 5.3 lists the maximum aspect ratio applicable to various structural systems in high-rise reinforced concrete buildings. Figure 5.5 suggests the stories applicable to a variety of structural systems in high-rise reinforced concrete buildings.

Table 5.4 lists the maximum height applicable to high-rise steel structural systems (m) in accordance with "Code for Seismic Design of Buildings" GB50011-2010 [33]. Table 5.5 is the maximum aspect ratio applicable to civil-use steel structures. Figure 5.6 is the stories applicable to high-rise steel structural systems.

Table 5.1 Structural systems and the characteristics of layouts commonly used in high-rise buildings

Structural system	Frame	Shear wall	Frame-shear wall	Framed tube	Tube in tube	Bundle tube
Structural floor layout (footprint)	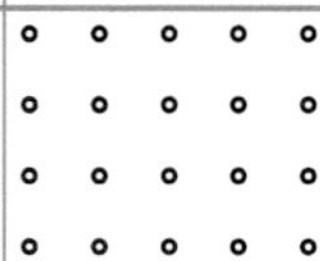	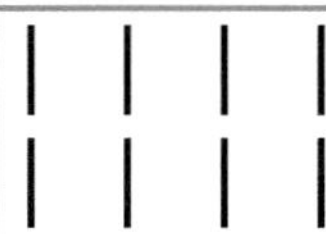	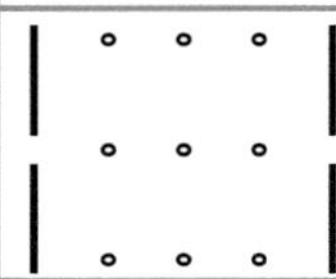	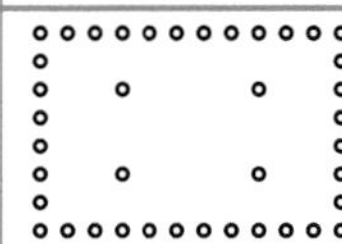	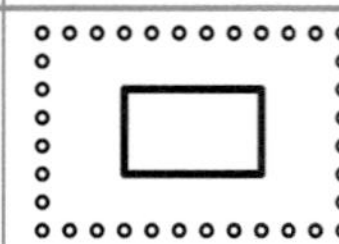	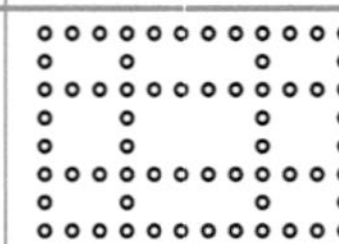
Room intersected layout	More flexible	Limited	Flexible	More flexible	Flexible	More flexible
Internal space	Larger space	Small space	Large space	Larger space	Large space	Larger space

Table 5.2 Maximum heights of varied reinforced concrete structural systems (m)

Structural system		Non-seismic design	Seismic fortification intensity			
			6°	7°	8°	9°
Frame		70	60	55	45	25
Frame-shear wall		140	130	120	100	50
Shear wall	All of the shear wall linked with the ground	150	140	120	100	60
	Part of framed shear wall	130	120	100	80	Not adopted
Tube	Frame core tube	160	150	130	100	70
	Tube in tube	200	180	150	120	80

Table 5.3 Maximum aspect ratio of various structural systems in high-rise reinforced concrete buildings

Structural System	Non-seismic design	Seismic fortification intensity		
		6°, 7°	8°	9°
Tube	5	4	3	2
Frame-shear wall	5	5	4	3
Cast-in-place shear wall	6	6	5	4
Tube in tube or bundled tube	6	6	5	4

Table 5.4 Maximum height applicable to high-rise steel structural systems (m)

Structural system	Seismic fortification intensity		
	6°, 7°	8°	9°
Frame	110	90	50
Frame-shear wall	220	200	140
Various tubes	300	260	180

Table 5.5 Maximum aspect ratio applicable to civil-used steel structures

Seismic intensity	6.7	8	9
The maximal aspect ratio	6.5	6	5.5

High-rise buildings develop rapidly in recent years, and the stories listed in Tables 5.2 and 5.4 might not cover all. To cope with that, China's Ministry of Construction issued No. 111 ministerial decree "Requirements of High-rise Construction for Seismic Resistance" [34] in 2010 to restrict the maximum height applicable to high-rise buildings. As for the high-rise buildings with excessive heights that exceed those prescribed in Tables 5.2 and 5.4, special censorship needs to be established and individual treatment is to be administered [35].

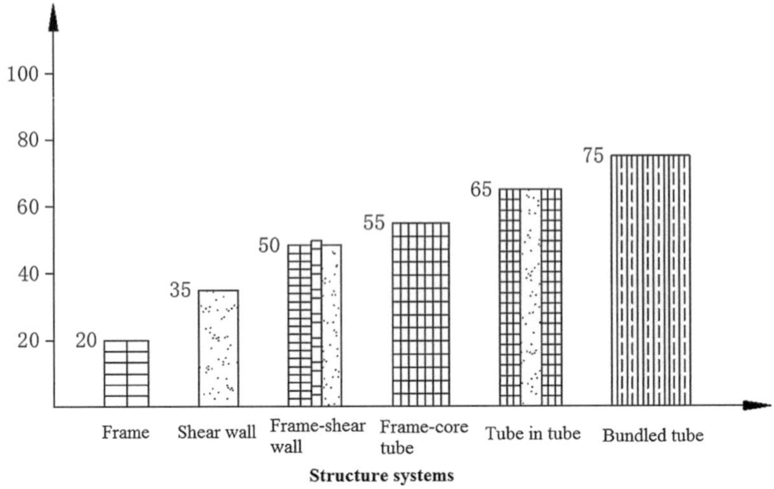

Fig. 5.5 Stories applicable to a variety of structural systems in high-rise reinforced concrete buildings

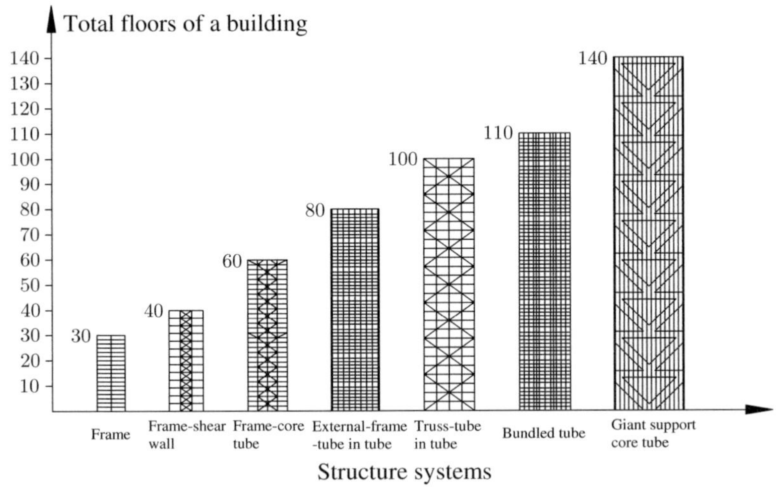

Fig. 5.6 Stories applicable to high-rise steel structural systems

5.1.2 Overall Analysis of Vertical Structures

As the height of the structure increases rapidly, horizontal stiffness and dynamic problems become prominent. The former is related to excessive horizontal displacement of the control structure under the action of wind loads, while the latter correlates to seismic response of the structure. Besides, the difference in temperature due to uneven sunlight exposure may also result in horizontal displacement. Steady wind loads can lead to horizontal displacement, while dynamic wind loads make the structure vibrate and generate additional horizontal displacement. From the perspective of structural functions, horizontal displacement derived from these factors must be properly restrained.

When seismic effects are taken into consideration, stimulation of the basic vibration mode of the structure is better avoided in order to avoid resonance. Since the frequency of the basic vibration mode usually is close to that of the earthquake and the displacement of the basic vibration mode is large and easily destructive, the basic vibration mode has to be controlled. Although some high-frequency components are entrained with seismic waves which might as well stimulate vibration modes of higher order, they do not serve as the decisive factors due to the small displacement they might cause.

It should be noted that as for high-rise buildings, the requirements of wind resistance and seismic resistance can be conflicting. Large stiffness might benefit wind resistance while penalize seismic resistance. On the contrary, when flexible structure is adopted, resonance with the earthquake might be small, while displacement will be large under the action of wind loads. Therefore, rational designs will make sure that the structure has sufficient stiffness to resist daily wind loads while having adequate flexibility and toughness when hit by major earthquakes.

As discussed in Chap. 3, detrimental vibration effects could be limited through active and passive control. For instance, an earthquake struck the city of Nicaragua, Managua, South America, in 1972. Among 511 neighborhoods of the city, houses destroyed reached a shocking number of 10,000, whereas the Bank of America, located at the epicenter downtown, stood in pride in the midst of the ruins. The Bank of America is the highest building in the city. Its resilient stance in this earthquake was attributed to the success in the application of an anti-seismic conception named "electric fuse" as well as the concept of passive control.

This building is a reinforced concrete tower. Four square tubes constituted a core tube through connecting beams to resist wind load and horizontal effects produced by earthquakes. Under normal wind loads, the core tube would not only resist the shear forces derived from the wind load but also provide a comparatively large stiffness to restrict horizontal displacement. When strong earthquakes strike, since connecting beams are relatively weak, seismic forces will first turn them into plastic state and then wreck them. As a result, restraints between four square tubes are lost (or weakened) and four individual tubes begin to work independently. Consequently, the stiffness of the entire structure plummets, its natural vibration period extends (according to an aftermath calculation, the natural vibration period extended from 1.3 s before the earthquake to 3.3 s after the earthquake.), and its resonance with the earthquake gets mitigated. Displacements of the structure at this time might increase, but, when faced with severe earthquake disasters, the priority is to ensure the structure would not collapse and guarantee the safety of personnel. Large deformation helps to consume earthquake energy and prevents the structure from collapsing. From the example of Bank of America, we know that through elaborate design, high-rise structures could not only limit displacements induced by wind loads in their normal service condition but also resist seismic forces in their ultra-limit state.

In the process of global urbanization, the construction of urban office building (such as the central business district) and a great many of residential buildings is in demand. Due to the limitation of urban land, the construction of high-rise buildings almost becomes the only solution. The overall mechanical state and deformation of a high-rise building are more complex than that of a low-rise building. As for the structural design of a high-rise building, people's attention will be shifted from strength problems derived from vertical loads to the limitation of horizontal displacement caused by horizontal loads. With the increment of height, the building's aspect ratio (*H/B*) also increases. The overturning moment generated by the wind load is proportional to the square of the height, while the horizontal displacement is proportional to the forth power of the height. Therefore, the displacement limitation becomes the major concern in the design of high-rise buildings. The purpose of all the research concerning overall mechanical state and deformation of a high-rise building is to design subsystems in a rational way, optimizing the composition of components so as to resist horizontal displacement induced by the wind load in normal service condition and to avoid collapse when major earthquakes strike.

In a high-rise structure, horizontal substructure (such as beam–slab system) is one of the major components that transfer vertical loads. As for this function, there is no significant difference between that in a high-rise building and that in a low-rise building. Vertical loads are transferred through beam–slab system to vertical substructure (such as columns, shear walls, and tubes) and then to the foundation. In a high-rise structure, horizontal loads are largely bore by vertical substructure. These substructural systems will bear not only all the vertical loads but also a majority of the horizontal loads. Since the vertical substructure selected has a direct impact on the overall mechanical state and deformation, the designation of a high-rise building often adopts the name of its vertical substructure, such as shear wall high-rise structural system, frame-shear wall high-rise structural system, frame-core tube high-rise structural system, and tube-in-tube high-rise structural system.

We will analyze mechanical states and deformation of varied high-rise structural systems and discuss how to limit horizontal displacement of high-rise buildings.

5.2 Design of Shear Wall Structures

The highlight in the design of a high-rise structure is the design of its vertical substructure. Fundamental vertical substructural system includes the shear wall structural system, frame structural system, and tube structural system, and each of them has its own mechanical characteristics. Often than not, many high-rise buildings would adopt a combined structural system with two or more structural systems integrated, such as frame-shear wall structural system and frame core-tube structural system.

As for a high-rise building, shear walls will not only effectively bear the vertical loads but also effectively resist horizontal loads. Generally speaking, shear wall structures are applicable to those high-rise buildings within 20 stories or so. If the shear wall is strong enough, it will also be applied to high-rise buildings with 30–40 stories.

The arrangement of shear walls should be symmetrical and bidirectional, so that they can effectively resist horizontal loads from different directions and possibly some additional torque. Normally, shear walls are collocated in pairs and results a gridlike formation. An arrangement like this will not only resist shear forces from different directions but also effectively counteract the effect of external torque.

The structure of a shear wall can be solid as concrete reinforced or hollow as in a steel truss. As for a steel truss, diagonal bars should be allocated in an "X" type or a "K" type to ensure the strength of the structure and to limit the deformation. Another problem is when holes have to be punched in order to install doors and windows, structural reinforcements have to be made to the surrounding areas of the holes. For instance, structural reinforcements have to be installed to strengthen the holes in a solid reinforced concrete shear wall. Local reinforced steel trusses have to be set in the opening of a steel truss shear wall so as to ensure the periphery of the opening has sufficient strength and avoids major deformation.

Tubes and elevator shafts composed by the shear walls could effectively resist the shear forces and the bending moment generated by horizontal and vertical loads. In many high-rise buildings, tubes and elevator shafts interconnect and demonstrate a synthesized effect in producing an enormous amount of resistance to horizontal loads. These shear walls better be arranged in a symmetrical manner in design to resist potential torque.

The following example is about the design or check of reinforced concrete lintels between the shear walls.

Example 1 Design for local reinforced concrete lintels between the shear walls in an n-story high-rise building bearing wind loads. The structure is shown in Fig. 5.7. The analysis is as the following:

1. A corridor is required on every floor and it is connected to two shear walls only through the lintel underneath the floor.
2. The overturning moment derived from the wind load in the shear wall is counteracted by a pair of force of reaction at the base.
3. The reaction T is the sum of all shear forces of lintels along the height of the building. Assuming that the shear forces are evenly distributed to each lintel, a share $V = T/n$ will be bore by each lintel. V produces shear deformation and bending moment M at the end of the lintel.
4. The bending moment M at the end of the lintel is produced by the shear force V in each lintel.
5. Sectional area and reinforcement arrangement of a lintel can be calculated through M.

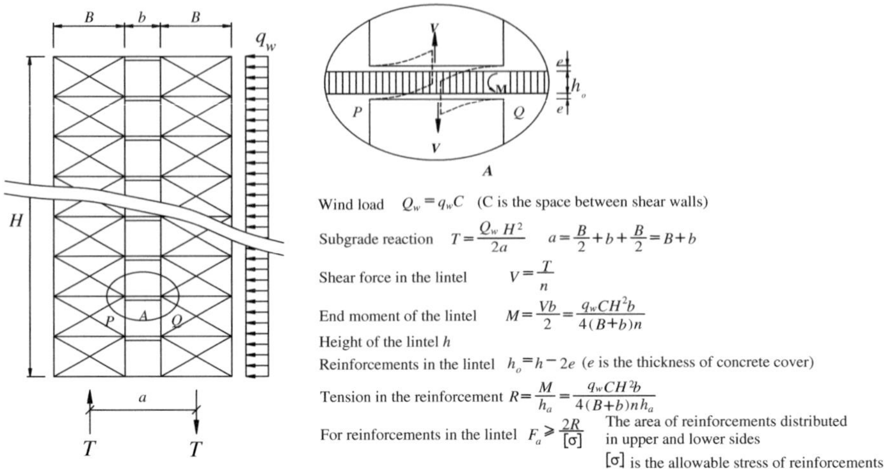

Fig. 5.7 Design of concrete lintels between the shear walls bearing wind load

The main part of the design or check of the lintels in this case is to determine the internal force of the lintels in light of the equilibrium of the shear wall. The calculation equations are:

$$\begin{aligned} M &= \frac{Vb}{2} = \frac{q_w C H^2 b}{4(B+b)n} \\ V &= \frac{T}{n} = \frac{Q_w H^2}{2(B+b)n} \end{aligned} \tag{5.1}$$

Sectional area of the lintels is determined with M and V, and sufficient tensile reinforcements and stirrups (sectional area) will be installed. To check the lintel of an existing building, sectional area of the lintels and safe operation of related reinforcements, stirrups under the action of bending moment M, and shear force V have to be guaranteed, as shown in Fig. 5.7.

5.3 Design of Frame Structures

Frame structure is often used in low-rise and moderate high-rise buildings. Integrated with shear wall or tube structures, it could also be applied to high-rise buildings. In comparison with shear wall structural system, frame structures have larger interior space which will facilitate arrangement and utilization with fewer limits on the installment of doors and windows. Another advantage of frame structural system would be the “rigid frame effect” generated by the points of inflection at the midpoints of beams and columns through a rigid connection between beams and columns. With this feature, frame structures are able to function as a whole and transfer horizontal and vertical loads to the foundation.

Compared with the shear wall, horizontal stiffness of the columns is relatively small. Under horizontal loads, horizontal displacement generated by frame structures in moderate high-rise buildings is far greater than shear wall structures. Therefore, frame structures are often considered as “ductile.” When an earthquake strikes, “ductile” structures will absorb considerably seismic energy due to deformation, which would prevent the structure from sudden collapse and protect personnel and property. Therefore, in seismic design of moderate high-rise buildings, reinforced concrete or steel frame structural system is often adopted.

With the addition of stories, excessive horizontal displacement will be produced by frame structures under service load. A combination of frame structure and shear wall structure will effectively limit horizontal displacement and enhance the strength and stiffness of the structure. The shear walls are usually collocated in pairs in a frame structure so as to bear a majority of horizontal loads and such arrangement often works.

Approximate analysis of high-rise frame structure under horizontal wind load and vertical loads can be conducted with the simplified calculation method introduced in Sect. 4.7. The conceptual design of high-rise frame structures under the action of seismic forces mainly concerns about the estimation of overturning bending moment inflicted by seismic forces upon the structure.

Example 2 A simplified model to calculate the seismic response of the frame structure in a high-rise building as shown in Fig. 6.8 is required. Major steps are as follows:

1. Calculate the deadweight W of the structure on each floor.
2. Seismic force inflicted upon each floor is estimated as 4 % of the deadweight W of each floor (refer to Sect. 2.3 and select in accordance with regional seismic fortification criterion); seismic effect on each floor is simply assumed to be identical; horizontal effect of the seismic force is imposed upon each floor. Shear force V at the base of the foundation and that distributed to each column $V_1 = V/2$ are determined from Fig. 5.8a. So can be determined the shear force of each column on each floor.
3. Assume the point of inflection of each column on each floor is at the midpoint of the column and the bending moment at the end of the beam in the second-floor frame structure is thus determined with shear force V_1 at the base, as shown in Fig. 5.8b. As for shear force of the columns on the upper floor V_2, it could be calculated as the principle indicates: The shear force induced by seismic force decreases as the story goes upward.
4. Calculate the bending moment induced by shear force V_1 and V_2 at the end of the beam M_3.

$$M_3 = M_1 + M_2$$

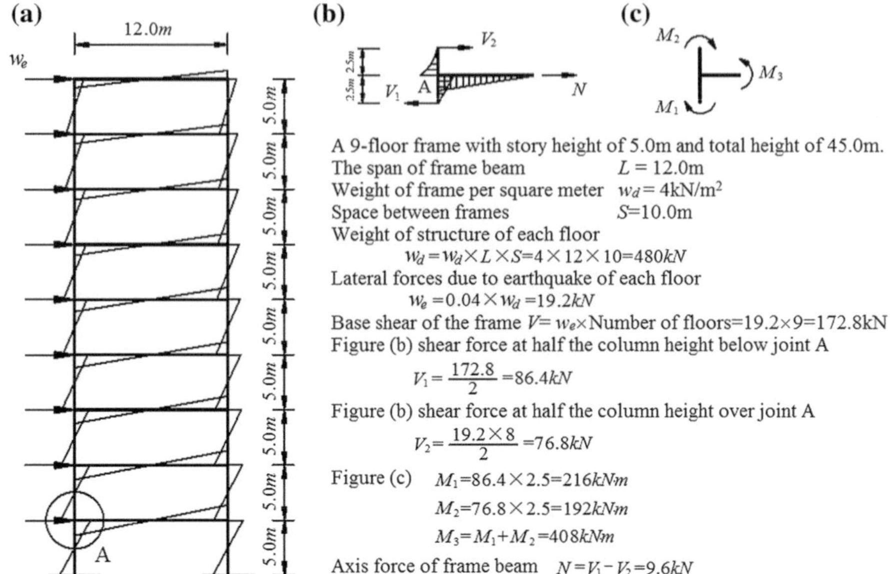

Fig. 5.8 Simplified calculation of earthquake on frame structures

5.4 Design of Tube Structures

The closed section serves as the key factor to resist buckling instability in a tube structure, which has also been proved by the evolution of natural selection of both animals and plants. For example, the bones of human or bamboos are hollow tubes, which is the result of biological evolution of millions of years. Similarly, tube structural system is commonly applied in high-rise structures ("Tube effect" has been introduced in Sect. 4.7 of this book).

A tube structure is comprised when four outer walls of the frame structure are to be connected. Figure 5.9 compares the footprint of a frame structure and that of a

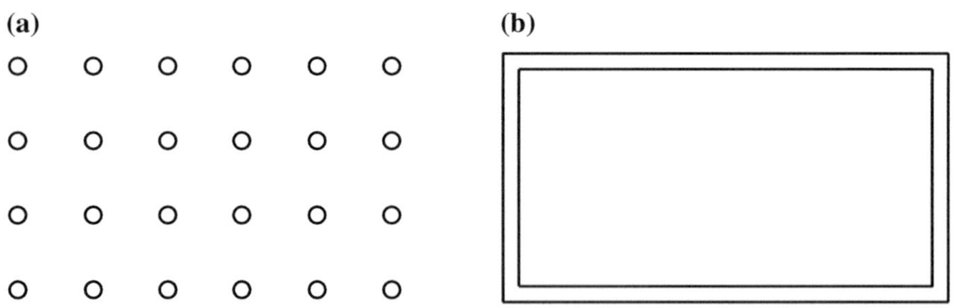

Fig. 5.9 Comparison of footprint of frame structure and tube structure. *Note* The overall "structural areas" are the same in (**a**) and (**b**)

tube structure with the same section area. From the figure, we could see that the moment of inertia of the tube structure is greater than that of the frame structure with the same section area, which in turn indicates that the tube structure provides a higher stiffness as well as strength compared with frame structure and shear wall structure. Since the section area of a tube structure is arranged away from centroid axis of the section, the surrounding walls can be utilized and provide maximum bending resistance arm. And the tube structure is therefore widely applied in high-rise or ultra-high-rise buildings. Tube-in-tube structure is also commonly adopted. The outer tube of a tube-in-tube structure is comprised of a frame tube or a truss tube, while the inner tube is a solid (or with openings like doors) reinforced concrete core tube, which under normal circumstances could (partly) serve as elevator shaft.

When the tube bends as a cantilever beam, walls will have to bear the shear force and overturning moment induced by horizontal loads. Since the tube usually has openings and local flection, the transmission of shear force is delayed. Shear lag is thus produced and a nonlinear redistribution of the stress within the wall occurs. As a result, the internal forces of the columns at outer corners are further augmented (approximately taken as 1.5 times average maximum stress), while that of the central columns are diminished (approximately taken as 0.5 times average maximum stress), as shown in Fig. 5.10.

If the opening area of the tube is considerably large (more than 50 % of the surface area), the tube structure will be regarded and calculated as a frame structure.

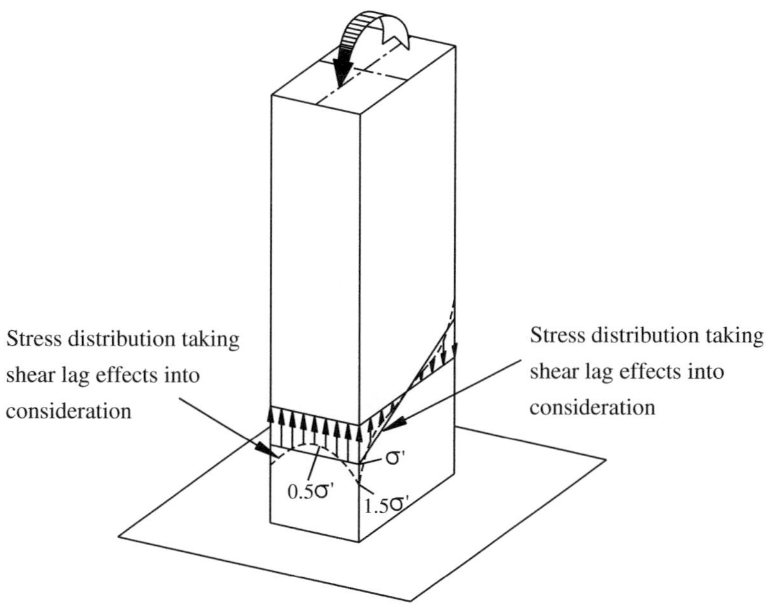

Fig. 5.10 Shear lag effects of framed tube

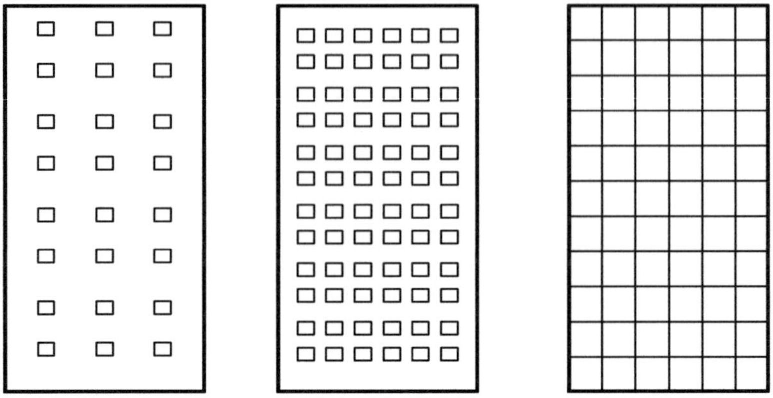

Fig. 5.11 Opening of tube affect the computational model

If the opening area is limited (less than 30 % of the surface area), the tube structure is more like a solid-web tube and will be calculated as an equivalent solid-web tube with a thickness reduction. When the opening area is between 30 and 50 %, shear lag effects need to be considered in the calculation of the internal forces, for there is a 50 % increase of the internal force for corner columns and it would be unsafe to exclude this effect in the calculation of internal forces (Fig. 5.11).

Obviously, the extent of the opening area on the tube walls has a major impact on the mechanical properties of the tube structure and the strength and stiffness will have to conduct corresponding reductions in light of the opening ratio of the walls. According to previous research, when the opening area is less than 30 %, the walls are devalued as equivalent solid-web tube walls with reduced thickness on the strength of the equivalent principle of shear deformation. Figure 5.12 illustrates the correlation between the opening ratio (of the shear wall) and the thickness of equivalent solid-web wall (of the shear wall). This curve is derived from finite element calculative analysis, and its theoretical base is the principle of same shear deformation between walls (in a shear wall structure) with varied opening ratio and their equivalent solid-web walls under the action of shear forces. The curve in Fig. 5.12 shows that with the upswing of the opening ratio (of a shear wall structure), the thickness of equivalent tube wall gradually decreases. When the opening area is less than 30 %, the tube can be devalued as equivalent solid-web tube. When the opening area is greater than 30 %, the devaluation is basically meaningless.

$$k = t/t^0 \tag{5.2}$$

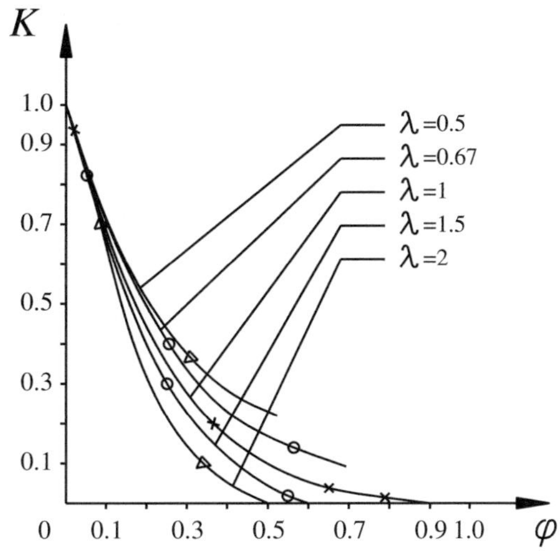

Fig. 5.12 Opening percentage of tube or shear wall and equivalent reduction coefficient of wall thickness

The t_0 in Eq. (5.2) represents the original thickness of the tube wall (or shear wall) with openings, while t is the equivalent thickness of the tube wall (or shear wall) with openings. K is the equivalent reduction coefficient and is in connection with the ratio and the shape of the opening, as shown in Fig. 5.12. The aspect ratio of the opening is $\lambda = \frac{a}{b}\left(\frac{l_2}{l_1}\right)$, in which equation a and b, respectively, represent the length and width of the opening and I_1 and I_2 are the length and width of the tube wall. The opening ratio of the tube wall is $\varphi = \frac{ab}{l_1 l_2}$.

Example 3 Calculate the equivalent wall thickness of a tube with openings. The original thickness of the tube wall is $t_0 = 0.20$ m, the length and width of the tube wall are $l_1 = 6.0$ m and $l_2 = 8.0$ m, and the length and width of the opening are $a = b = 2.0$ m. Then,

$$\varphi = \frac{ab}{l_1 l_2} = 2.0 \times 2.0/6.0 \times 8.0 = 4.0/48 = 0.08$$

$$\lambda = \frac{a}{b}\left(\frac{l_2}{l_1}\right) = 2.0 \times 8.0/2.0 \times 6.0 = 1.33$$

In light of Fig. 5.12 and the interpolation method, $K = 0.73$. Since $K = t/t_0 = 0.73$, the equivalent thickness of the tube wall with opening is $t = 0.146$ m.

If calculation is directly conducted with volume deduction of the tube wall, then the deduced thickness after the punctuation is $t = (0.2 \times 6 \times 8 - 0.2 \times 2 \times 2)/(6 \times 8) = 0.183 \text{ m} > 0.146 \text{ m} \approx 0.8 \times 0.183$ m.

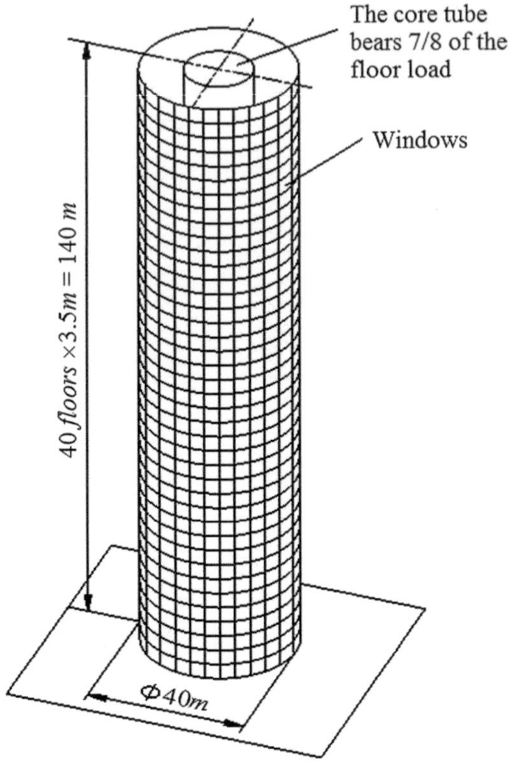

Fig. 5.13 Simplified calculation of reinforced concrete tube-in-tube structure

Hereby, we could see that the deduced thickness of the tube wall, which is calculated by the principle of same shear deformation of the tube wall and the equivalent solid-web tube wall, has more reserve for safety.

Example 4 Figure 5.13 is a 40-story reinforced concrete tube structure. The height of each story is 3.5 m and the total height is 140 m. The diameter of the outer tube is 40 m, and the openings are as illustrated in Example 3. 30 MPa concrete is applied in the outer tube wall. In simplified design, it is assumed that 5/8 of the vertical loads are bore by the inner tube and 3/8 of the vertical loads and all horizontal overturning moment are bore by the outer tube. With all the information available, we will determine the thickness of the outer tube wall under given horizontal and vertical loads. Calculation is as follows:

1. The shape factor of wind load is taken as 0.6 according to "Load code for the design of building structures" GB50009-2001. The wind stress factor of Shanghai is 0.55 kN/m^2. Assuming the wind load is evenly distributed along the height of the structure, the uniformly distributed line load along the height of the structure is: $P_w = 0.6 \times 1.4 \times 0.55 \times 40 = 18.48$ kN/m.

2. Calculate the wind-induced overturning bending moment at the foundation:

$$M_w = 18.48 \times 140 \times 140/2 = 181{,}104\,\text{kN m}$$

3. Assuming that the ultimate thickness of the outer tube is $t_0 = 0.20\,\text{m}$, the deduced thickness after punctuation (calculated in Example 3) is t = 0.146 m.
4. Determine the moment of inertia produced by the outer tube at the foundation:

$$I_t = [3.14 \times 40^3 \times 0.146]/8 = 3667\,\text{m}^4$$

Determine the maximum tensile or compressive stress in the outer tube wall generated by the overturning bending moment at the foundation

$$\sigma_w = M_w \times D/(2I)_t = 181{,}104 \times 20/3667 = 988\,\text{kN/m}^2$$

5. The deadweight (the effects of the openings are taken into consideration) of the outer wall is estimated as q = 2 kN/m^{-2}, and therefore, the deadweight per unit length along the periphery of the foundation is

$$2\,\text{kN/m}^2 \times 140\,\text{m} = 280\,\text{kN/m};$$

Compressive stress produced by the deadweight along the periphery of the foundation is

$$280/0.146 = 1918\,\text{kN/m}^2$$

6. The summation of the dead and live load of the floor is assumed to be q_2 = 8 kN/m^2. On the strength of the distribution assumption of vertical loads mentioned above, 3/8 of the vertical loads on the floor is bore by the outer tube and that would be bore by per-unit-length wall along the periphery of the foundation is 40(8 × 3/8) × 3.14 × 20 × 20/(3.14 × 40) = 1200 kN/m. Corresponding compressive stress is 1200/0.146 = 8219 kN/m^2
7. Determine the maximum and minimum stress induced by vertical loads and overturning moment (horizontal loads) at the foundation
σ_{max} = 1918 + 8219 + 988 = 11,125 kN/m^2 = 11.1 MPa (compressive stress) <30 MPa concrete design strength = 14.3 MPa
σ_{min} = 1918 + 8219 − 988 = 9149 kN/m^2 = 9.15 MPa (compressive stress).
8. The thickness of the tube wall should make sure that compressive stress is maintained within the wall; in other words, there will be no tensile stress, so that the equilibrium requirement of the structure will be met. In this previous example, both the maximum and the minimum stress of the tube are compressive stress and the maximum stress is less than the designed strength of the concrete. Thus, the selection of the thickness t = 0.20 m is viable.

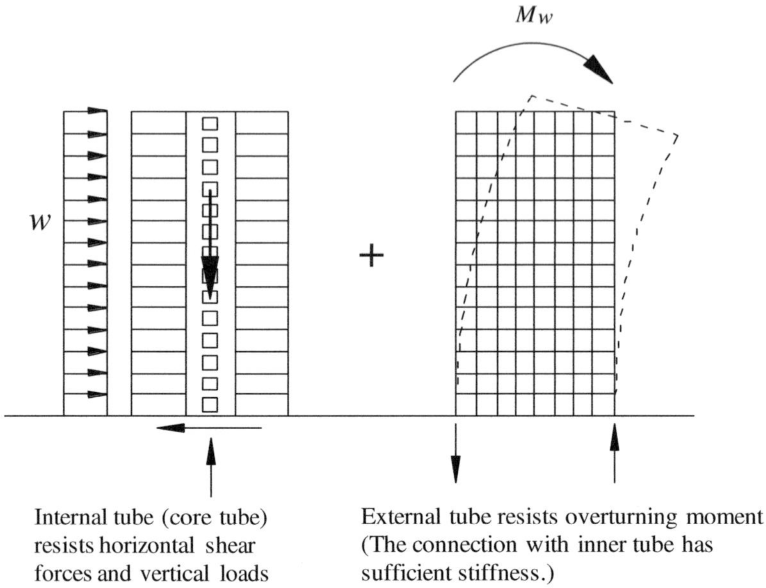

Fig. 5.14 Loading features of inner and external tubes of the tube-in-tube high-rise structure

The essence of the tube-in-tube high-rise structure is the utilization of vertical stiffness and shear resistance of the inner tube and the bending stiffness of the outer tube. The shear force induced by horizontal loads and a majority of vertical loads will be bore by the inner tube, while the overturning moment derived from horizontal loads will largely be bore by the outer tube through bending stiffness. The inner and outer tube will not only function individually but also work in cooperation (as shown in Fig. 5.14).

5.5 Design of Mixed Vertical Structures

To meet the operational requirements of structures with interior wide span or ultra-high-rise structures, extra consideration sometimes has to be made for high-rise structures. When it is necessary, the combination of the aforementioned structural systems would give rise to some mixed structural systems, such as skip-floor truss structural system, giant truss structural system, connecting-tower structural system, framed tube structural system, suspended core-tube structural system, and inclined column truss structural system. These mixed structural systems could meet the requirements of strength and stiffness for horizontal load transmission in high-rise structures in more economical way compared with single structural system.

5.5.1 Skip-Floor Truss Structural System

In order to meet the operational requirement of internal wide span, skip-floor truss (or skip-floor deep beam) structural system can be adopted. First, we will analyze this multi-story frame structure. Columns and beams are bending components. When the span is extended, the bending moment of the beam will multiply in accordance with the square of the span. Beams are affected the most, while the columns are not so much influenced. Thus, to avoid overstrengthening of the beams when the span is extended, connecting the upper and the lower beams with diagonal bars and making a truss out of it could be a practical choice.

Figure 5.15 presents a multi-story frame structure. The upper and the lower beams of every other story are connected with diagonal bars to form trusses, which in other words is the "skip-floor" truss wall. Such a "skip-floor" truss structure will not only provide large internal space with wide span but also significantly increase the stiffness of the structure. That is because every beam in this frame structure is embedded in and fixed by floor slabs of two adjacent stories and a truss with enormous stiffness within its own plane is constructed. Under the action of horizontal loads, the shear force between skip-floor trusses will mainly be bore by the trusses; as for the unfortified upper and lower stories, since they are the skipped stories between trusses and thanks to the horizontal stiffness provided by skip-floor trusses, a majority of the shear forces between floors would be bore by them. Thereby, we believe in the multi-story frame structure as shown in Fig. 5.15 that the shear forces between floors are all but bore by skip-floor trusses and columns basically do not function to resist shear forces. Knowing that trusses could only have considerable horizontal stiffness within its own plane and to attend to the randomicity of horizontal wind loads, skip-floor trusses are also to be installed along the side perpendicular to that allocated with trusses.

The structure shown in Fig. 5.15 does not only meet the operational requirement of internal large span but also guarantee that the structure would not produce excessive horizontal displacement under the action of wind load or seismic forces due to the enormous horizontal stiffness of the trusses. Moreover, the mechanics and the path of force transmission of this structure are clear and justified.

Fig. 5.15 Skip-floor truss structure

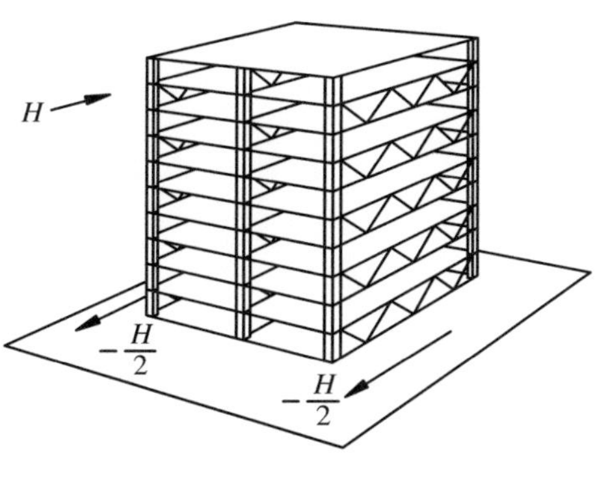

Although "skip-floor" trusses and deep beams are two different structures, their mechanical features are quite similar. Thereby, frame structure with skip-floor trusses installed has the same function as that with skip-floor deep beams installed. As for reinforced concrete high-rise structures, "skip-floor" deep beams can easily be made into quadrilateral vierendeel trusses. The middle vertical bars in a vierendeel truss without supporting diagonal bars generally have sufficient stiffness to bear the shear forces between floors; as for the upper and lower chords, they could be designed through the simplified calculation of tensile and compressive stress necessary to resist maximum bending moment bore by the vierendeel truss.

5.5.2 *Giant Frame Structures*

With the multiplication of the stories in high-rise buildings, the enhancement of horizontal stiffness and the control of horizontal displacement have become the major concerns. Since the overturning moment is mainly balanced by the countermoment derived from the axial forces of the surrounding columns, the fortification of side columns has become a necessity. However, the enhancement of side columns alone will not be able to elevate the horizontal stiffness of the whole structure. To connect giant columns on both sides with high-strength beams or truss beams and bring forward a giant frame system to resist horizontal loads would be a very effective solution (as shown in Fig. 5.16). And that would be giant frame structural system.

Figure 5.16 displays a 50-story high-rise building. Since the structure stands high up in the sky and has insufficient horizontal stiffness, one-story or two-story layer of trusses or that of horizontal beams could be installed on mechanical floors every 15–20 stories. Rigid connection between the layer of trusses or that of horizontal beams and side columns will contribute to construct a giant frame, which

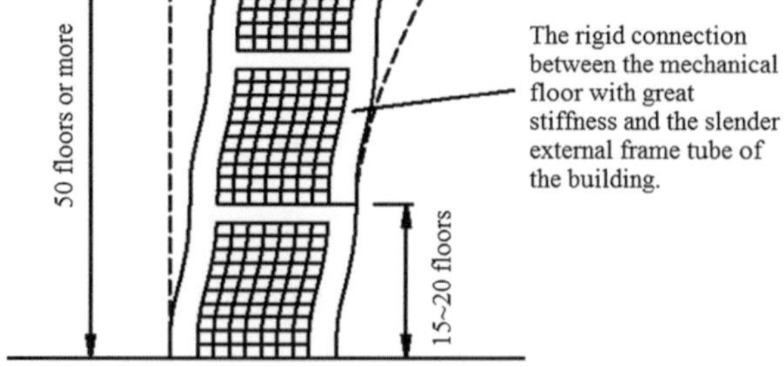

Fig. 5.16 Mega-frame structure

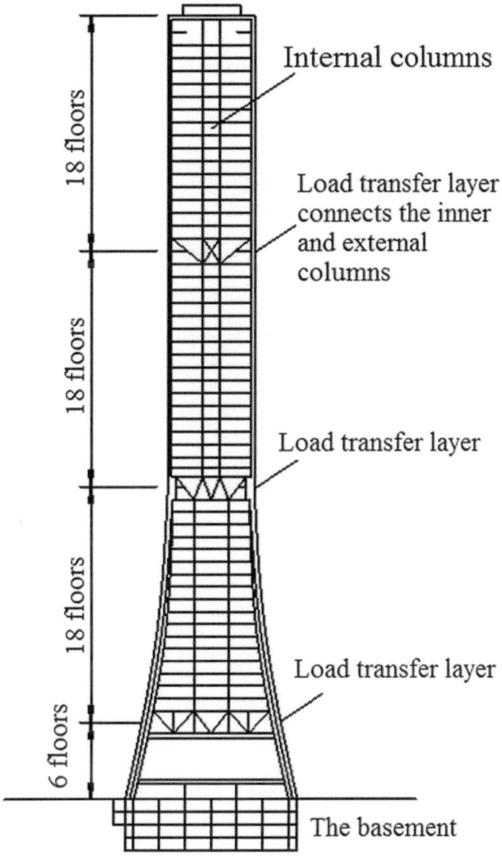

Fig. 5.17 Load transfer layer acts as mega-frame structure

will in turn greatly enhance the overall horizontal stiffness and restrict horizontal displacement. Conventional inner frame structure as shown in Fig. 5.16, although serving as a secondary frame, strengthens the outer giant frame structure. Figure 5.17 is a diagram of the National First Bank of the USA in Chicago. This building is 243.8 m tall, and it has 60 stories above the ground. Sectional area of the foundation is about 5110 m^2 and that of the top is about 2649 m^2. The first floor has no columns, providing an open space of more than 5000 m^2 for the business of the bank. Aside from 4 stories underground, a two-story layer of trusses is installed between 5th floor and 7th floor to serve as a layer of force transmission and transformation and it is rigidly connected with four high-strength outer columns in the formation of a giant frame structure. All the horizontal and vertical loads from 6th floor to 60th floor are transferred via this layer of trusses from the columns on one side to those on the other side without outstanding horizontal displacement. The 2nd layer of trusses is installed between 23rd floor and 25th floor, and it transfers horizontal and vertical loads from 23rd floor to 60th floor. Similarly, horizontal and vertical loads from 41th floor to 60th floor are transferred via the layer of trusses installed between 41th floor and 43rd floor.

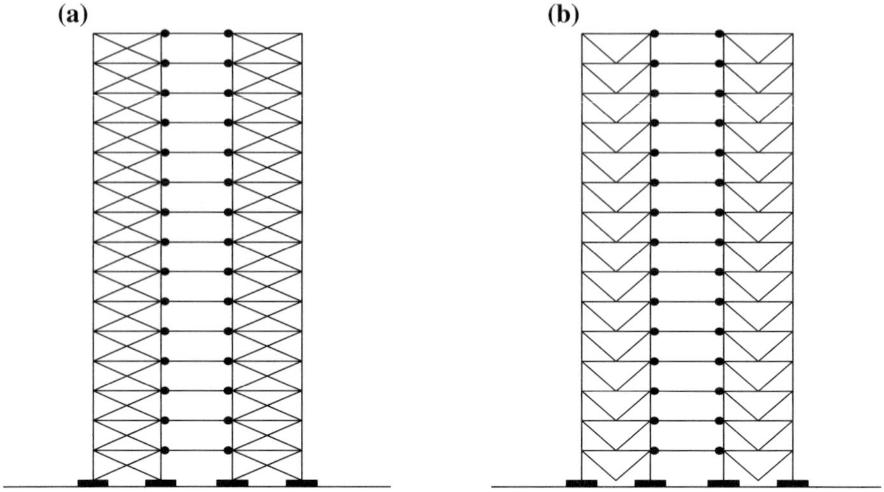

Fig. 5.18 Connecting-tower structural system. **a** X-type. **b** K-type

5.5.3 Connecting-Tower Structural System

Figure 5.18 is a frame structure comprised of two independent towers. Each one of them is able to bear corresponding vertical and horizontal loads, and they are rigidly connected with a horizontal connecting beam so as to make the structure work as a whole. When the high-rise structure is under the action of horizontal wind load or seismic load, an integrated structure with tremendous rigidity derived from two independent towers will counteract the overturning moment generated by horizontal loads and limit horizontal displacement to a significant extent.

The tower as shown in Fig. 5.18a adopted X-type crossed truss in its structure. Since the diagonal bars are great in length, buckling instability will be prone to happen in a steel structure. K-type truss tower as shown in Fig. 5.18b has short diagonal bars, and buckling instability is less likely to happen in a steel structure.

As for the connection between the connecting beam and two towers, rigid connections are to be employed.

While hinged connection could connect two towers with beams, the two towers would still work individually with their own rigidity and the horizontal deformation of the whole structure would be the same with structures with only one tower. If the beams are rigidly connected, "rigid frame effect" will significantly enhance the horizontal rigidity of the structure. In seismic design, connecting beams are often designed to be destructible. Under the action of service load, the rigid connection between two towers would limit horizontal displacement of the structure. When the structure is hit by major earthquakes or strong winds, the connecting beams would be destroyed first or turn into a plastic hinge. Without a rigid connection between two towers, the overall rigidity of the structure declines, natural vibration frequency is changed and seismic, or wind response of the system is modulated. This is also an example of the application of the concept "structural fuse" discussed in Sect. 4.7.

5.5.4 Framed Tube Structural System

Circular framed tube structure as shown in Fig. 5.19 is comprised of horizontal curved beam and columns allocated at the outer ring through rigid connection. Mechanical features of this structure are similar to that of a cylinder with openings. When the columns at the outer ring are sufficiently allocated (the opening ratio is less than 50 %), the structure could be regarded as a tube structure and has considerable bending rigidity which is often adopted in high-rise structures. Moreover, according to previous research [6, 9], the circular shape will reduce 20–40 % wind load, which is another outstanding merit circular tubes have. Therefore, circular framed tube is often adopted in high-ride structures.

Rectangular framed tube structure, as shown in Fig. 5.20, is divided into the outer tube and the inner tube. The outer tube provides an overall bending rigidity and counteracts the horizontal overturning moment, while the inner tube, elevator shafts included, bears the majority of vertical loads and shear forces (Fig. 5.14). Since such a tube-in-tube structure is succinct in design and explicit in component functions and load distribution, it is one of the most adopted structural forms in high-rise buildings.

5.5.5 Suspended Core-Tube Structural System

Suspended core-tube structural system, as shown in Fig. 5.21, is ideal for high demand in operational open space on the ground floor. The hallmark of this

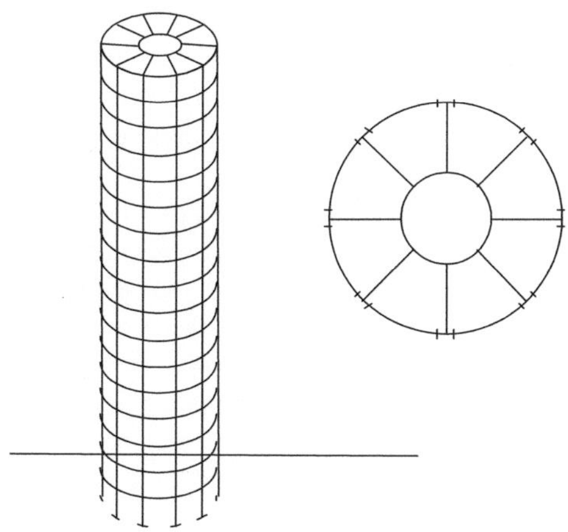

Fig. 5.19 Circular framed tube structure

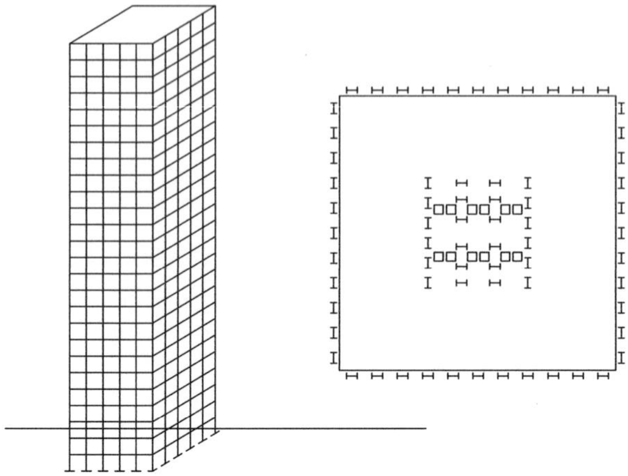

Fig. 5.20 Rectangular framed tube structure

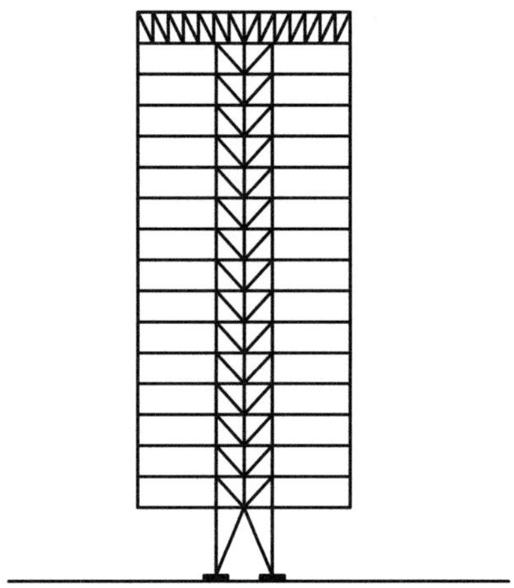

Fig. 5.21 Suspended core-tube structural system

structure is the high-rigidity cantilever trusses, allocated at the top of an independent core tower, and each and every floor is suspended by these trusses. The core tower, as an independent load-bearing system, bears all vertical and horizontal loads.

If columns at the foundation are sufficient in rigidity, the structure, as shown in Fig. 5.21, will effectively restrain the horizontal displacement under the action of horizontal loads. Otherwise, the bending deformation of columns at the foundation might give rise to considerable horizontal displacement. In that case, a suspended core-tube structure, as shown in Fig. 5.22, can be adopted, for it extends wall columns on both sides to the ground floor and then to the foundation. Meanwhile, since vertical columns are connected with the cap trusses, the core tower will

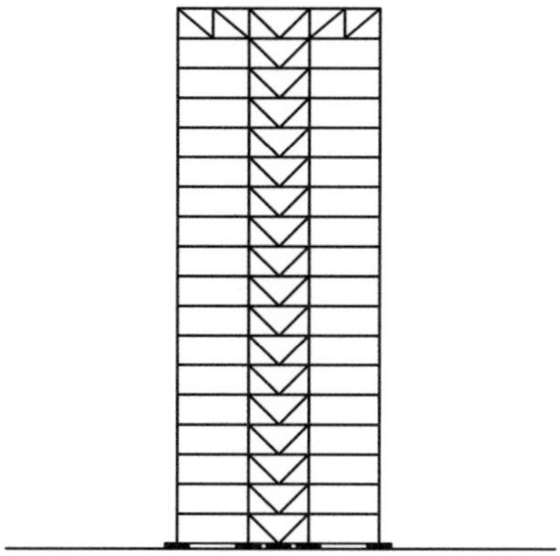

Fig. 5.22 Suspended core-tube structural floor system

produce a point of reflection under the action of wind loads, which in turn will substantially reduce sideway displacement at the top of the structure.

Further developing the concept of suspended core-tube structural system, we will have strengthened core-tube frame structural system, as shown in Fig. 5.23. Several outrigger trusses are added to the central core tube. These outrigger trusses would not only serve as the layer of equipment shift but also connect the inner core tube with the outer giant frame, which will effectively elevate the overall strength and rigidity of the structure. Figure 5.24 is the application of this type of structure in Shanghai Jin Jiang Restaurant.

5.5.6 Inclined Column Truss Structural System

Due to increasing aspect ratio, the horizontal displacement under wind loads and seismic loads could be difficult to control. In case like that, the adoption of inclined column structural system would be an effective solution. As for the structures with a narrow top and a wide bottom, the horizontal displacement is often effectively restricted. In actual constructions, such a structure system could serve as a candidate design to reduce horizontal displacement under the action of horizontal loads. Inclined column truss tube system, as shown in Fig. 5.26, is the extension of conventional truss tube system, as shown in Fig. 5.25. Through the adjustment of the distance between columns, the type of truss, and the inclined angle of columns, mechanical performance of the structure under static and dynamic loads would be controlled.

Pyramidlike structure (Fig. 5.27) is one of the variant of inclined column truss tube system, as shown in Fig. 5.26. The feature of this type of structure is the convergence of all columns at the top. This structural form will generate enormous rigidity and considerably restrain horizontal displacement. Nowadays, this concept of design has further developed for some architects as the basis of their design for ultra-high structures or even some world-class skyscrapers. Figure 5.28 is the

Fig. 5.23 Strengthened core-tube frame

schematic design for the Millennium Tower intended to be built in Tokyo, Japan. Figure 5.29 is a contemporary community building which adopts inclined column truss tube system.

The Millennium Tower in Tokyo, as shown in Fig. 5.28, is a conic pyramidlike structure. The tower is 800 m high and the diameter at the bottom is about 180 m. The tower is designed to accommodate fifty thousand people. In the conceptual design of this structure, the flow-splitting and wind load-reducing effects of the shape of a conic pyramid are taken into consideration. Thereby, some outstanding advantages of this structure are as follows: light wind load, comparatively low point of application of wind load and seismic forces, and relatively small overturning moment. Cross-giant space truss tube that unfolds surface-wise the cone is adopted as the major structural system to resist horizontal loads; some fortified "restraining trusses" are installed in the central locale along the height, which will effectively bear horizontal loads, help with the ventilation, and further reduce the wind load.

Fig. 5.24 Sketch of Shanghai Jin Jiang Hotel

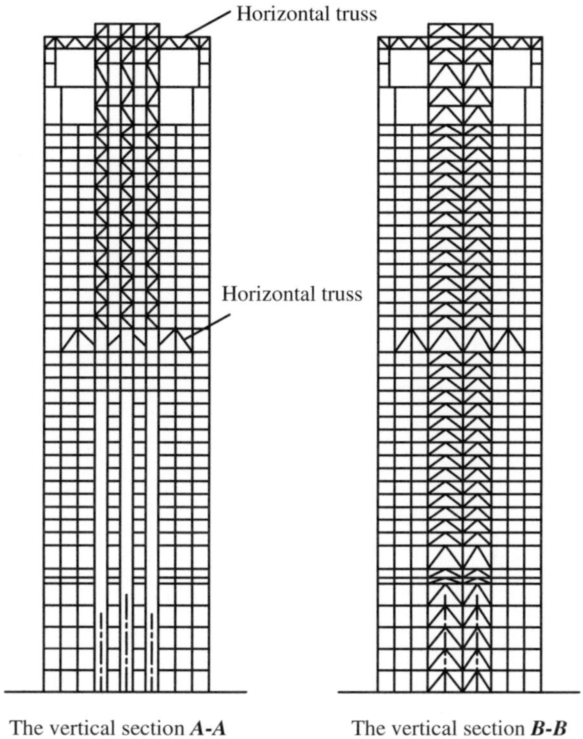

Fig. 5.25 Trussed tube system

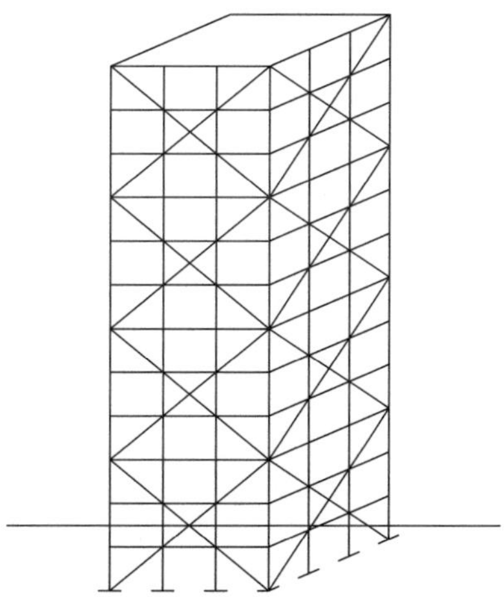

To deal with the soaring population in urban cities, architect Stanley Tigerman proposed a design of so-called community building, as shown in Fig. 5.29. Two symmetric 180-m-high triangle buildings incline toward each other and finally converge at the top, while rivers or highways would be bridged over underneath the

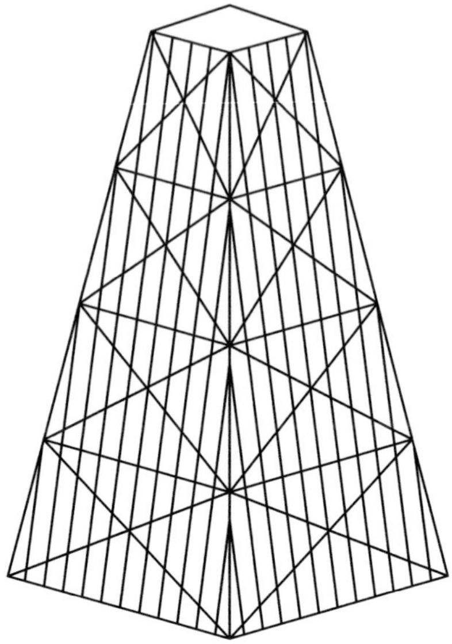

Fig. 5.26 Inclined column trussed tube system

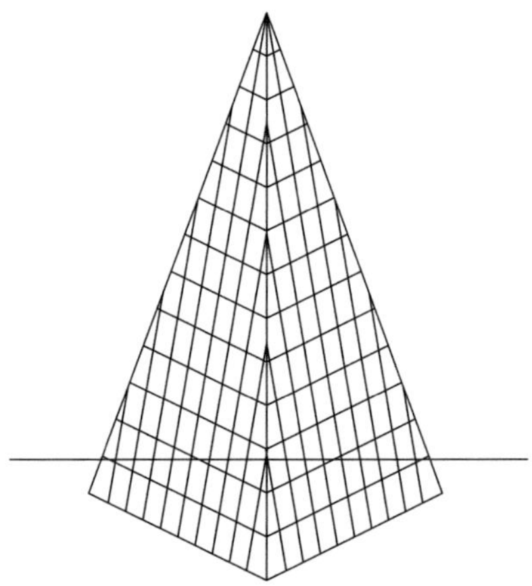

Fig. 5.27 Pyramidlike structures

structure. Geometric space effect was considered in the conceptual design, which lowers the position of the resultant force of wind loads or seismic forces, transforms the overturning moment generated by the wind loads at the foundation into lateral push of these two triangle buildings, and reduces horizontal displacement of the whole structure.

In previous sections, mechanical features, structural forms, and some simplified calculation methods of high-rise structures have been discussed. Four typical high-rise structures will be analyzed and discussed in details next.

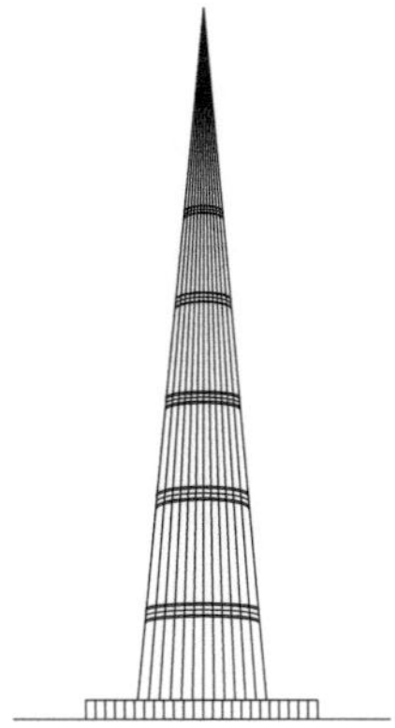

Fig. 5.28 Schematic design for the Millennium Tower intended to be built in Tokyo, Japan

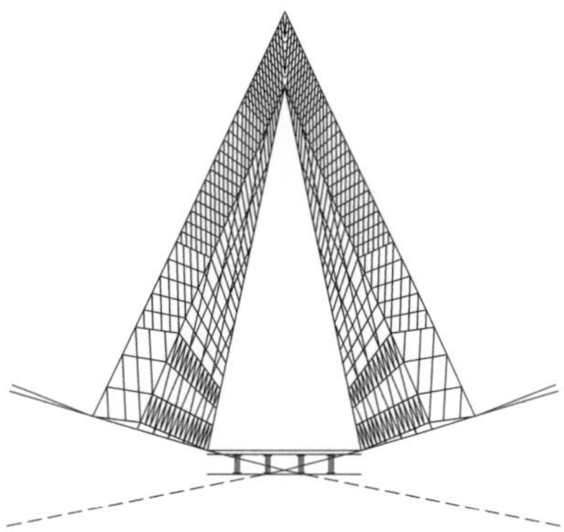

Fig. 5.29 Contemporary community building

5.6 Comparative Analyses of Milestone High-Rise Buildings

5.6.1 Outer Trussed Tube-in-Tube System in Shanghai World Financial Center

Project background

Shanghai World Financial Center is located in Lujiazui, Pudong, Shanghai [36]. The central construction area of this project takes up about 30,000 m^2 and the gross building is up to 381,600 m^2. Specifically, the building area of the main building is about 252,935 m^2, that of the annex building is 33,370 m^2, and that of the underground is 63,751 m^2. There are 101 stories above the ground and 3 stories under the ground. The height of the main building amounts to 492 m (about 70 m higher than Shanghai Jin Mao Tower), and the project reached its final completion in 2008. Shanghai World Financial Center was the highest building at its

Fig. 5.30 Model of Shanghai World Financial Center

completion in China. Shanghai World Financial Center has become a new tourist attraction, due to its panorama of urban Shanghai at the top of the building.

The major part (7th to 77th story) of the building provides office space, with an average coverage of nearly 3300 m^2 per story. It also provides commercial facilities, multifunctional conference halls, hotels, and sightseeing facilities. A 55-m-long sightseeing corridor at the height of 472 m is higher than the glass floor at 447 m of CN Tower in Toronto, Canada. The shape of the wind tunnel on top of the building was adjusted from circular to inverted-trapezoidal, which did not only lower the difficulty of construction and the costs but also shorten the construction period and display a more succinct appearance (Fig. 5.30).

Structural system

The structural system of Shanghai World Financial Center is a tube-in-tube structure, comprised of reinforced concrete core tube and outer steel truss tube (Fig. 5.31). Structural deadweight and vertical live loads are jointly bore by the inner and outer tube; as for horizontal overturning moment and shear forces generated by wind load or seismic forces, they will, respectively, be sustained by the outer truss tube and inner core tube.

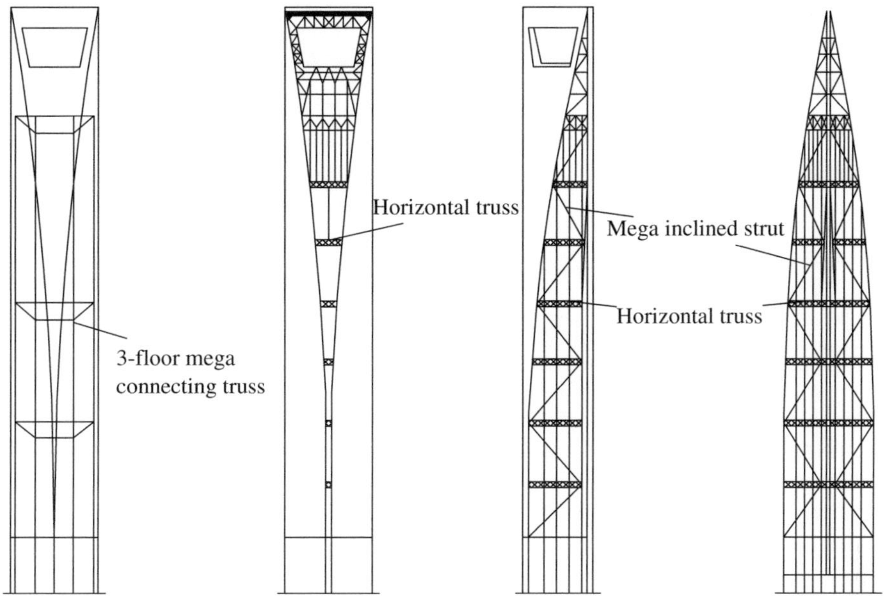

Fig. 5.31 Structure system of Shanghai World Financial Center

1. Outer steel truss tube

Outer steel truss is composed of surrounding giant columns, giant diagonal braces, and giant horizontal truss beams, which will take a minority of vertical loads and a majority of the bending moment induced by horizontal loads.

Giant columns are welded steel box. Under the action of wind loads or seismic forces, these columns will play an important role in resisting the overturning moment inflicted upon the structure. In addition, these giant columns will also handle the vertical loads transferred from truss beams, while these truss beams provide support for the deadweight passed on from surrounding minor columns.

Giant diagonal braces are welded steel box, which will enhance not only the shear rigidity but also the bending resistance of the structure. They connect surrounding giant columns with horizontal truss beams which are constructed along the height so as to constitute an integral outer truss tube structure.

2. Reinforced concrete core tube

Reinforced concrete core tube is responsible for a majority of vertical loads and shear forces induced by horizontal loads. To strengthen the connection between the core tube and the outer truss tube, steel trusses are preinstalled within reinforced concrete core tube at the locale reserved for connecting beams between the inner and the outer tube (as shown in Fig. 5.32).

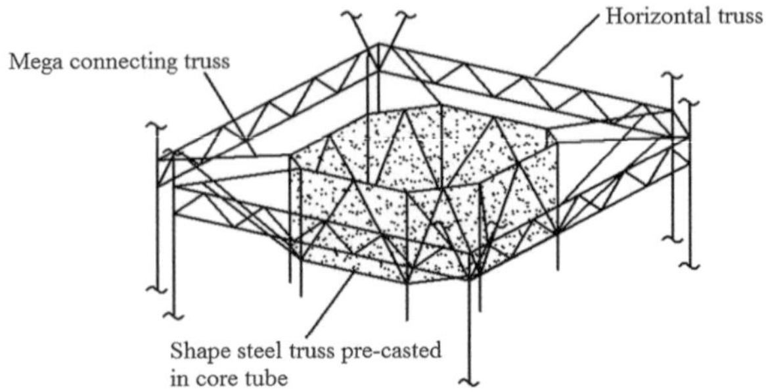

Fig. 5.32 Vertical substructure system of Shanghai World Financial Center

3. Giant connecting trusses

Three giant connecting trusses fasten the inner core tube and the outer truss tube together so that the structure will act as a whole in its resistance of horizontal loads and substantially reduce horizontal displacement. These giant connecting trusses are three-story high (as shown in Fig. 5.32). According to the design conception of "structural fuse," giant connecting trusses will be the first to wreck which will separate the inner and the outer tube, reduce the overall rigidity of the structure, and prevent the major structure from collapse.

4. Primary and secondary beam–slab structural system

Horizontal substructure, comprised of primary and secondary beam–slab system, sustains the deadweight and live loads (as shown in Fig. 5.33). The path of vertical load transmission is from the floor to secondary steel beams to primary steel beams and further to the inner core tube and the outer truss tube and eventually to the foundation.

As for the foundation, the structure takes up reinforced concrete raft structure supported by pile groups. And the substructure underground adopts the cast-in-place reinforced concrete diaphragm wall system, which includes the surrounding wall, through wall, and wing wall (as shown in Fig. 5.34).

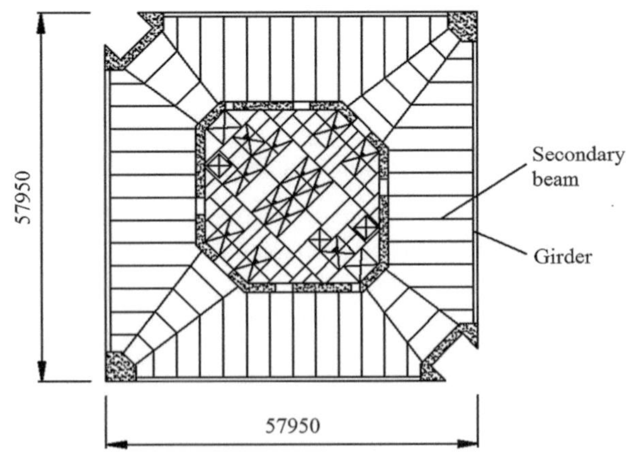

Fig. 5.33 Horizontal substructure system of Shanghai World Financial Center

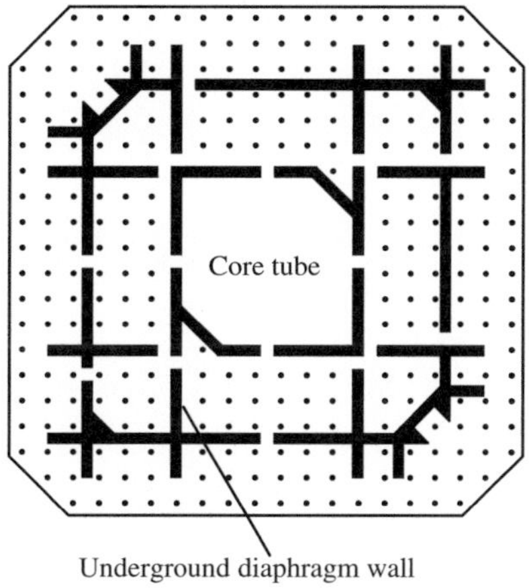

Fig. 5.34 Foundation layout of Shanghai World Financial Center

Another feature of the structural system of this building is its inverted-trapezoidal wind tunnel, which will considerably reduce the horizontal loads induced by wind at the top of the building and thereby lessen the bending moment and torque applied to the foundation. Meanwhile, the installation of integrated steel space truss system around the inverted-trapezoidal wind tunnel and the rest area on top of the building practically makes a high-rigidity cap truss out of the structure, which will not only strengthen the connection between the inner and the outer tube and immensely reduce the horizontal displacement at the top of the building but also lighten the deadweight of the structure.

Key points in preliminary design

1. Load conditions
   Deadweight: 3–5 kN/m^2, with reference to operational functions of different stories, take 4 kN/m^2 as average;
   Live load: 2–5 kN/m^2, with reference to operational functions of different stories, take 4.8 kN/m^2 as average;
   Wind load: 100-year design reference period (in normal service condition) (Wind load condition. 3.86 kN/m^2);
   200-year design reference period (in ultimate limit state).

2. Design conditions
   Seismic design: Fortification intensity is 7, class B structure, foundation grade is IV.
   Displacement restrictions: vertical displacement in the middle of the beam $\leq L/500$ or 20 mm,
   horizontal displacement between stories $\leq H/450$,
   inelastic horizontal displacement between stories $\leq H/100$.

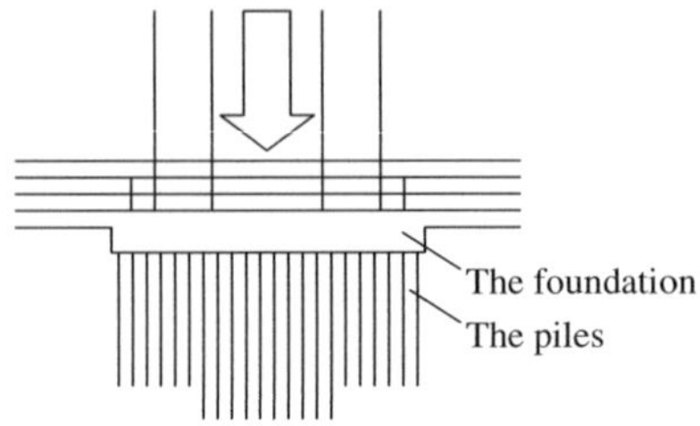

Fig. 5.35 Under vertical loads

3. Simplified calculation of pile foundation [36]

   (1) Under vertical load (as shown in Fig. 5.35), total vertical load is calculated as:

$$W = 381{,}600\,\mathrm{m}^2 \times \left(4\,\mathrm{kN/m}^2 \times 1.2 + 4.8\,\mathrm{kN/m}^2 \times 1.4\right) = 4{,}396{,}000\,\mathrm{kN}$$

Since the groundwater (depth: 10 m) provides buoyancy 6200 $\mathrm{m}^2 \times$ 98.1 $\mathrm{kN/m}^2$ = 608,220 kN
Average pressure on each pile among 1177 piles is:
P = (4,396,000,608,220)/1177 = 3218 kN < 4300 kN (designed bearing capacity of a single pile)

   (2) Under horizontal and vertical loads at the same time (as shown in Fig. 5.36), total bending moment generated by horizontal wind load is:

$$M = 492 \times 60 \times 3.86 \times 492/2 = 28{,}000{,}000\,\mathrm{kNm}$$

Bending resistance arm of force of the pile groups is estimated as: $B = 34 \times 17/2 \times 2/3 \times 60$ m = 11,560 m
Maximum pressure on one pile in 1177 piles is:
$P' = p + M/B$ = 3218 + 28,000,000/11,560 = 5640 kN < 1.33 × 4300 kN = 5719 kN.

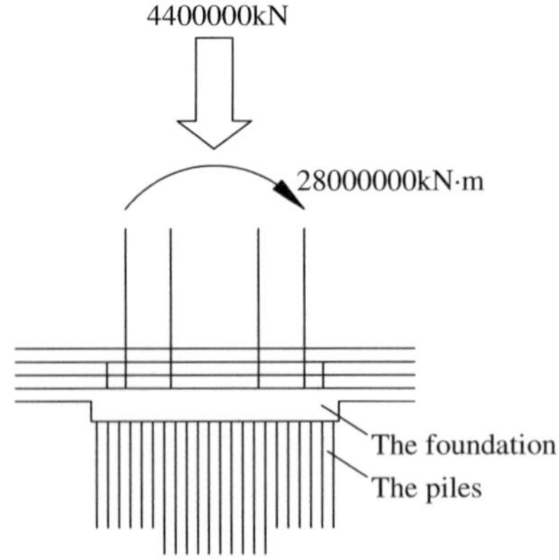

Fig. 5.36 Under vertical and horizontal loads

Table 5.6 Forces resisted with major external loads [36]

Type of loads	Vertical reaction (kN)	Horizontal shear (kN)	Overturning moment (kN m)
Total vertical load	4,420,000	0	0
Wind load (service condition)	0	77,400	23,493,000
Wind load (ultimate condition)	0	89,400	27,059,000
Seismic response (response spectrum method)	0	61,392	9,820,000
Seismic response (time history analysis)	0	53,400	7,351,000

The equation above meets the requirement for bearing capacity of one pile; 1.33 is the bearing capacity enhancement coefficient of one pile under horizontal dynamic load, stipulated in relevant US norms [31].

4. Estimation of major external loads

Table 5.6 [36] lists the forces generated by major external loads in preliminary design. It is the basis of static and dynamic calculation of the structure and the reference of further detailed design.

From Table 5.6, we can see that the most unfavorable horizontal load of this project is the wind load, to which enough attention should be paid. The effect of seismic load in Shanghai is only fraction of wind load. Nevertheless, dynamic response under seismic forces of this ultra-high structure has to be checked in the later stages of detailed design.

Summary

The structural system that Shanghai World Financial Center took is the tube-in-tube structure, which is quite suitable for ultra-high structures. The inner tube of this structure is concrete core tube and the outer tube is steel truss tube. Three giant horizontal trusses integrate the concrete core tube with the steel truss outer tube and enable the structure to resist horizontal and vertical loads as a whole. Meanwhile, horizontal displacement is reduced and strength requirements as well as deformation restriction are met. The inverted-trapezoidal wind tunnel will considerably reduce horizontal loads generated by the wind on the top of the building and thereby deduct the bending moment and torque bore by the structure. The integrated steel space truss system installed around the wind tunnel and the rest area on top of the building has made a high-rigidity cap truss out of the structure, in which case the connection between the inner and the outer tube is strengthened and horizontal displacement is substantially lessened.

As for the construction, major challenge comes from the height. To ensure a smooth construction, the builder preordered two mega-cranes with giant crane arms that can reach the altitude of 500 m. The height of lift of these two cranes has also been the highest in construction in China so far. The steel used in this structure amounts to 62 thousand tons, which made the building No. 1 individual construction in steel consumption in China.

Discussion

1. The outline of Shanghai World Financial Center is designed to be a bottom-up toothpaste tube. What are the characteristics of such a design in regard to force bearing? What controversial symbolic meaning does it represent when it comes to aesthetic expression of the architecture?
2. How were vertical and horizontal substructures put together? What is the principle of arrangement when three-story high giant connecting trusses were installed?
3. Could you analyze the process of force transmission regarding horizontal and vertical loads with reference to the force-bearing structural system and load conditions of this building?
4. Would you be able to make estimations over components and major structural systems after knowing the force-bearing system and load conditions in this case?
5. Openings on top of high-rise buildings reduce the wind load, while the cap truss on the top strengthens the structural integrity and limits horizontal displacement of the top. Would you be able to make qualitative analysis over what is mentioned above? How should cap trusses be arranged so as to lessen horizontal displacement on the top?
6. How was the construction of Shanghai World Financial Center unfolded? Which de facto checking of the structure has to be taken into consideration in construction design? Since the core tube would display compressive deformation during construction, does the outer steel truss tube have to be coordinated with this?
7. The interior of Shanghai World Financial Center was once on fire before the building came to service. Please analyze precautions against fire that steel structures should take.
8. Shanghai World Financial Center installed on 90th floor an active vibration control. Would you please explain the mechanism of this active control?

5.6.2 The World Trade Center in New York, USA

Basic analysis of structural systems

Prior to the final design decision of a structural system, simplified analysis of the structure should be conducted to check whether strength, rigidity, and stability meet all requirements. Only then, further detailed analysis is to be conducted.

Feasibility study of the structural design has to be simple, ready, and precise to certain extent so that the designers will have an easy grasp on the subject. World Trade Center is taken as an example of complete estimation of structural system to demonstrate how simplified models are established and how approximate analysis is conducted when high-rise structures are approached.

1. Basic structural information
   Stories: 110
   Total height: 412 m
   Dimension of the foundation: 63.5 m × 63.5 m
   Structural system: Tube-in-tube

The outer tube is a framed tube structure comprised of dense columns (as shown in Fig. 5.37). Nineteen steel box columns were aligned on each side of the frame tube with a space of 3.05 m between each one of them. The sectional area of the column is 0.263 m^2. The inner tube is a core tube composed by steel columns and concrete [32].

2. The assumption of simplified model for structural analysis
   (1) The whole structure is simplified as a cantilever beam with the end embedded in the ground;
   (2) As for a tube-in-tube structure, compared with the outer tube, the inner tube has a minor role to play in the resistance of overturning moment. Thus, in approximate analysis, the outer tube is regarded as major structural system to resist horizontal loads;
   (3) As for the dense columns installed in the outer tube, they are rigidly connected to the beams at the top, which strengthened the integrity of the structure.

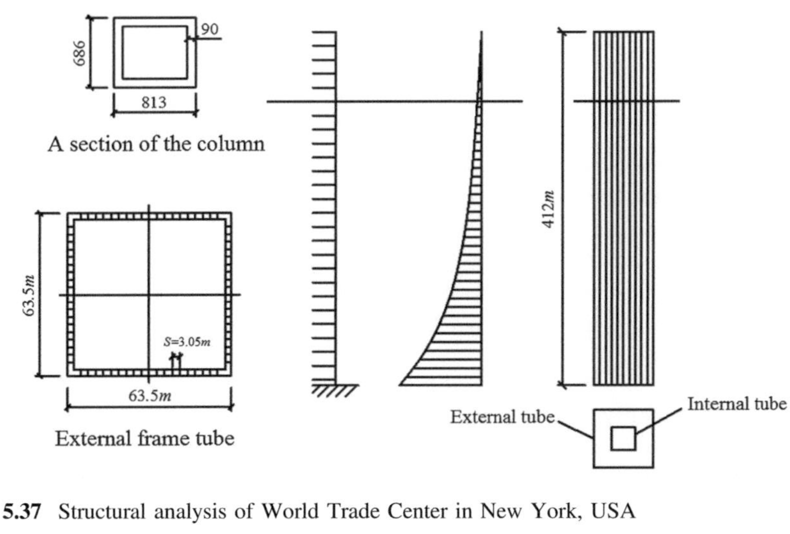

Fig. 5.37 Structural analysis of World Trade Center in New York, USA

(4) Columns on each floor are restricted by floor systems above and below. As for floor system itself, one end is to be placed on the short cantilever beam of the outer column, while the other connects with the core tube and transfers vertical loads on the floor to the inner and outer tube.
(5) To simplify calculation and analysis, sectional area of all columns in the outer tube is assumed to be the same and wind load is the only factor to be considered in the analysis during conceptual design. Since wind load is the control load, seismic load is ignored.

3. Internal force analysis [37]

(1) The overturning moment generated by the wind load at the foundation is

$$M_w = 0.5\, q_w\, BH^2 = 0.5 \times 2.2 \times 63.5 \times 412^2 = 11.86 \times 10^6\,\text{kN m}$$

q_w is wind pressure. With reference to the aspect ratio of the structure (*H/B* = 412/63.5 = 6.49) and North America and the Atlantic Wind Load Table, we have 2.692 kN/m^2 as the wind load above 30 m high and the equivalent of its average wind pressure as 2.2 kN/m^2.
(2) To simplify the calculation, dense columns of the outer frame tube have to be converted to solid tube with an equivalent wall thickness of t in light of the sectional area.
Equivalent wall thickness t = sectional area of one column/the distance between two columns = 0.263/3.05 = 0.0862 m.
(3) The moment of inertia of the outer tube is simply calculated as:

$$\begin{aligned} I &= 2 \times 0.0862 \times 63.5 \times (31.75)2 + 2 \times 0.0862 \times 63.53/12 \\ &= 11{,}036 + 3679 = 14{,}715\,\text{m}^4 \end{aligned}$$

(4) The converted stress of the outer tube induced by the overturning moment from the wind load is

$$\sigma_w = M_w y/I = 11.86 \times 10^6 \times 31.75/14{,}715 = 25{,}590\,\text{kN/m}^2$$

The axial force of each column in the outer tube derived from the overturning moment from the wind load is

$$N_w = A_c \times \sigma_w = 0.263 \times 25{,}590 = 6730\,\text{kN}$$

4. Displacement analysis

The displacement of a cantilever beam under evenly distributed horizontal load is $\delta = q_w BH^4/8EI$. Since in this case, sectional area of the box column decreases to the top, displacement should be greater than that of box columns with identical sectional area. Thus, the value is doubled in approximate calculation, that is,

$\delta = q_w BH^4/4EI$, so we have $\delta = q_w BH^4/4EI = 2.2 \times 10^3 \times 63.5 \times 412^4/(4 \times 2 \times 10^{11} \times 14715) = 0.342\,\text{m} = H/1205$.

With reference to Uniform Building Code in USA, allowable displacement of high-rise buildings under horizontal loads is $[\delta] = H/500$, and thereby, horizontal displacement mentioned above $\delta = H/1205 < [\delta] = H/500$ meets the requirement of design.

Further displacement analysis indicates that horizontal displacement is highly affected by the aspect ratio of the structure. When the width of the structure B is kept constant and the height H is increased, horizontal displacement on top of the structure will be augmented to the fourth power of the height. However, when the height is kept constant and the width is decreased, horizontal displacement will also see a swift upswing for the moment of inertia in the "footprint" downscales to the square of the width. Therefore, attention must be paid to the selection of aspect ratio in conceptual design. With currently available material and technique, the aspect ratio of conventional high-rise structures should not be greater than 10; otherwise, displacement requirement could be rarely met.

Background of the collapse

It is well known that on September 11, 2001, World Trade Center in New York suffered terrorist attack and soon collapsed. This unexpected incident shocked the world and left many matters for people to ponder upon. Apart from political issues, from a technical perspective, problems remain to be deeply thought.

Why would these two skyscrapers literally crumble to the ground after the collision with the plane? Obviously, this catastrophe resulted in severe loss of life and property and about 3000 lives were lost in this attack. It has given rise to a matter of "robustness" or "vulnerability" of the structure. In other words, how sensitive the structure would be to external destruction when sudden wreck strikes (see more in Sect. 3.7).

Since the twin towers collapsed after the plane crash, we know that the structures themselves were in fact vulnerable. In this section, we will apply the design concept of "vulnerability" to analyze the composition and formation of the structure, illustrate the structural layout, and explain why the twin structures would collapse in the flames after they were hit by the planes. This example will help us to understand the mechanism of vulnerable structures and the importance of "robustness" of a structure in conceptual design.

Analysis of the collapse of world trade center

The reason of the collapse of World Trade Center after the attack has its reason in the design and construction of the structure. By way of analyzing the cause, beneficial lessons could be learned from this tragic incident and people will realize the importance of considering the "robustness" in early structural design. Reasons that led to World Trade Center's collapse are as follows:

1. Columns on each floor are restrained by floor systems above and below, which are comprised of truss and primary and secondary beam–slab systems. One end of the floor was placed on the short cantilever beams of the columns in the outer frame tube, while the other is allocated in the inner core tube, transferring vertical loads to the outer and inner tube. Meanwhile, columns in the outer frame tube are continuous columns with each floor serving as their support and the span of horizontal load is the height between floors. Such a structural layout made the structure vulnerable.
2. When the building was hit by the hijacked aircraft, columns in the outer frame were destroyed which severely affected the force-bearing capacity of the structure. In the meantime, the plane exploded and the fuel oil instantly set fire to the building. Under the action of long-time overheat, horizontal substructure (primary and secondary beam–slab system) was breached and floors collapsed which resulted in sequential destruction of the columns. Therefore, the breakdown of the whole structure was the result of the interplay of these following factors:

   (1) Beam–slab system was wrecked due to the crash and fire;
   (2) Columns in the outer frame lost support and were unstable. Since the outer columns were continuous columns with each floor serving as their support, the buckling of columns on one floor would inevitably lead to the instability of columns on the upper and the lower floors which would keep spreading in the same way, that is the so-called buckling propagation of columns on adjoining floors.
   (3) Floors overhead tumbled in the manner of a free fall and the acceleration of gravity gave rise to enormous impact force which smashed the lower floors directly to the ground.

As a matter of fact, vertical collapse took the failure mode with minimum damage, and casualties were confined to that within the building. If the structure was unsymmetrical and tumbled sideways, a densely populated area within the radius of 300–400 m would be affected and casualties and property loss would have been more severe.

Summary

World Trade Center was built in New York in the 1970s. The design concept was to make full use of a frame tube formed by densely allocated columns in the outer tube and contributed to a tube-in-tube structure in which primary and secondary beam–slab system served as its horizontal substructure. Since it has a clear path of force transmission, the structure was rewarded with a light deadweight and scored a balance between economy and practicality. Back then, it was an advanced and well suited design. However, the designer did not take precaution against the "vulnerability" of the structure in his design. Having considered collision with small planes, the design team did not conduct serious analysis and argumentation regarding the composition of the structure. When the incident took place, a ready collapse of the twin towers was not predicted by any specialist.

The tragedy of 911 taught us that structural design should be executed through the perspective of "vulnerability" of the structure. Composition of the structure should be thoroughly studied so as to find "implicit damage chain" and avoid interlocking destruction. Thereby, whether there is "implicit damage chain" and whether the damage of one component on the chain would result in that of the next component in the direction of force transmission have become the key to the problem.

"Implicit damage chain" did exist in World Trade Center. There were 110 stories in the building. When the primary and secondary beam–slabs on the ith floor were compromised, columns in the outer tube on the $(i - 1)$th floor would suffer and tumble. In that case, loads from all the above stories would be imposed upon the $(i - 1)$th floor which would in no time be wrecked and lead to the breakdown of columns on the $(i - 2)$th floor. Following this chain of deduction, the structure would have no choice but to cave to the ground. Nonetheless, when the "chain" was not triggered, the structure itself might as well function without error.

In sum, conclusion could be reached that the internal cause of this incident was the existence of "implicit damage chain" in the composition of the structure and the external cause was the collision with the planes in 911 attacks which triggered the "chain." After the 911 event, Oral Buyukozturk, professor of civil engineering at Massachusetts Institute of Technology (MIT), pointed out that redundancy of the structure should be promoted in future design of skyscrapers which would prolong the time of collapse and provide more time for people to escape. Hence, in conceptual design of the structure, designers are recommended to start from the "redundancy" and "vulnerability" of the structure and avoid "implicit damage chain" in structural composition to ensure "redundancy" and safety.

Discussion

1. From the analysis mentioned above, we know that only when there existed both the "implicit damage chain" in the structure and the trigger could motivate a Domino effect. Without a trigger like 911 attacks, World Trade Center might as well function with the "chain" inside. The avoidance of "implicit damage chain" in early structural design becomes the key to prevent the structure from sequential collapse. In the meantime, attention needs to be paid to four measures to improve upon the "vulnerability" of the structure introduced in Sect. 3.7, which are to alternate load path, to strengthen key components, to install energy absorber, and to control the incident.
2. Even with a trigger like 911 attacks, if there existed a mechanism to break the "chain," Domino collapse would have not taken place. The question is whether there could be such a chain-breaking mechanism? The answer is positive. By prior design of such a system, Domino collapse could be avoided and damage could be confined to limited extent. For instance, subsea pipeline under

hydraulic pressure in offshore development is regarded as one of those structures with "implicit damage chain." Once local buckling occurred under hydraulic pressure, cylindrical shell of the pipeline would stretch the dent and give rise to long-range bucking propagation. The mechanism to stop "implicit damage chain" in subsea pipelines emerged in the form of buckling arrestor in the 1970s. This buckling arrestor is comprised of numerous ring supports that are allocated at certain distance between each other. Once local buckling occurred, it would be confined to a limited range.

3. The mechanism to stop "implicit damage chain" is a new subject of theoretical research that needs long-time investigation and efforts. Moreover, mechanisms to stop "implicit damage chain" vary when it comes to different structures. Corresponding mechanism to arrest the "chain" in different structures would undoubtedly display distinct characteristics. For instance, vertical high-rise structures could stop the "chain" by installing structurally transitional story, while horizontal multi-span structures could substitute consecutive structures with simply supported structures or install "buckling arrestor" in consecutive structures as the subsea pipelines.

For reasons above, in the early stage of design, the priority should be to avoid "implicit damage chain" in the structure. If that is not avoidable, interruption mechanism is recommended to be installed to the structure designed and attention should be paid not to trigger the "chain."

5.6.3 Beijing CCTV Tower

Project background

The new China Central Television (CCTV) tower is located in the newly established CBD (central business district) in the eastern third-ring road in Beijing, as shown in Fig. 5.38. The peculiarity of its extrinsic appearance and the whimsicality of the conception have given rise to controversy. The analysis and discussion of this example would not only introduce structural characteristics of this building but also provide readers a chance to think about some controversial issues regarding architectural and structural design.

Architecture design and arrangement

In accordance with the operational requirements of CCTV, activities are to be integrated within the building and therefore, the architectural conception has to discard conventional arrangement of internal division and centralize all personnel and activities involved in the operational process such as the production of news and TV programs, and the organization of studio and media broadcast into one three-dimensional consecutive tubelike structure [39].

Fig. 5.38 Model of China Central Television (CCTV) tower [38]

The building is 234 m high (56 stories) and takes up a land of 187,000 m^2 and a floor area of 550,000 m^2. There are 3 stories underground, 9 stories in the bottom above the ground (lower convex section), superstructure (inclined pipe column), and 48 stories and 36 stories overhanging upper convex section in the main building. The whole structure is a continuous hollow pipe column tube that is comprised of four sections: lower and upper convex sections, inclined column 1 and 2 with a relative angle of inclination of 6 toward each other. And it was twisted into a three-dimensional continuous cylindrical tube, as shown in Fig. 5.39. With 6 degrees of angle of inclination toward each other, the pair of inclined pipe columns diminished the eccentricity of the upper convex body relative to the lower convex body (Fig. 5.40).

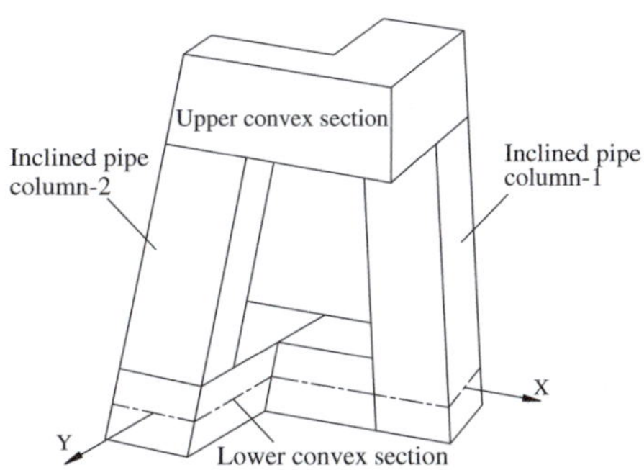

Fig. 5.39 Components of CCTV tower

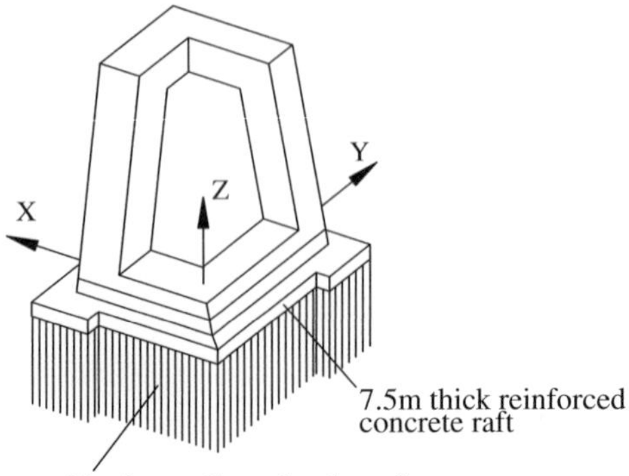

Fig. 5.40 Foundation of CCTV tower

Structural design and arrangement

The main building of CCTV adopted the structural form of connected inclined two towers. Seismic fortification intensity is 8°, and designed service life is 100 years. It is classified as complex, overlimit, and ultra-high connected structure. The main structure employed high-rigidity outer tube supporting system. The originative structural style brought new subjects of research and difficulty upon the conception as well as the construction. The technical difficulty encountered in construction was successfully solved by theoretical analysis and experimental study. Key research and technologies include the integration of performance-based seismic design, construction simulation, high-steel-content reinforced concrete columns, elastic–plastic time history seismic analysis under major earthquake, the application of novel nodes, and the mechanism to stop consecutive collapse [33].

The major concern in the structural arrangement of CCTV building is the resistance of enormous deadweight of the structure itself, in-service live loads, horizontal wind load, and seismic load. Extra attention is needed for that distortion and inclination of such an arrangement had given rise to colossal additional eccentricity and torque which multiplied the burden on the structure.

As for ultra-high connected structures like this, pipe tube structural system is well suited. The main body of this building consists of a frame tube, comprised of long and hollow column section and connecting beams, and a continuously distributed diagonal connecting bar system. The outer pipe tube is shrouded by tubelike curved surfaces and has excellent bending and torque resistance. Diagonal connecting bar system is applied to resist shear forces derived from additional eccentric bending moment and torque. Meanwhile, since these connecting bars are distributed consecutively, they will effectively strengthen the turning locales of the structure, resist stress concentration at these points, and guarantee a reliable transfer of the abovementioned disadvantageous loads to the foundation.

The foundation of this building is located on reinforced concrete raft through 3 stories of basement. The floor plan of the raft ("footprint") is identical with that of the lower convex section. To obtain a uniform reaction from the bottom of the raft, the thickness of the raft has been increased to 7.5 m. Meanwhile, stressed area has been enlarged to sustain the concentrated force at the lower end of the inclined pipe column, as shown in Fig. 5.40. In order to diminish the non-uniformity of reaction at the foundation induced by eccentric loads, the application point of resultant force of the deadweight of the upper structure (the upper convex section, inclined pipe column 1 and 2) and the loads is designed to approximate the centroid of the lower convex body in the "footprint" of the raft. Foundation pile groups of 1.2 m in diameter and 56 m in length piles are stricken into the bottom of the raft so as to solve the non-uniform distribution of reaction at the foundation under the action of horizontal loads.

The structural system of CCTV building is a three-dimensional closed bent pipe tube. In fact, all four parts of the structure, namely the lower and the upper convex section and 6-degree inclined pipe column 1 and 2, are nevertheless comprised of vertical load-bearing columns and horizontally supporting beams and floors. All the vertical loads, including the floor system of the upper convex section (also named as the overhanging layer) from 36th to 48th floor, are bore by vertical load-bearing columns. However, the vertical load-bearing columns of 6-degree inclined pipe column 1 and 2 could not possibly be kept within the same vertical line from the top to the bottom. To solve this problem, a two-story high layer of transitional truss is needed to be installed at certain height of the structure so as to transfer the loads from the upper columns to the lower ones.

The conceptual design of foundation

Due to the complexity of the structure, the simplified calculation of the main body above the ground is omitted, while foundation raft is taken as an example in the explanation of simplified calculation of complicated structures in which force transmission system serves as the key factor and needs to be figured out.

The principle in the design of a foundation raft is to check the reaction of the foundation under given dimensions to make sure there is no tensile force. Procedure of design generally involves preselecting dimensions of the raft based on practical experiences and conducting checking calculations. Until all requirements are met, modifications would iteratively be made.

When it comes to this project, with reference to previous experiences, preselection of the thickness of the raft is 7.5 m, that of the width is 60 m, and the enlarged area at the end of two inclined tube columns is 80 m × 80 m. The thickness of the raft is designed to be 7.5 m so as to guarantee enough rigidity to uniformly distribute loads transferred from the upper structure to the foundation and the piles.

As shown in Fig. 5.41, the centroid of the foundation in the "footprint" is R' and its coordinates are (53, 53, 0) in coordinate system $O'X'Y'Z'$. Therefore, $R'O' = 53/0.707 = 75$ m, $R'L_o = 75 + 30/0.707 = 117$ m, $O'J = 110 - 30 + 40 = 120$ m, $O'L_c = 75 + 40 \times 0.707 = 103$ m.

In accordance with the plane arrangement as shown in Fig. 5.41, the moments of inertia along the neutral axis of n–n and m–m in the "footprint" are (readers could do the simplified calculation). $I_{nf} = 38{,}770{,}408$ m^4 and $I_{mf} = 187{,}807{,}686$ m^4. From

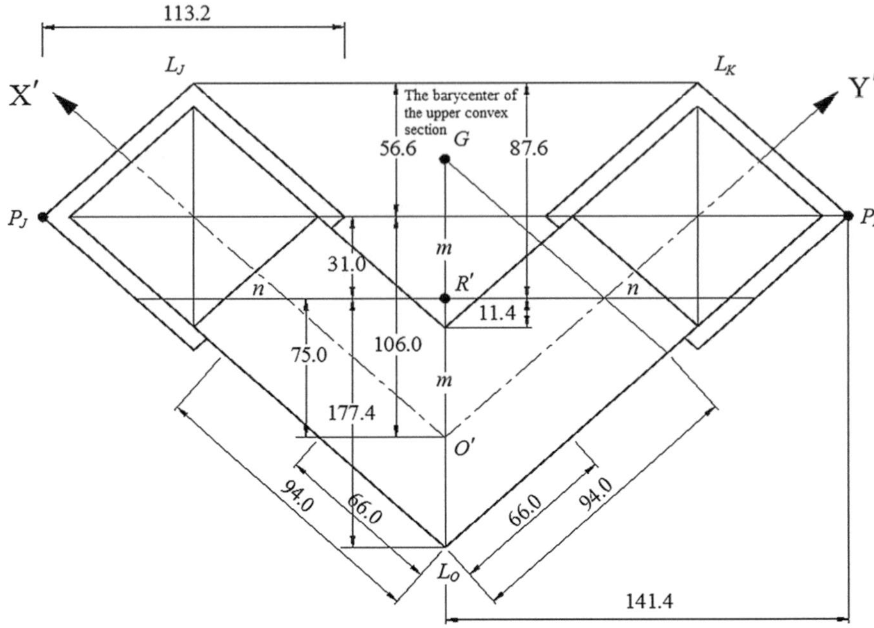

Fig. 5.41 "Footprint" of CCTV tower

the results, we notice that the direction *n–n* axis represented is comparatively weak in force bearing. The deadweight of the structure would give rise to additional eccentric bending moment in this direction, not to mention that the bending moment generated by the wind load will be added too. Therefore, it is emphasized that checking calculation of force bearing has to be conducted in *n–n* axis direction. The simplified calculation is executed as the following:

It is known that the deadweight of the upper convex section G_{TT} = 514,800 kN and its centroid coordinates are (64, 64, 188.5). The deadweight of the lower convex body G_{TC} = 367,200 kN and its centroid coordinates are (36, 36, 0). The deadweight of the inclined tube column G_{TC} = 367,200 kN, that of the raft G_{RF} = 4,425,000 kN, and that of the basement G_{BS} = 162,000 kN. Total eccentric bending moment generated by eccentric deadweight of the upper convex body and two inclined tube columns along *n–n* axis is

$$\begin{aligned} M_{nR'1} &= G_{TT}(64-36)/0.707 + 2\times G_{TC} \\ &= 514{,}800\times 39.6 + 2\times 367{,}200\times 31 = 43{,}154{,}510\,\text{kN m} \end{aligned}$$

Thus, total eccentric bending moment revolving *n–n* axis is:

$$M_{nR'} = M_{nR'1} + M_{nR'2} = 43{,}154{,}510 + 1{,}053{,}430 = 44{,}207{,}940\,\text{kN m}$$

Assume that the deadweight of the structure distributes evenly at the foundation through underground layers and the raft, deadweight-induced compressive stress at the foundation will be:

$$\begin{aligned}R_w &= (G_{TT} + G_{BT} + 2G_{TC} + G_{BT} + G_{BS} + G_{RF})/A_{RF}\\ &= (514{,}800 + 2 \times 367{,}200 + 486{,}000 + 162{,}000 + 4{,}425{,}000)\\ &\quad /(5 \times 60 \times 60 = 351.2\,\text{kN/m}^2)(\text{compressive})\end{aligned}$$

In the expression, A_{RF} represents the bottom area of the raft.

Eccentric bending moment M_{nR}-induced stress R_{nM} at the foundation is:

$R_{nM} = M_{nR'}h/I_{nf}$ (h is the distance from the measuring point to the neutral n–n axis in the footprint at the foundation)

As for point L_o in Fig. 5.41

$$R_{nM} = 44{,}207{,}940 \times 177.4/38{,}770{,}408 = 202.3\,\text{kN/m}^2\ (\text{tensile})$$

As for point L_g and point L_J in Fig. 5.41

$$R_{nM} = 44{,}207{,}940 \times 56.6/38{,}770{,}408 = 64.5\,\text{kN/m}^2\ (\text{compressive})$$

As for point P_g and P_J in Fig. 5.41

$$R_{nM} = 44{,}207{,}940 \times 31.0/38{,}770{,}408 = 35.3\,\text{kN/m}^2\ (\text{compressive})$$

As shown in Table 5.7, under the action of the wind load, tensile stress does not manifest at the foundation, which means preliminary selection of raft dimensions is reasonable. Stress data of the foundation shown in Table 5.7 could serve as the basis for further division of the foundation and design of the pile groups.

Discussion

Ever since the construction of CCTV building originated in Beijing, people began to talk about it and never stopped. Some praised it for its novelty, while the other

Table 5.7 Total stress at bottom of the raft (refer to Fig. 5.41)

Control point	Generated by R_W	Generated by R_{nM}	In total R_{mM}
L_o	361.2 kN/m^2 (compressive)	202.3 kN/m^2 (tensile)	148.9 kN/m^2 (compressive)
L_J	361.2 kN/m^2 (compressive)	64.5 kN/m^2 (compressive)	410.9 kN/m^2 (compressive)
L_K	361.2 kN/m^2 (compressive)	64.5 kN/m^2 (compressive)	420.5 kN/m^2 (compressive)
P_J	361.2 kN/m^2 (compressive)	35.3 kN/m^2 (compressive)	378.6 kN/m^2 (compressive)
P_K	361.2 kN/m^2 (compressive)	35.3 kN/m^2 (compressive)	394.4 kN/m^2 (compressive)

disapproved of it for its peculiarity. To sum up, the debate over the construction of CCTV building followed the questions emerged:

1. What principle and policy should be promoted as the guidance regarding urban construction nowadays?
2. How to appraise this building from an architectural perspective?
3. What is the evaluation of the structure of this building? Does there exist any potential structural security issue?
4. How to assess the economy of this building? Does obvious waste exist?

Time will tell the settlement of this dispute. However, as a structure engineer or an architect, we should participate in the discussion in terms of professional civil engineering practice. What kind of a structure is economically advisable? The optimal structure should have short and smooth path of force transmission and therefore has the lowest cost. Specifically, the internal forces will be transferred in the shortest and smoothest way to the foundation. From the perspective of energy, we know that internal forces, when transferred, will produce deformation to the path of force transmission, which means that work has been done by the internal forces to the path and energy has reserved in the structure. As the path of force transmission prolongs, energy reserve multiplies. Economically advisable structure has a comparatively low energy reserve. Unnecessary path results in unnecessary energy reserve and cost increase. Such a structure will not only increase material consumption but also decrease the operational efficiency.

From a load-bearing perspective, we can see that the overhanging upper convex body is the origin of enormous additional eccentric bending moment and torque. As shown in Table 5.7, additional eccentric bending moment M_{XH} at the joint of the lower convex body and the tube columns amounts to 8,733,100 kN m and additional torque is up to 21,621,600 kN m. As a result, the internal forces at joints are more complicated, stresses are more concentrated, and economically advisable material amount has been surpassed, which brings more difficulty to construction and maintenance afterward.

It is noted that CCTV building still has possible potential structural security issue. Given that it was built in the area in which the seismic fortification intensity is 8°, the building will have to face the challenge from seismic forces. The price it has to pay might give rise to the irrationality of high costs and unsound functionality. In comparison with the problems posed by additional bending moment and torque induced by constant loads, seismic effect is another factor that is worth serious consideration in structural design.

Professionals generally agree that the pursuit of visual effect in architecture will easily lead to the waste of resources and hidden defects as well. Does it agree with the local strategy of sustainable development in construction industry? These questions call for people's reconsideration.

Summary
Vertical high-rise structures have become the most common and the fastest-growing structural form in modern engineering projects. On the one hand, available urban land resources are quite limited; on the other hand, the economy of modern metropolis has a rapid development which entails a high concentration of space and personnel. These are all reasons behind the development of high-rise structures.

In this chapter, we started from the characteristics of high-rise structures and simplified the structure as a cantilever beam with one end embedded in the foundation when we calculate. Apart from vertical deadweight, the main loads that the structure bears would be horizontal wind load and seismic load. Therefore, the enhancement of structural strength and rigidity against bending moment and shear forces are emphasized. Six measures of such kind have been summarized in this chapter, such as the addition of "enclosing" giant frames, "three-dimensional" structural members, utilization of diagonal supporting system to enhance horizontal rigidity, marginalized section in the "footprint," the fortification of critical structural locales, and the increment of pre-load on the columns. Readers could comprehend these methods from the perspective of structural features and conduct overall analysis of actual high-rise structures.

Three fundamental vertical structural systems were introduced in this chapter, namely shear wall structure, frame structure, and tube structure, as well as mechanical properties and simplified analysis of major parameters of the above-mentioned structural systems. To meet the needs of internal wide span and ultra-height in high-rise structures, vertical mixed structural systems were introduced, such as skip-layer truss structural system, giant frame structural system, connecting-tower structural system, framed tube structural system, overhanging core-tube frame structural system, and inclined column truss structural system. Readers will be able to see efforts from architects to restrain horizontal deformation of high-rise structures. Reflection is needed in regard to mechanical connotation and the effects that corresponding structural measures achieved so as to better the understanding of structures.

Demonstration of structural concepts with example has been most effective in the improvement of comprehension, which has also been highly recommended in this book. Simplified analysis of Shanghai World Financial Center, the Sears Tower in Chicago, New York World Trade Tower, and Beijing CCTV tower has been provided, in which mechanical properties, how the structural concept was applied as well as how some of the special requirements were met through the management of structural design, were all covered.

The structural system of Shanghai World Financial Center has been the tube-in-tube structure, comprised of a concrete core tube and an outer steel truss tube. Three giant connecting trusses integrated them all so that the structure could bear an enormous amount of shear forces and bending moment induced by vertical and horizontal loads.

New York World Trade Center was a tube-in-tube structure, comprised of a concrete core tube and an outer frame tube. In 911 terrorist attacks, the twin tower collapsed in the blaze. We analyzed the cause of the collapse in detail and discussed issues such as “redundancy” and “vulnerability” of the structure. Concepts such as “implicit damage chain” of vulnerable structure and the development of the “mechanism to arrest implicit damage chain” have been proposed, which will help readers in further consideration and research of these issues.

Chicago Sears Tower is a bundled tube structure, in which varied sections of this high-rise structure were designed to bear different bending moment and shear force. Nine bundled tubes with divergent length were combined together, which has become a newly emerged effective ultra-high-rise structural system.

The Beijing CCTV tower has been in the spotlight and the center of attention for the peculiarity of the appearance. A preliminary estimation and some data were released for readers’ reference so that further consideration over controversial issues regarding architectural design raised by this building would be possible.

Chapter 6
Comparative Design and Analyses of Large-Span Structures

6.1 Characteristics of Large-Span Structures

Large-span structures conventionally take the form of long horizontal strip, such as in bridges, portal rigid frames, and large-span stadiums. As for these structures, vertical load is the control load. Special attention needs to be paid to systematic design of horizontal substructure when it comes to the conceptual design of large-span structure. Specifically, the force-bearing capacity of vertical load and the control upon vertical displacement shall be enhanced so that the structure has sufficient strength and stiffness to resist vertical load as a whole.

Bridge system is a typical horizontal large-span structural system [40]. Therefore, the classification of horizontal structures is also applied to bridges. Horizontal large-span structural systems are not confined to conventional bridges of all kinds but have been applied quite often to large-span industrial workshops and large-span stadiums and exhibition centers of various kinds.

Horizontal large-span structural systems are comprised of simply supported beam system, continuous beam system, portal rigid frame system, arch system, suspension cable system, cable-stayed system along with diversified combinations of aforementioned structural systems, such as the composite system of arch and suspension cable structure, suspension cable and cable-stayed structures, arch and cable-stayed structure, that of arch, suspension cable, and cable-stayed structured, etc. Figure 6.1 shows the diagram of various horizontal large-span structural systems.

1. Simply supported beam system. It is mainly through the bending of beam that such a system transfers the external force upon the beam to the supports. Since the section of beam is not fully stressed when it is bent, the effectiveness of material could not be fully utilized, and therefore, simply supported beam system is not a structural system with high economic efficiency. Nonetheless, the construction of such a system is quite simple; it has clear-cut path of support and force transfer; the internal stress of beam is rarely affected by the differential

S. Lin and Z. Huang, *Comparative Design of Structures*,
DOI 10.1007/978-3-662-48044-1_6

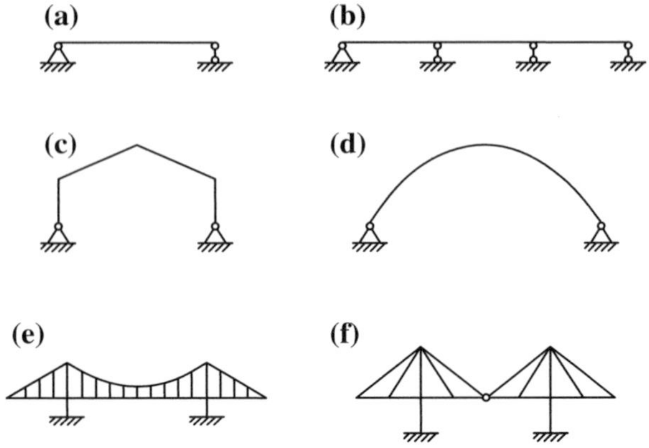

Fig. 6.1 Horizontal large-span structural systems. **a** Simple beam. **b** Continuous beam. **c** Portal frame. **d** Arch. **e** Suspension. **f** Cable stayed

settlement of the supports. All these advantages make simply supported beam system as one of the most frequently applied bridge structural systems. For the moment, ultra-large-span Shanghai Donghai Bridge has a total length of 34 km. Except for the main bridge which took the form of cable-stayed system, the rest parts of the bridge employed simply supported beam or continuous beam as the structural system.

2. Continuous beam system. Similar with simply supported beam, continuous beam system transfers the external force upon the beam to the supports through the bending of beam as well. The difference is that there exists a mid-support negative bending moment, and therefore, maximum positive bending moment of the beam is reduced for continuous beam system. Hence, compare to simply supported system, continuous beam system is more economic and therefore also frequently applied. And the only drawback is that the differential settlement of the supports of this system significantly affects the performance of continuous beam system. Most elevated freeway in downtown area of Shanghai takes the form of continuous beam system.
3. Portal rigid frame system. This is an improved edition of beam system, which as a matter of fact is an arch system. Portal rigid frame system employs the side thrust within its own supporting system to balance part of the bending moment within the rigid frame. Under the action of identical vertical load, maximum bending moment within a portal rigid frame is smaller than that of a beam. Such a system is often applied to large-span steel factory building, and the achieved largest span is up to 80 m, such as in large-scale aircraft maintenance hangar of Beijing Airport.
4. Arch system. This is a refined edition of portal rigid frame system. Arch system turns polygonal path of forces of portal rigid frame system into a pure compression structure form, which makes the paths of forces more streamlined and smooth. Arches barely bear any bending moment but only uniform pressure, which could well utilize the effectiveness of material. Arch system is of high efficiency when force bearing is concerned but note that relative side thrust shall

appear in the application of arch structure. The span of an arch bridge could be hundreds of meters, such as in Shanghai Lupu Bridge, the span being 550 m. It was the arch bridge with the longest span in the world at completion.

5. Suspension cable system. It is a "conjugated" version of arch system by turning it upside down. Suspension cable system employs catenary to determine the configuration, at the completion of which tension shall be the only force without a single trace of bending moment. Since such a system could fully utilize the effectiveness of material, which makes such a structure system highly efficient. High-strength steel cables are often employed to make suspension cable system. Since buckling instability is not a concern when the material is pulled, this system has more economic efficiency than the arch system and therefore the span of suspension cable system is of much greater than that of arch system and that of beam system. For instance, the span of Japan Akashi Kaikyo Bridge is 1991 m and it is currently the world's longest suspension bridge. Jiangyin Yangtze River Bridge (with a span of 1385 m), discussed in Sect. 6.5.3 in this chapter, is also a typical suspension cable bridge.
6. Cable-stayed system. Unlike suspension cable system, the catenary of which is pulled, cable-stayed system employs and symmetrically places high-strength steel cables to stabilize the whole structure. High-strength cable and concrete are often combined in cable-stayed bridge so that the high compressive resistance of concrete could be utilized in the tower and deck of bridge, while the high-tensile strength of steel cable contributes to a bridge with economic efficiency and reliability. For instance, the aforementioned Shanghai Nanpu Bridge, Yangpu Bridge, and Donghai Bridge take the form of cable-stayed structural system. For the moment, Russky Bridge (completed in 2012) in Russia is the cable-stayed bridge with the longest span in the world, the span of which is more than 1100 m.
7. A composite system of arch and suspension cable. In order to balance the enormous arch-induced side thrust, a composite system of arch and suspension cable is employed, in which case the arch-induced side thrust is balanced by suspension cable-generated lateral tension. Minneapolis Federal Bank is a renowned example of practical application of such a composite system [1, 2].
8. A composite system of arch and cable-stayed system. Such a composite system is sometimes needed in the construction of an arch structure with ultra-long span. The advantage of arch structure and that of cable-stayed structure are fully utilized in this combination, which is often applied to cable-stayed structure. A well-known practical application is Guangxi Nanning Bridge, the double-saddle inclined arch bridge discussed in Sect. 6.5.2.
9. A composite system of suspension cable and cable-stayed system. Two or three structural systems could be combined in practical application so as to skillfully utilize the advantage of suspension cable and that of cable-stayed system. A noted practical application is Gibraltar Strait Bridge, the cable-stayed suspension cable bridge discussed in Sect. 6.5.4 in this chapter.

In sum, among horizontal structural systems, the arch system, suspension cable system, and cable-stayed system are the ones with relatively large span. In the following sections, mechanical characteristics and key points in design of these horizontal structural systems shall be introduced in detail; further analysis of a large number of representative practical applications will help readers understand to a deeper level the mechanical concepts and the application of these structural systems.

6.2 Design for Arch Structural System

Arch structural system has been employed in bridge structures from long ago. The existent Zhaozhou Bridge in China (as shown in Fig. 6.2) is a perfect demonstration of the wisdom of the ancient Chinese people. Zhaozhou Bridge has also been known as Anji Bridge, and it was built in Sui Dynasty by a well-known craftsman Li Chun from 605 to 618 A.D. The length of the bridge is 64.4 m, and the span is 37.0 m. It is a single-span open-shoulder stone arch bridge with the longest span and the longest history of construction in the whole world. Since there are two openings on each shoulder of the bridge, the bridge is not entirely solid and therefore designated as a bridge with open shoulder (those without opening on the shoulder are named bridges with full shoulder or solid shoulder), with the help of which floods could be discharged when catastrophic floods occur.

It has been 1400 years since Zhaozhou Bridge was built. It has remained intact after ten floods, eight wars, and numerous earthquakes, especially the one that happened in Xingtai in 1996 which was measured 7.6 on the Richter scale. In September 1991, Zhaozhou Bridge was selected as an "International Historic Civil Engineering Landmark."

Figure 6.3 shows the calculation and comparison between the arch and beam system. As for beam system, the resisting bending couples (*T* or *C*) could only be regulated through addition to the height of beam (*h*) or to the width of beam (*b*). To control vertical deflection of beam, more material is needed in the section. As for arch system, the resisting bending couples (*T* or *C*) could be controlled through an

Fig. 6.2 Hebei Zhaozhou Bridge

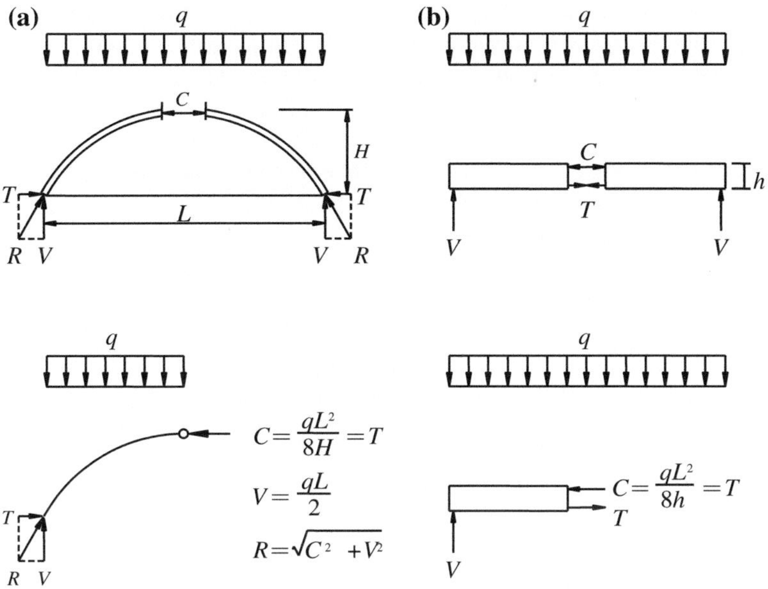

Fig. 6.3 Comparison between the stress state of an arch and a beam. **a** Arch structure. **b** Beam structure

overall enhancement of the rise of arch (H) instead of an addition of the height of the section of arch (h). The rise of arch (H) generally is much larger than the height of the section (h). When the rise of arch is up to a certain amount, the side thrust and the bending moment of the arch would be reduced to the minimum. Thereby, under the action of identical load upon a span of identical length, material employed in an arch system is less than that employed in a beam system.

Quite a number of types exist for arch structures. Classified by composition and way of support, there are three hinged arch, two hinged arch, and hingeless arch, as shown in Fig. 6.4. Three hinged arch is statically determinate, and there is no bending moment at the hinge. The stress analysis of such an arch could be conducted by the equation as shown in Fig. 6.3. Two hinged arch or hingeless arch is statically indeterminate and therefore bears more redundancy. Hence, practical construction projects employ more of two hinged arch and the hingeless arch.

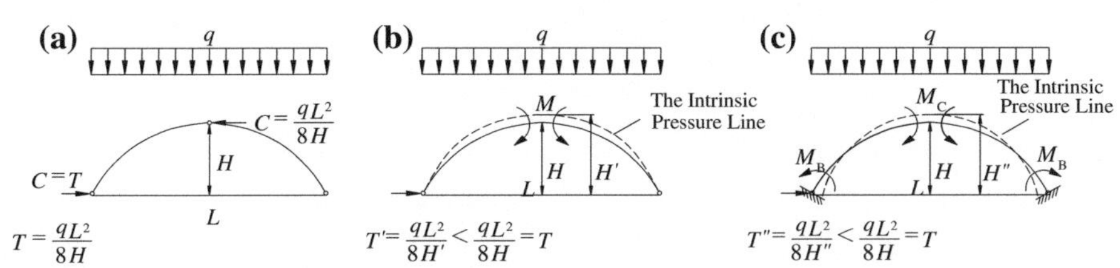

Fig. 6.4 Types of arch. **a** Three hinged arch. **b** Two hinged arch. **c** Hinged arch

Precise calculation of stress of these two types of arch is rather complicated, and a simplified calculation as followed could be adopted in the early stage of design.

Under the action of uniform symmetrical load, three hinged arch, or two hinged and hingeless arch, the axis of which approximates to the ideal pressure line, as shown in Fig. 6.4, could be calculated through static equilibrium, in which case half of the arch is taken as the isolated body. It is assumed that there is no bending moment or shear force in the middle of the arch. Calculate the bending moment with support of the arch as the center:

$$C \cdot H = qL^2/8 \tag{6.1}$$

$$C = T = \frac{qL^2}{8H} \tag{6.2}$$

The resultant force at the point of support is

$$R = \sqrt{V^2 + T^2} \tag{6.3}$$

where, q refers to uniform load; L to the span of the arch; H to the rise of the arch; C to the compression on the top of the arch; T to the side thrust; and R to the resultant force at the support.

When it comes to actual structures, the arch might be affected by asymmetrical non-uniform live and constant loads of all kinds and the axis of the arch therefore deviates from the intrinsic pressure line, which leaves part of the arch with some amount of bending moment (as shown in Fig. 6.5). Even when the axis of two hinged arch and that of hingeless arch approximate to the intrinsic parabola of the arch, deformation shall invite the appearance of the bending moment and the internal force and the reaction force at the support shall be thereby different from those when the aforementioned three kinds of arches are in their ideal state. However, since the height of the section is much smaller than the rise of the arch, the bending moment induced by the deformation of the arch is usually quite small. Hence, in the stage of preliminary design, the aforementioned equation could still be employed in the calculation of the internal force and the reaction force at the

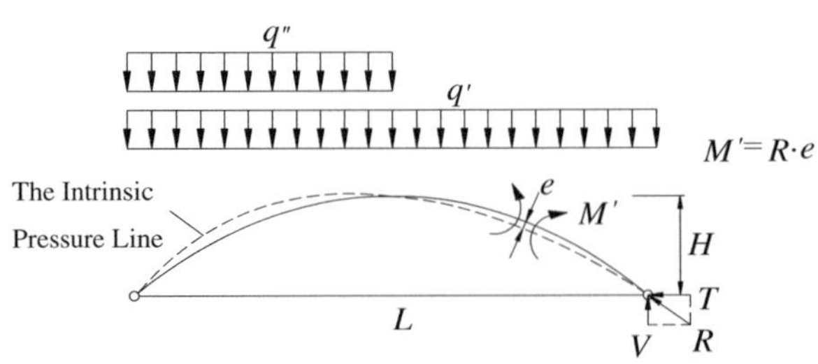

Fig. 6.5 Pressure line and simplified calculation of the bending moment of the arch

support with only a slight regulation in accordance with the proportion of the action of the bending moment and that of the arch. For instance, a simplified calculation as shown in Fig. 6.5 could be used to calculate the bending moment within the arch.

The outlook of the intrinsic pressure line of the arch is directly related to the load (as shown in Fig. 6.6). When the load is uniform and perpendicular to the surface of the arch, the ideal pressure line of the arch is a circular arc. When the load is vertical uniform load, the ideal pressure line of the arch is a parabola. When the load is vertical eccentric load, the ideal pressure line of the arch is an eccentric parabola. When the load is concentrated load acting upon the top of the arch, the ideal pressure line of the arch is a triangle-shaped inclined straight line. When the pressure line of the loaded arch deviates from the axis of the arch, bending moment shall be acting upon the arch rib. In practical construction projects, the structure has to bear various kinds of loads; there is uniform load and there is concentrated load; the possibility that there is only one kind of load for the structure to bear is quite small. Thereby, it is rather difficult to determine a perfect axis to adapt to the action of loads of all kinds. Usually, a major load (such as a uniform constant load) is employed to determine a relatively ideal axis of the arch, which shall keep the local bending moment at a small amount. As shown in Fig. 6.6, the pressure line of the arch under the action of uniform constant load is a parabola and that is why an arch more than often takes the form of a parabola. As for the shield segments built in the

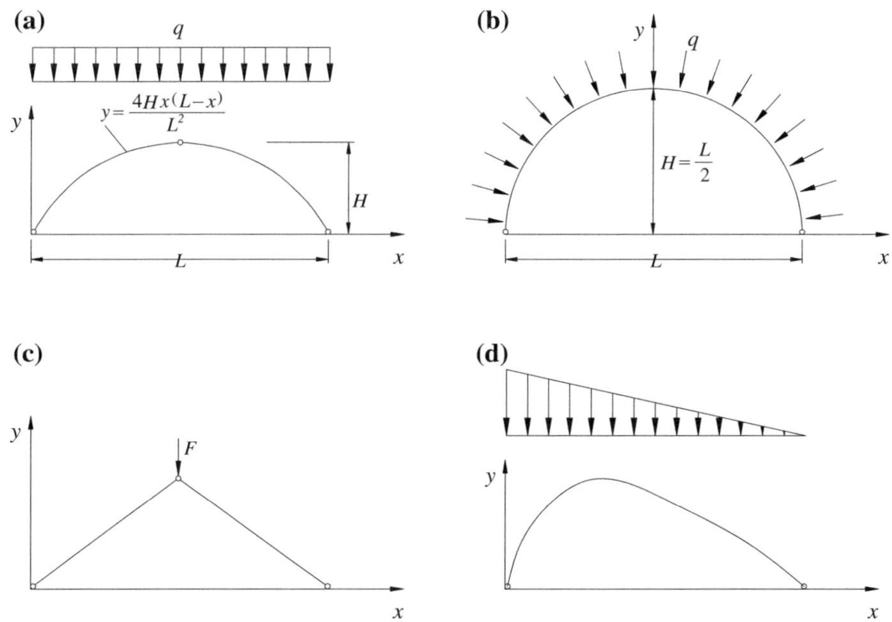

Fig. 6.6 Intrinsic pressure line of the arch. **a** The vertical uniformly distributed load. **b** The uniformly distributed load perpendicular to arch axis. **c** Central concentrated load. **d** The vertical eccentrically distributed load

tunnel, they are largely under the action of uniform water and soil load that is perpendicular to the surface of the arch and their pressure line takes the form of a circle, which serves as the reason why the shield segments in the tunnel are usually round or oval-shaped.

Arch structure is self-adaptable to construction error. Many a designed axis of the arch goes through changes after the completion of the structure and that is because the assumed foundation and support are not exactly the same with the actual foundation and support that the actual construction project is faced with. Arch structure often adjusts its own axis and the angle of support to achieve a more balanced structure. From certain perspective, arch structure figures the shape out through self-modification. Naturally, careful design and calculation of the axis and support of the arch are indispensable in schematic design, detailed design, and its required calculation.

1. **Design for the axis of an arch**

A reasonable determination of the axis of the arch is the main quest in the design of the axis of the arch. The purpose is to prevent all or a major proportion of the bending moment from the arch.

Generally, the larger the rise of the arch is, the smaller the side thrust and local bending moment within the arch are. When aesthetics, economy, and practicality are synthetically considered, it is ideal if the ratio between the span and the rise of the arch is between 5 and 10. And it is quite proper if the ratio between the height of the section and the span of the arch is about 1/50. By comparison, the ratio between the height of the section and the span of a beam is usually between 1/10 and 1/20, in which case the effectiveness of material of an arch is much greater than a beam.

The axis of an arch is usually designed to approximate to the intrinsic pressure line of an arch when it is under the action of vertical uniform load, which is by the way a second power parabola (as shown in Fig. 6.6a). In this way, relatively minor local bending moment shall be generated under the action of various non-uniform loads. And the equation for the axis of the arch is:

$$y = 4Hx(L - x)/L^2 \tag{6.4}$$

where x is the horizontal coordinate of the axis of the arch; y is the vertical coordinate in the direction of the rise of the arch; H is the rise of the arch; and L is the span of the arch.

Note that arch structure is expected to undertake certain amount of the bending moment and the bending design of an arch is a complex issue. In the early stage of design, an approximate calculation as shown in Fig. 6.5 could be employed to calculate the bending moment within an arch. In particular, when the uniform constant load is relatively of a large number and the non-uniform live load is relatively of a small number, maximum bending moment could be approximately derived from the section that is farthest from the pressure line. Checking could be made to see whether the section meets the requirement of bending resistance.

2. **Design for the section of an arch**

When it comes to the section of an arch, it could be made into solid rectangle. As for the material, stone, brick, and reinforced concrete which have sufficient compressive strength are frequently employed and an example is the stone-made Zhaozhou Bridge. The section of the arch could also be made as box-shaped, I-shaped, T-shaped, or lattice-shaped; as for the material, steel or reinforced concrete could be adopted in structures, such as the steel box-sectioned arch in Shanghai Lupu Bridge. Modern arch structure basically employs reinforced concrete or steel in construction, for that these materials could handle not only compression but also tension and provide resistance to potential local bending moment within the arch. Hence, the span of the arch is improved, construction more economized, and safety more secured.

Uniform section is always there to choose when the body of the arch is concerned and so it is with a tapered section which is better in terms of material efficiency. Since the internal force of the arch gradually increases from the top to the springer, a relatively small section could be for the top while a relatively large section for the springer, which reduced the material employed.

3. **Design for the support of an arch**

Arch structures have to undertake thrust. A skillful counterbalance to the side thrust remains a key issue in arch design. Generally, feasible constructional measures are listed as followed (also see Fig. 6.7):

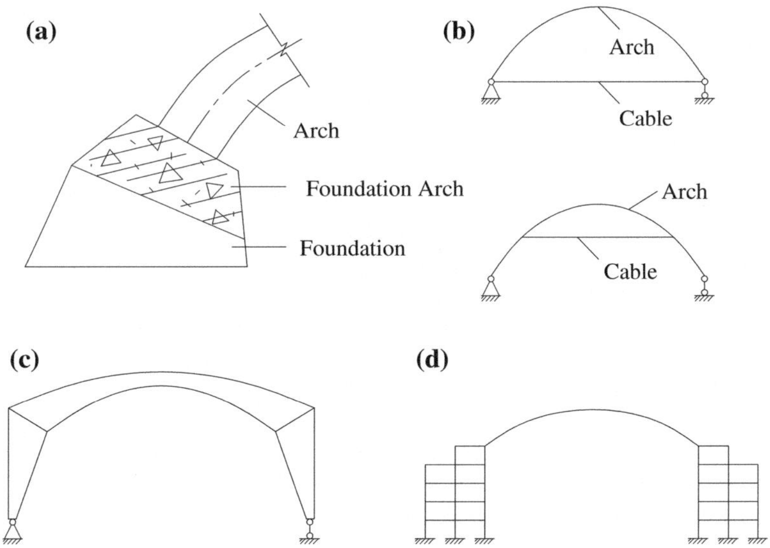

Fig. 6.7 Measures designed for the support of arch. **a** Directly support. **b** Tension brace. **c** Stiffness enhancement supports on edges

(1) Direct support. As for ground-based arch structure, when the side thrust is of small magnitude or the foundation is in a sound condition, the thrust could directly be bore by the supports. When this method is adopted, the size of the support has to be relatively large and material applied is of larger amount. To counteract the side thrust more effectively, supports are often designed to be slantwise so as to be perpendicular to the axis of the arch, as shown in Fig. 6.7a.
(2) Continuous arch. A frequently adopted method is to replace a single large arch with numerous continuous small arches to reduce the side thrust.
(3) Tension brace. The addition of tension members between the supports of the arch or within the body of the arch could counterbalance all or part of the side thrust at supports (as shown in Fig. 6.7b). Take Shanghai Lupu Bridge for instance, it is a half-through arch deck bridge and to counteract the enormous side thrust induced by ultra-large-span giant tension members are added to the deck.
(4) Stiffness enhancement at supports. As for non-ground-based arch structures, horizontal stiffness of supports on the edge could be enhanced. Specific measures include the following: the adoption of rigid frame structure, that of tube structure, that of frame structure, and that of shear wall structure as supports. In this way, the side thrust could be undertaken by these horizontal-stiffness-enhanced supports (as shown in Fig. 6.7c).

4. **Stability design for an arch**

When the arch structure is designed to be slender, having a large span to the height of the section ratio, local buckling instability shall manifest, which in the essence is an issue of the instability of compression members. Measures to tackle this problem are listed as follows (also see Fig. 6.8):

(1) Increase section height. To add to the height or the width of the section could directly solve the problem of buckling instability. However, this solution will also result in more materials consumption.
(2) Add braces to arch rib. To avoid in-plane stability of the arch, the arch could be designed as a truss arch and latticed arch or to be added with stiffening steel plate to the arch ribs, which shall vertically stiffen the arch ribs (as shown in Fig. 6.8a). Besides, the arch ribs could be horizontally stiffened, which means to install horizontal support between two main arches so as to prevent out-of-plane instability of the arch (as shown in Fig. 6.8b).
(3) Adopt groove plate arch. When it comes to large-span-arched roofing, groove plate arch or wave-shaped arch is frequently adopted, in which case buckling instability that easily occurs to the surface of thin-walled arch could be significantly reduced (as shown in Fig. 6.8c).

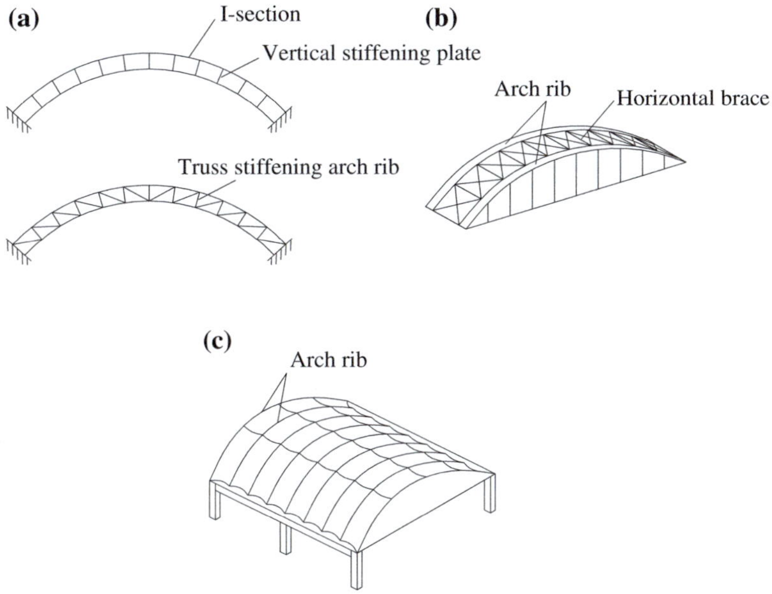

Fig. 6.8 Measures designed for the stability of arch. **a** Vertical stiffening arch rib. **b** Lateral stiffening arch rib. **c** Wave form stiffening arch rib

5. **Practical construction of arch structures**

Some practical constructions of steel arch structures shall be introduced next [41, 42].

(1) National (Tianjin) Tennis Training Center (with the span being 42 m), China

Type of structure: large-span rigid steel arch frame; no internal column within the span, as shown in Fig. 6.9.

Fig. 6.9 National (Tianjin) Tennis Training Center, China

Enclosing system: pressed metallic composite board roof; bricks wall.

Dimensions of the structure: span being 42 m; length being 104 m; eaves elevation being 6.7 m; the rise of the arch being 7 m.

Gross floor area: 4300 m^2.

Features of construction: Arched rigid frame is employed to solve large-span-induced mechanical issues and to display a reasonable light steel structural system. This would not only utilize mechanical properties of material, economizing upon steel as well as the cost of construction, but also provide an aesthetically graceful outlook for the structure.

(2) Xi'an Zhongti Water Park (with the span being 56 m), China

Type of structure: large-span-arched steel rigid frame; no internal column within the span, as shown in Fig. 6.10.

Enclosing system: insulated pressed metallic composite board roof and wall.

Dimensions of the structure: the span is 56 m; the length 90 m; eaves elevation 26 m; the rise of the arch 9 m.

Gross floor area: 5400 m^2.

Features of construction: Arched rigid frame is employed to solve large-span-induced mechanical issues and to provide a majestic outlook for the structure. Compared with the proposal with spatial truss, one-third cost of construction was saved. Another noteworthy point of this construction project is that an integrated sliding installation of the arched rigid frame is employed so as to prevent damage to the costly imported facilities that had already been installed, which adds to the difficulty in construction.

(3) Shenyang Auto Trading Center (with the largest span being 71 m), China

Main exhibition hall: arched steel rigid frame; net span is 71 m; column spacing 9 m; length 120 m; the entrance is installed with threefold steel frame; eaves elevation 22 m.

Trading hall: arched steel rigid frame; net span is 50 m; column spacing 9 m; length 315 m; eaves elevation 13 m; the rise of the arch 4 m (as shown in Fig. 6.11).

Hall of Science and Technology: circular threefold steel frame; the diameter is 30 m; eaves elevation 22 m.

Fig. 6.10 Xi'an Zhongti Water Park in construction, China

Fig. 6.11 Shenyang Auto Trading Center in construction, China

Wing exhibition hall: arched steel rigid frame; the net span is 40 m; column spacing 9 m; length 235 m; eaves elevation 13 m; the rise of the arch 3.6 m.

Wing trading area: arched steel rigid frame; the net span is 39.6 m; column spacing 9; length 432 m; eaves elevation 9 m; the rise of the arch 3.6 m.

In sum, the structural design for arch construction projects could be summarized as follows:

1. Design for the axis

Second-degree parabola is taken as the shape of the axis of the arch, and specific parameters are determined by the net height and span of the arch designed.

2. Design for the section

I-shaped section is chosen for the arch, and the material of selection is steel, which undertakes not only compression but also tension and potential local bending moment as well.

The main body of the arch employs a differential section. Since the internal force of the arch gradually increases from the top to the bearing and reaches the maximum at the bearing and at the top of the column, a smaller section is designed for the top of the arch and a larger section for the bearing. Round holes are often engineered in the web of the I-sectioned arch ribs so as to further lessen the weight of the arch, to economize upon the steel employed, and to facilitate the installation of equipment and facilities.

3. Design for the support

The side thrust of the arch is undertaken by the large-stiffness rigid frame on the edge.

4. Design for the stability

The addition of rib stiffeners to the webs at the largely stressed springing of the arch could instantly solve buckling instability of the arch. Thereby, vertical stiffeners are to be welded to the I-sectioned arch ribs and to be increased in largely stressed areas so as to prevent in-plane instability of the arch. Horizontal stiffening of the arch ribs through purlins and diagonal tension members prevents out-of-plane instability of the arch. Meanwhile, the installation of angle supports at certain spacing could prevent local instability of webs and wings of the arch ribs.

6.3 Design for Suspension Cable Structural System

Suspension cable is a "conjugated" system of the arch (as shown in Fig. 6.3). Since suspension cable determines its outline through catenary, tension is the only force that the system has to take. Steel cables with high tensile strength are usually employed to for suspension cable system. Since there is barely any issue of instability when the material is pulled, suspension cable system has a better economic efficiency than the arch system and it spans over the longest distance in all large-span structural systems.

Compare the stress system of suspension cable as shown in Fig. 6.12 with that of beam as shown in Fig. 6.3b. We could see that the needed anti-bending force couple could be regulated through mid-span internal force *H* and the sag of suspension cable system *y* instead of through the addition to the diameter of the section of suspension cable *d*. As for the stress system of beam, the needed anti-bending force couple could only be regulated through adding to the height *h* or to the width *b* of the section. When it comes to the regulation of vertical deflection, more material is needed for the expansion of the section. Since the sag of suspension cable *y* is usually much larger than the height of the section of beam, under the action of identical load upon identical span, material employed by suspension cable system is much less than that employed by beam system. Furthermore, suspension cable undertakes only uniform tension and leaves no room for buckling instability which

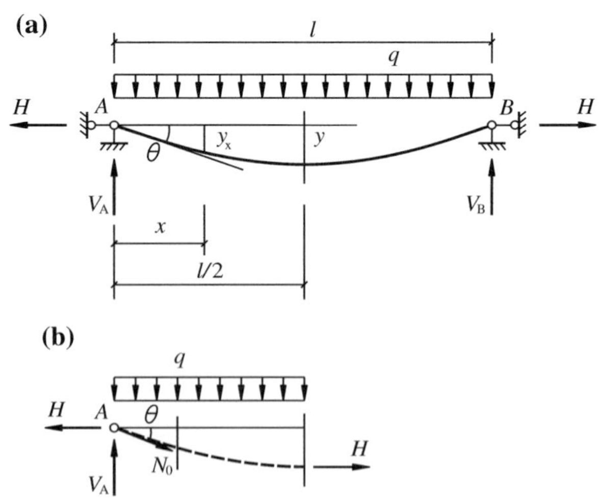

Fig. 6.12 Stress analysis of suspension cable

could occur to arch structures. Therefore, suspension cable system has an advantage of less material employed over the arch system.

The calculation of suspension cable is quite similar to that of the arch structure. As shown in Fig. 6.12, suspension cable is under the action of symmetrical uniform load q. Half of the suspension cable could be taken as the isolated body to conduct static equilibrium calculation. In this case, there is no bending moment and shear force at mid-span and the bending moment relative to the support of suspension cable is:

$$Hy = qL^2/8 \tag{6.5}$$

$$y = \frac{qL^2}{8H} \tag{6.6}$$

The resultant force at the support of suspension cable is

$$R = \sqrt{V^2 + H^2} \tag{6.7}$$

where q is uniform load; L the span of suspension cable; y the sag of suspension cable; H horizontal tension at the bottom of suspension cable; V vertical force of reaction at the support of suspension cable; and R the resultant force at the support of suspension cable.

In actual construction, the ratio between the span and the sag of suspension cable is usually between 10 and 20. Although there needs to be no concern of buckling instability, the stiffness of suspension cable is to be fortified so as to prevent relatively large deformation and vibration induced by potential dynamic load or local concentrated load. On the other hand, although suspension cable in general adopts high-strength steel cable which could not undertake bending moment, it has in theory the same issue of intrinsic pressure line to deal with. Therefore, the fortification of stiffness of suspension cable is of significance for when it is necessary it could undertake local bending moment.

Suspension cable is also a self-adaptive structure. Slender cable could adapt to external loads through instant changing of the shape of the axis. When cables are not stiffened, they will vary curvature and deformation to the condition of load and coordinate the stress of the structure. From certain perspective, suspension cable determines its own shape through self-regulation. Naturally, when it comes to the design and calculation of suspension cable structure, attention is needed in the design for the axis and support of suspension cable. Extra attention is to be paid to lateral tension in the same process.

1. **Design for the axis of suspension cable**

A rational determination of the axis of suspension cable so that it bears no or minor bending moment serves as the major concern in the design for suspension cable.

Generally speaking, the larger the sag of suspension cable is, the smaller lateral tension is. When aesthetics, economy, and practicality are synthetically considered, it is quite ideal if the ratio between the span and the sag is between 10 and 20. In certain cases, this ratio of suspension cable in large-span roofing system could be raised to about 50.

The axis of suspension cable is in general designed to approximate to the intrinsic pressure line of suspension cable when it is under the action of vertical uniform load, namely second power parabola, so that the suspension cable generates only a relatively small lateral tension under the action of various loads.

$$y = 4y_0x(L - x)/L^2 \tag{6.8}$$

where x is the horizontal coordinate of the axis of suspension cable; y the vertical coordinate in the direction of the sag; y_0 the sag of suspension cable; and L the span.

2. **Design for the support of suspension cable**

Suspension cable has to deal with lateral tension. A skillful counterbalance of lateral tension is a key issue in the design for suspension cable structure. In general,

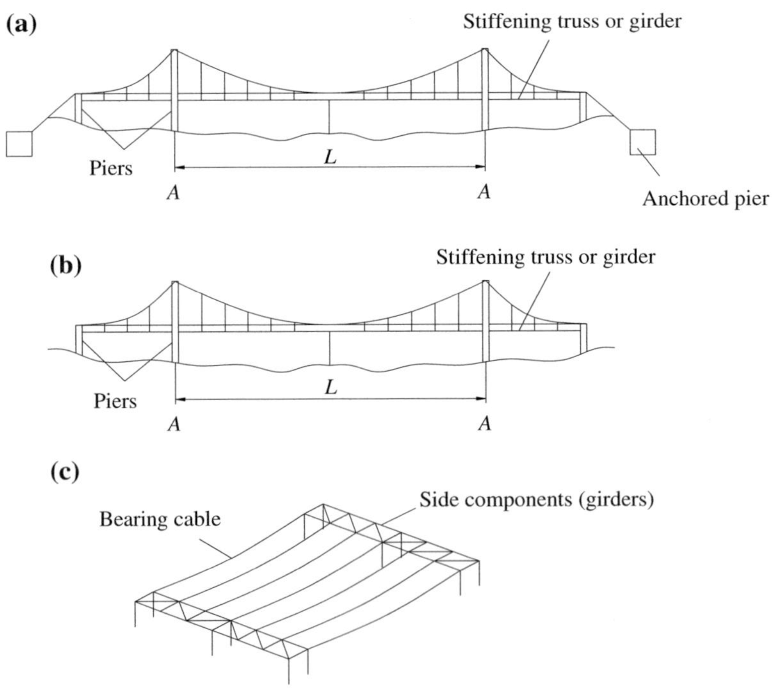

Fig. 6.13 Measures designed for the support of suspension cable structure. **a** Support anchoring. **b** Self-anchoring. **c** Stiffness enhancing of side-supported structure

measures as follows (as shown in Fig. 6.13) could be employed to solve the problem:

(1) Support anchorage. As for ground-based suspension cable, when the lateral tension is not of large amount or the foundation is in good condition, the lateral tension could be dealt with through simple anchorage of the support (as shown in Fig. 6.13a). When this approach is adopted, the dimensions of support will be relatively large; more material will be employed; and it has to be anchored under the ground. There are gravity anchored support and embedded anchored support.
(2) Self-anchorage. To counteract the tension that suspension cables bear through the compression upon horizontal rigid support is frequently adopted in the design for self-anchorage suspension cable bridge (as shown in Fig. 6.13b).
(3) Stiffness enhancement at supports. As for non-ground-based suspension cable, especially for suspension cable roofing, horizontal stiffness of side support could be enhanced in ways such as the adoption of reinforced concrete circular beam or that of inclined arch so that the lateral tension that suspension cables has to bear could be undertaken by large-horizontal-stiffness supports (as shown in Fig. 6.13c).

1. **Examples of suspension cable structures**:

   Following are the famous suspension cable bridges from Source: http://en.wikipedia.org/wiki/List_of_longest_suspension_bridge_spans.

- Akashi Kaikyō Bridge (Japan), 1991 m—1998
- Xihoumen Bridge (China), 1650 m—2009
- Great Belt Bridge (Denmark), 1624 m—1998
- Yi Sun-sin bridge (South Korea), 1545 m—2012
- Runyang Bridge (China), 1490 m—2005
- Humber Bridge (England, United Kingdom), 1410 m—1981 (longest span from 1981 until 1998)
- Jiangyin Suspension Bridge (China), 1385 m—1997
- Tsing Ma Bridge (Hong Kong, China), 1377 m—1997 (longest span with both road and metro)
- Verrazano-Narrows Bridge (USA), 1298 m—1964 (longest span from 1964 until 1981)
- Golden Gate Bridge (USA), 1280 m—1937 (longest span from 1937 until 1964).

Major applications of suspension cable in roof system are:

- Cable-suspended roofing of Beijing Workers' Stadium. The plane layout is a circle with a 94 m diameter.
- Cable-suspended roofing of Zhejiang People's Stadium. The plane layout is elliptical with the diameter of the long axis being 80 m and that of the short axis 60 m.

6.4 Design for Cable-Stayed Structural System

Cable-stayed structural system could be regarded as a variant of suspension cable. Both of these two systems employ cables as their major force bearing and transfer system. Since there is no issue of instability when the cables are pulled, cable-stayed structural system could fully utilize the high tensile strength of cables and combine them and concrete to materialize the cable-stayed bridge. Compressive strength of concrete is utilized in the tower, while the high tensile strength is embodied through cables, in which case an economically efficient and reliable structure is constructed.

Cable-stayed structural system is frequently applied to large-span bridges or large-span roofing systems. The distance it spans over is only second to suspension cable and could be up to 2000 m in theory. Cable-stayed bridges generally span over a distance of 500–1000 m, and bridges of such kind include Shanghai Nanpu Bridge, Yangpu Bridge, Xupu Bridge, and Suzhou-Nantong Yangtze River Bridge. All roofing systems employed in large-span stadiums have resorted to cable-stayed structural system for the good returns from the aspects of technology and economy. One example is the Huanglong Stadium in Hangzhou, China.

In suspension cable structural system, the main cable is curved in the fashion of a catenary. In cable-stayed structural system, all cables are straight. Cable-stayed structural system is a semi-self-balancing system that is comprised of a set of stay cables, compressed towers, and compressed decks or roof panels, and it is quite similar to a doubled cantilever when it comes to the feature of the structure as a whole (as shown in Fig. 6.14). Unlike suspension cable structural system, the deck or the roofing of cable-stayed structure is compressed by the horizontal component force of stay cables. When buckling instability occurs to the deck or the roofing, the entire cable-stayed system shall fail and this is the major cause behind the inferiority of cable-stayed structural system in regard to distance spanned in the presence of suspension cable structural system.

The stress analysis of cable-stayed system could be conducted through approximate calculation as shown in Fig. 6.15. Assume that total vertical load stay cables have to bear is W and the inclined angle between stay cables and horizontal deck or roofing is θ. Then, the calculation equation of the tensile force within stay cables R or the compressive force within horizontal deck or roofing H could be derived from the equilibrium state of joint (as shown in Fig. 6.15).

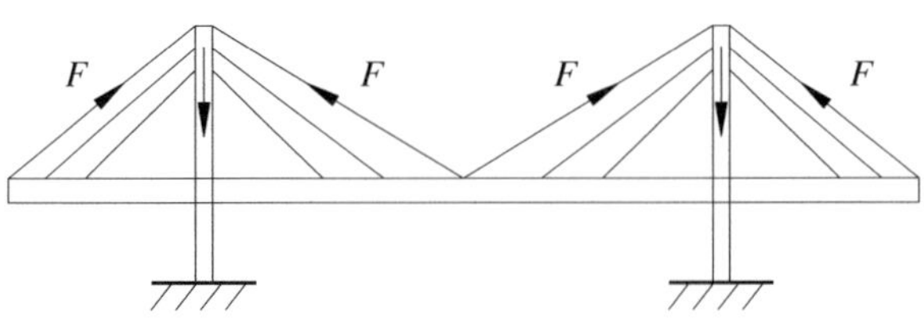

Fig. 6.14 Cable-stayed structural system

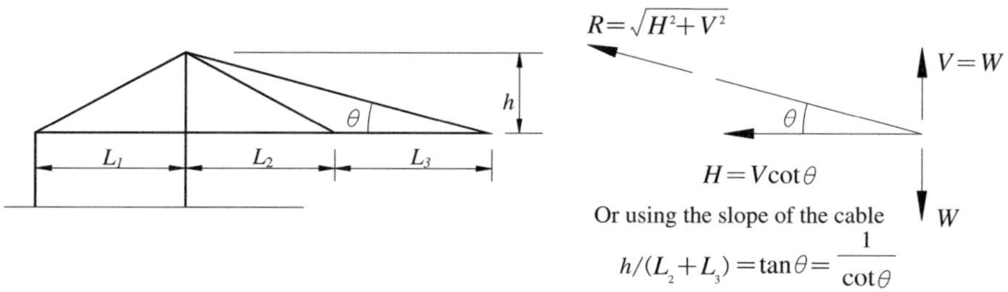

Fig. 6.15 Stress analysis of cable-stayed structural system

When it comes to the design for cable-stayed system, the priority has been given to the determination of the inclined angle, the internal force of cables, and the stress upon the deck. The inclined angle has a direct impact upon the height of the tower, the internal force of cables, and the compression upon horizontal structures such as the deck or the roofing. When the inclined angle is small, the height of the tower will be relatively small, but the internal force of cables as well as the compression upon horizontal structures will be relatively large, which requires that the deck or the roofing has sufficient compressive strength to undertake compression without having to yield to the failure of buckling instability. Otherwise, when the inclined angle is large, the internal force of cables and the compression upon horizontal structures are relatively small, but the height of the tower will be comparatively large. A reasonable determination of the inclined angle is in need of a synthetic analysis, in which case requirements from technology, economy, and aesthetics are balanced and a comparatively optimal design is derived.

Example

The structural system and geometric dimensions of a cable-stayed bridge have been simplified as shown in Fig. 6.16. It is known that the sum of the deadweight of the deck and the equivalent live load is q = 100 kN/m. Please carry out a simplified calculation of the internal force of major components in this system.

1. To determine the vertical load that each cable bears

As for cable AF, $W_1 = 0.5 \times 15 \times 100 = 750$ kN
As for cable BF, $W_2 = 0.5(15 + 15) \times 100 = 1500$ kN

Fig. 6.16 Calculation of cable-stayed system

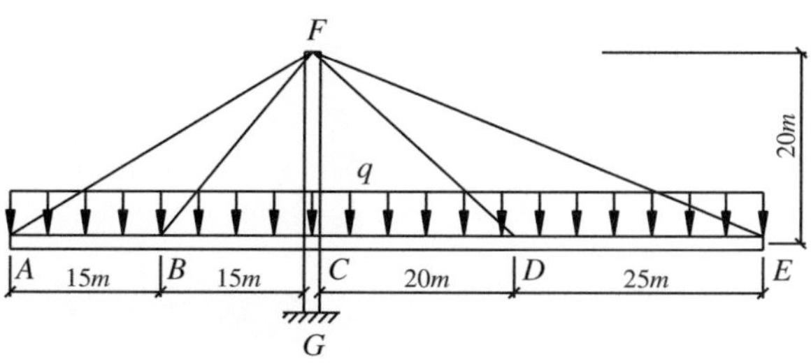

As for cable *DF*, $W_3 = 0.5(20 + 25) \times 100 = 2250$ kN
As for cable *EF*, $W_4 = 0.5 \times 25 \times 100 = 1250$ kN.

2. In light of the equilibrium state of each joint,

Compression upon beam *AB*, $H_1 = W_1(30/20) = 1125$ kN
Compression upon beam *BC*, $H_2 = H_1 + W_2(15/20) = 2250$ kN
Compression upon beam *ED*, $H_4 = W_4(45/20) = 2810$ kN
Compression upon beam *DC*, $H_3 = H_4 + W_3(20/20) = 5060$ kN
Tension of cable *AF*, $R_1 = 1350$ kN
Tension of cable *BF*, $R_2 = 1880$ kN
Tension of cable *DF*, $R_3 = 1770$ kN
Tension of cable *EF*, $R_4 = 3080$ kN
Compression upon column *EF*, $Z = 100(15 + 15 + 20 + 25) = 7500$ kN
Compression upon column *FG*, $M = 0.5(45 \times 45 - 30 \times 30)100 = 56{,}300$ kNm.

6.5 Comparative Analyses of Well-Known Large-Span Structures

6.5.1 Shanghai Lupu Bridge

1. **Project background [42]**

Lupu Bridge was constructed in 2003, and it was the first all-steel arch bridge over the Huangpu River, serving as the connection between Pudong and Puxi. And it is also the 4th bridge over the Huangpu River within urban Shanghai area. The other three bridges are Yangpu Bridge, Nanpu Bridge, and Xupu Bridge. Lupu Bridge (as shown in Fig. 6.17) has a length of 3900 m. Its main bridge is 750 m long and

Fig. 6.17 Shanghai Lupu Bridge [42]

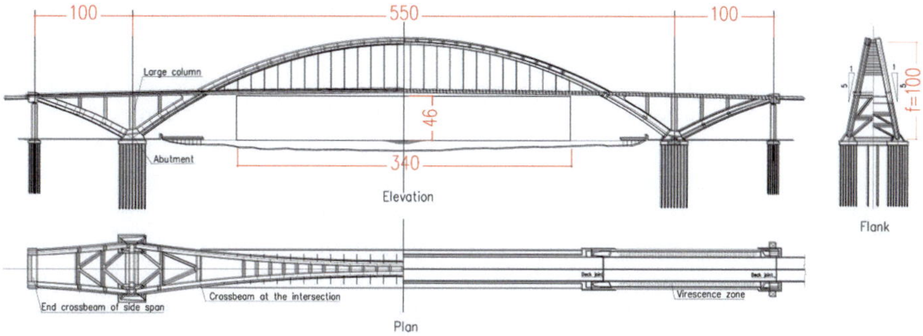

Fig. 6.18 Layout, elevation, and side view of Shanghai Lupu Bridge

28.7 m wide. The net height above the water is 46 m, and the deck provides service for two-way six lanes. The bridge displays a graceful arc over the Huangpu River.

Shanghai Lupu Bridge has the longest span in the world among box-shaped steel arch bridge, which spans over 550 m (as shown in Fig. 6.18). During the construction of this bridge, the lifting weight of a single structural member reached 8600 kN and the sum of maximum lifting weight of the middle span was 4800 kN, both of which are still No. 1 in the world. The construction of Lupu Bridge integrated the construction technology for three different types of bridge: the arch bridge, suspension bridge, and cable-stayed bridge. Steel consumed in the construction of the main bridge was over 35,000 tons and that employed in construction measures in the same process was more than 11,000 tons. The 16 horizontal tension cables applied in the construction of this bridge are currently the longest (760 m) with the largest diameter (18 cm), the heaviest weight of a single cable (1100 kN) as well as the highest tensile strength of a single cable (17,000 kN) in the world.

2. **Structural system**

Shanghai Lupu Bridge is a large-span box-sectioned steel arch bridge. The design and construction of this bridge integrated that of three types of bridges: the arch bridge, suspension bridge, and cable-stayed bridge [42].

(1) **Steel arch bridge system**

The main span of Lupu Bridge spans over a distance of 550 m. The two solid-web box-sectioned steel ribs of the main arch of Lupu Bridge are 5 m wide, 9 m high at the feet of the arch, which gradually changes to 6 m high at the top of the arch. The bridge has the world's largest steel box-sectioned arch rib. To expedite the construction as well as to achieve economic efficiency and aesthetics, Lupu Bridge was welded together instead of being bolted together, the difficulty of which was unprecedented in the history of bridge construction.

(2) **Horizontal cable system**

Shanghai has a typical soft soil foundation. The construction of an arch bridge upon foundation of such kind has always been a headache for engineers, let alone an arch with ultra-large span. In total, 16 horizontal cables of 760 m long each effectively counterbalanced the enormous side thrust upon the arch and restrained the lateral displacement of the bridge.

(3) **The construction system of Cable-stayed bridge**

Due to the imitation of construction conditions, cable-stayed bridge construction system had to be adopted. Two 128-m-high all-steel cable towers were temporarily erected on both sides of the Huangpu River. The cable tower was employed to take a hold of the main arch ribs through the construction technique of cable-stayed bridge. The ribs were welded together and became a whole arch structure at the completion of the closure (as shown in Fig. 6.19).

(4) **Wind and seismic resistance system**

The analyses of wind and seismic resistance of Lupu Bridge was carried out prior to the actual construction. Corresponding anti-wind/seismic vibration control measures were taken so the bridge could resist wind of 12th grade and earthquake of 7 fortification intensity. To prevent wind-prompted vortex-induced resonance in the stage of construction and after closure, numerous membrane structures were installed at the top of the arch to direct wind. Meanwhile, to guarantee that there is no collision of deck boards from both sides of the expansion joints, when the bridge is under the action of horizontal wind, seismic force, and the braking force of vehicles, four "damping safety belts" were installed in the bridge. Specifically, two 3-m-long viscous dampers were placed to each of the two mid-span beams of the bridge. Once the bridge began to vibrate severely, this installation should instantly generate enormous amount of damping force and serve as a buffer so as to prevent serious mid-span collision and vibration of decks from both sides.

Fig. 6.19 Cable-stayed construction of Lupu Bridge [42]

3. **Construction technology**

(1) **Forced closure**

The key in the construction of large-span bridge lies in the closure. Prior to the closure, the arch ribs of Lupu Bridge varied in length under the action of thermal expansion and contraction. When the temperature befitting the closure was reached, the screw holes of steel arch ribs were aligned perfectly. When this occurred, builders shot the positioning pins in the middle, took steps to install high-strength bolts, and completed the initial and final tightening procedure. In accordance with previous calculation and measurement, Lupu Bridge should be closed at 19 °C. However, the closure day was too hot so that the originally determined "natural cooling closure" did not turn out quite as expected. The forced closure was applied to accomplish the mission. First step was to adjust the internal force of cables. The arch ribs were elevated by a few centimeters so that the arch ribs were distanced from the section of closure and extensive bolt connection was facilitated since the bolt holes were able to be aligned. As for the next step, eight jacks with 50-tons jacking force were employed in thrusting so that ribs were contracted as required and forced closure was completed (as shown in Fig. 6.20).

(2) **Switching control of structural systems**

In the construction of Lupu Bridge, multiple switches among the technologies of bridge construction and that among the stress systems are quite necessary. Besides, due to the ultra-long span of the bridge, the stress state of the bridge varied to some considerable extent and therefore real-time controls and regulations were indispensable throughout the process of construction. To solve these problems, computer software was needed in the modeling of structural system so as to conduct tracking calculation in real-time construction and to guarantee a safe switching between structural systems.

4. **Summary**

Shanghai Lupu Bridge is a half-through steel arch bridge with an ultra-large span. In the early stage of design, controversy arose over the plan of steel arch bridge.

Prior to the construction of Lupu Bridge, in Shanghai metropolitan area, Yangpu Bridge, Nanpu Bridge, and Xupu Bridge have been successfully constructed. All three bridges adopted the well-matured and economically efficient form of

Fig. 6.20 Lupu Bridge prior to closure [42]

cable-stayed bridge. To construct an ultra-large-span steel arch bridge upon soft soil foundation would not only encounter considerable difficulty in technology but also be faced with high cost of construction, which from either perspective was unfavorable. Nonetheless, as the landmark structure of Shanghai and the main structure in Shanghai World Expo, Lupe Bridge had to adopt a more aesthetically agreeable form of arch bridge. To achieve such a goal, designers and builders overcame numerous difficulties in technology, originated quite a number of techniques in design as well as in construction and ultimately made Lupu Bridge a success.

4. **Discussions**

(1) To provide a quality view of urban Shanghai, Lupu Bridge took the challenge from soft soil foundation and made its way to "the world No. 1 arch." Which aspects do you think that attention should be paid to when technology is concerned?
(2) The main arch of Lupu Bridge employed 35,000-odd tons of steel, while the construction measures around the main tower consumed 11,000-odd tons of steel. In your opinion, is it worth it?
(3) Readers have known that the construction of Lupu Bridge employed 16 horizontal tension members, the diameter and the internal force of which were respectively 18 cm and 17,000 kN. Would you be able to do a reverse calculation over the magnitude of the uniform load upon the arch and conduct a conceptual design for the steel section of the main span?

6.5.2 Guangxi Nanning Bridge

1. **Background of the project**

Guangxi Nanning Bridge is located in the southeast of Nanning City. It spans over the Yong River, and the length of the bridge is 1314.77 m. It serves as the main channel between Nanning downtown and the circular highway of southeastern Nanning. Two circularly curved saddle-shaped inclined arches were engineered to the sides of the bridge, which from a distant point of view look like a connected pair of harps. This bridge ever since its completion has become a new landmark of Nanning.

The construction of Nanning Bridge began from the end of 2004. A large-span bridge with asymmetrical arch ribs, proposed by Professor T.Y. Lin, was adopted [43] (as shown in Fig. 6.21). The deck provides service for two-way six lanes, and the deck system is composed of in-plane arch-shaped steel box-sectioned beams, which were supported by the inclined arch ribs on both sides with uneven heights and non-identical inclined angles via dozens of inclined tension hangers. The approach bridge is curved posttensioned box-sectioned concrete beam, being supported by concrete pier bases. The 300 m span of the bridge provides sufficient navigation clearance, while the bridge itself connects with the existent urban streets.

Fig. 6.21 Double-saddled sideways-suspended arch design for Guangxi Nanning Bridge [43]

Two inclined arches provide support for a horizontally curved deck system so that the in-plane and out-of-plane bending moments the bridge has to undertake are of very small amount.

Nanning Bridge is the first large-span slantwise-suspended bridge with an in-plane sinuous deck in the world. This bridge is of unique originality and carries itself with magnificence. Considering the stress state of the structure is quite complicated, there was great difficulty in construction and the cost of construction was quite high. The realization of large-span asymmetrical structure is a breakthrough in international bridge design theories. To balance requirements from suitability, aesthetics, reasonability, effectiveness, and feasibility of construction has brought a sense of originality to the history of bridge construction worldwide.

2. **Structural system**

The schematic design of Guangxi Nanning Bridge is the last piece of work conducted by Professor T.Y. Lin himself for a significant project of construction. The design of this "double-saddled sideways-suspended arch bridge" is based on an almost perfect understanding of the integration of structure and architectural aesthetics, which enables this high-rising, graceful, and large-span sideways-suspended arch structure to meet the challenges from complicated loads, seismic force, and hurricane.

From the perspectives of architecture and structure, the feature of this bridge shall be the winding main arch which spans over a distance of 300 m. This main span is supported by two asymmetrical box-sectioned inclined arch ribs; no connecting members were installed between the upper part of these two arch ribs; and the surface of the arch converges below the deck. These two pieces of arch ribs have non-identical height and top so as to facilitate the regulation of the internal force flow within the structure. As for the deck, it is hung over the arch ribs like a crib.

The saddle-shaped sideways-suspended arch ribs of the main span are composed of box-sectioned steel arches and are supported by reinforced concrete piers bases and pile foundation. Scientific experimental research has been conducted over six areas including wind tunnel, seismic resistance, wind resistance, box-sectioned steel arch, adjustable/changeable tension members, and tie bars. All these research effort laid a sound foundation for the success of the project.

The adoption of such a structural system could simultaneously meet the requirements from architectural aesthetics and structural mechanics. Architectural aesthetics has been well expressed by the novelty of these saddle-shaped sideways-suspended arch ribs, which have been in perfect harmony with the surrounding environment. The feature of structural mechanics goes to the horizontally winding steel bridge supported by sideways-suspended tension cables (as shown in Figs. 6.22 and 6.23). This curved steel bridge has uniform internal stress and a minimum amount of bending moment. The tensioning of multiple high tensile strength sideways-suspended tension cables gives a full play to the advantage of material employed. These sideways-suspended tension cables act upon the planes created by saddle-shaped inclined arches so that the inclined arch ribs are compressed within the plane and the action of torque is reduced. Besides, the box-shaped section of the arch ribs could fully utilize its capacity of instability resistance. Furthermore, the adoption of differentially sectioned box-shaped arch provides an enlarged section of arch ribs at the support, and the installation of rib stiffeners prevents instability induced by compression and the additional bending moment. In addition, the horizontal component of multiple tension cables counterbalances the horizontal loads upon the winding deck and this horizontally curved

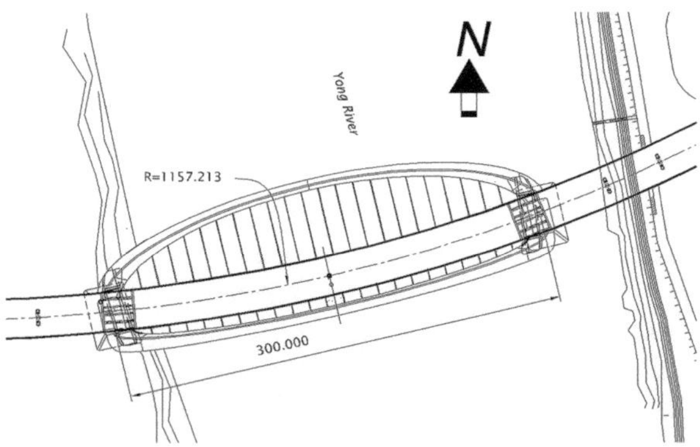

Fig. 6.22 Layout of Nanning Bridge

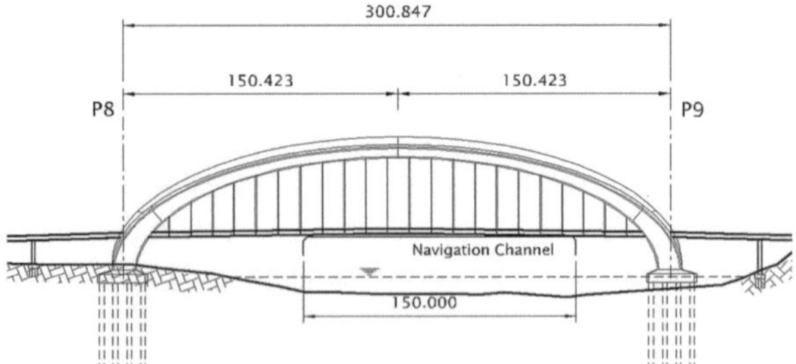

Fig. 6.23 Elevation of Nanning Bridge

deck system therefore plays the role of the third arch (an arched deck). In this way, the weight of the sideways-suspended arch ribs and that of the arched deck could support and depend on each other. From the analysis, we could see that the design of this project of construction is highly originative and is a success in the integration of architectural aesthetics and structural mechanics.

This bridge employed curved tube-shaped concrete pier bases. Thanks to the geometrically curved shape, these pier bases could smoothly pass the load from the above to the foundation. The design for the pier bases has to consider the connection between the pier bases, the arch ribs, the primary beams of the deck structure, the foundation, and the elevated surface of the approach bridge. Meanwhile, it has to provide anchorage for the posttensioned tie bars of the arch ribs located within the primary beams of the deck. Thereby, the pier bases become the connection and turning points of all parts of the bridge.

The arch ribs and the primary beams were designed to be closed box-sectioned so as to meet the requirements from strength, stiffness, and instability resistance and to integrate the bridge as a structural entity. A series of sideways-suspended steel cables were employed to support the weight of the deck. And the elevated approach bridge was connected to both ends of the arch. The primary beams of the approach bridge adopted continuous posttensioned reinforced concrete vierendeel box-sectioned beams. A continuous structure from one pier base to another is composed by the span of the arch and the elevated approach bridge.

3. **Design for the axis of the arch ribs**

The principle of design concerning the geometric shape of the arch ribs is to approximate the pressure line of the arch ribs when they are under the action of deadweight to the axis of the arch ribs and its geometric shape shall guarantee that a major proportion of the deadweight is undertaken by the compression of the arch ribs so as to lessen the amount of bending moment and torque within the arch ribs. Specific principles of design are as follows:

(1) The constant load upon the primary beams of the deck is undertaken by the arch ribs via the suspenders and therefore the bending moment that the deck has to bear is relatively small;
(2) The arch ribs undertake the dead load of the deck, thrust, and a lesser amount of bending moment and torque synthetically induced by the end-of-the-arch constraints and the primary beams of the deck;
(3) Vertical reaction force and side thrust of the base of the arch shall be close to the center of the base so as to minimize the overturning moment and torque;
(4) A certain level of balance shall be reached between the height of the arch ribs, the thrust that arch members have to resist, the stiffness of the structural system, and the amount of steel employed;
(5) The configuration of asymmetrical arch ribs and suspenders shall meet the requirement from architectural aesthetics; and
(6) A simple curvature was designed for steel arch ribs so as to facilitate construction.

To meet the criteria of design when the axis of the arch ribs was being conceived, computer software specializing in shape optimization was employed and the results were modified with reference to the actual construction site so that the requirements from aesthetics, operation, and the feasibility of construction were gradually met with. The actual axis of the arch ribs was derived from multiple iterations.

4. **Design for the arch ribs**

Research has proved that an optimized shape of the arch rib has following characteristics (as shown in Figs. 6.24 and 6.25):

(1) Each piece of the arch ribs is within an inclined plane, the angle of which is determined by the weight of the arch rib and that of the deck;
(2) The sideways-suspended arch rib that is outside the curvature of the deck has a 5° angle from the perpendicular line and therefore the angle of the suspenders is relatively small, too;
(3) The sideways-suspended arch rib that is inside the curvature of the deck has a 30° angle from the perpendicular line, and the horizontal component force of suspenders induced by this inclination angle shall give rise to an eccentric torque upon the arch rib;
(4) The inside-the-curvature arch rib is higher than the outside-the-curvature one so as to reduce the eccentricity of the deadweight and side thrust from the foundation; and
(5) Parabola is the optimal curve within the plane of the arch rib.

The arch rib took the form of steel box-shaped cross section. The exterior was constructed with steel plates; the interior was stiffened by longitudinally reversed trapezoid-shaped steel plates; as for the suspenders, they were enhanced by horizontal stiffeners. The sectional area of the closed box-shaped section at the top of

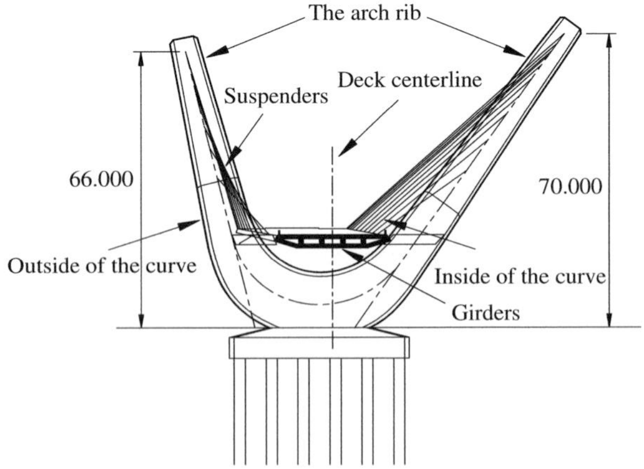

Fig. 6.24 Cross section of the bridge

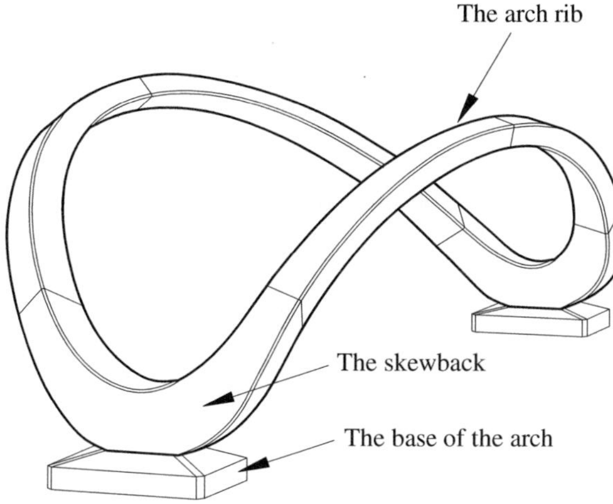

Fig. 6.25 Model of the arch rib

the arch is about 5 m^2 and that at the bottom of the arch is gradually increased to 7 m^2. The box-shaped section removed the four angles to reduce the wind load and to add to the graceful looks (as shown in Fig. 6.26). The arch ribs are directly connected with the abutments, which are fixed with steel plates to the concrete base of the arch. High-strength bolts were employed to reinforce the fixation of the abutments so as to guarantee a smooth transference of side thrust from the arch ribs to the abutments and to the foundation.

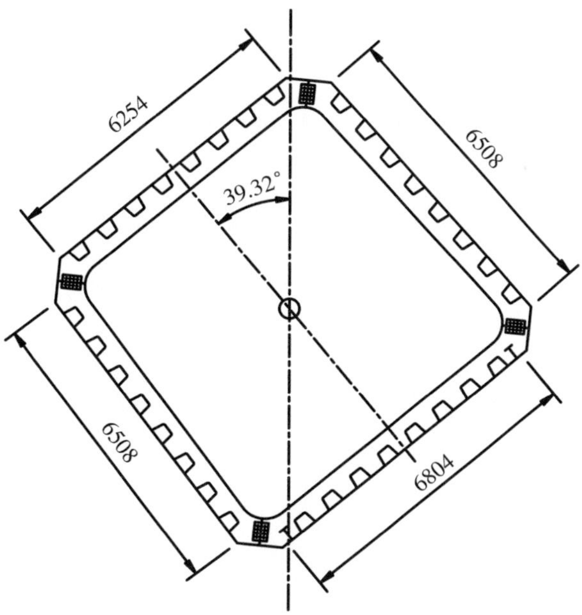

Fig. 6.26 Typical section of the arch rib

It is necessary to conduct a checking of compressive strength, bending strength, and shear strength of box-shaped steel plates and horizontal and longitudinal stiffeners. As for the arch rib as a whole, a checking of structural stability concerning the overall structure and local components has to be conducted with the help of linear and nonlinear buckling theories.

5. **Design for the primary beam of the deck**

The primary beam of the deck is an orthotropic box-shaped steel beam (as shown in Fig. 6.27). The height is about 2.7 m, and the width is about 27 m so there is sufficient space for all the traffic lanes. There has to be sufficient horizontal stiffness and aerodynamic stability guaranteed for the primary beam to resist wind load. Within the stretch of the entire span, the horizontally curved primary beam of the deck is continuous without the installation of expansion joints and extends directly to the concrete abutments at both ends.

The primary beam of the deck is composed of longitudinal roof plates, bottom plates, inclined side plates, and stiffeners. The primary beam is hanged to the sideways-suspended arch at a distance of 12 m via the suspenders. Attention shall be paid to the connection between the primary beam and the lifting points of the tension cables. Stress concentration of the lifting points shall be improved; vertical shear resistance and horizontal bending resistance shall be guaranteed for those lifting points. Besides, to prevent low-cycle fatigue, the constitution of the roof plates, bottom plates, inclined side plates, and stiffeners shall be as simple as possible, and attention shall be paid to the requirement of fatigue resistance of the welding seam.

6. **Design for the base of the arch**

As shown in Fig. 6.28, the abutment is the major weight-bearing component of the arch and it provides support for the steel arch rib, the steel primary beam of the deck, and the concrete pier base of the approach bridge. The approach bridge is comprised of reinforced concrete deck plates and posttensioning force. The exterior of the abutment is a curved surface and transfers the side thrust from the arch rib to the foundation. The abutment as a matter of fact is part of the base of the arch, and it serves as a transition member between the box-shaped primary beam of the main span and posttensioned reinforced concrete approach bridge.

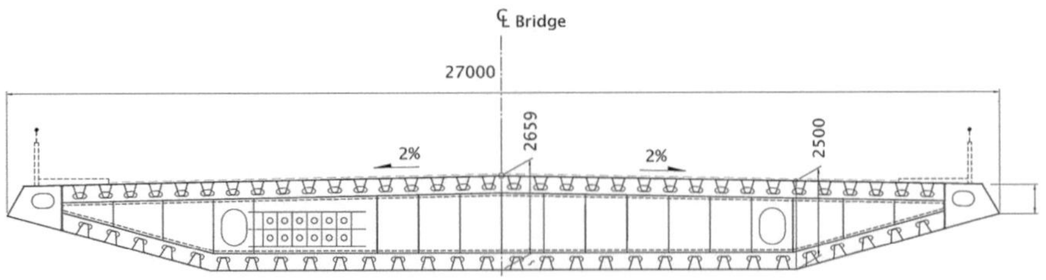

Fig. 6.27 Typical cross section of beam of the deck

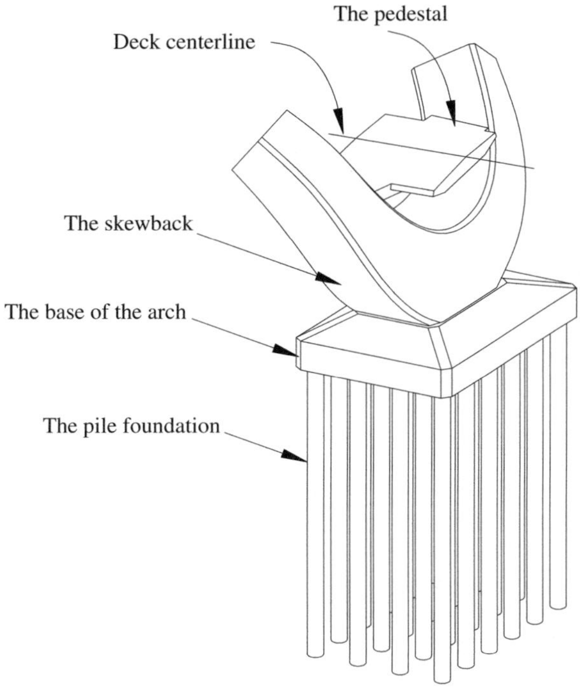

Fig. 6.28 Pedestal, the abutment, the base of the arch, and the pile foundation

The base of the arch is supported by reinforced concrete pile cap and concrete steel pipe pile. Square pile cap is enhanced by reinforcement so as to transfer vertical and horizontal load to the pile foundation. To provide a pleasing outlook and to facilitate water discharge, the top of the square pile cap is tilted. The steel pipe pile was driven to the bearing stratum. Soil was unearthed; reinforced cage was put in place; and concrete was poured in.

7. **Design for the viaduct of the approach bridge**

The span of the viaduct of the approach bridge varies from 30 to 50 m to the variations of terrain. The primary beam of the viaduct adopted posttensioned vierendeel concrete box-shaped continuous beam (as shown in Fig. 6.29) and was supported by the concrete pier base of the pile foundation. The bending style of the exterior of the viaduct tallies with that of the primary beam of the arch. When it comes to the design of the viaduct, factors which should be taken into consideration include economy, aesthetics, convenience of construction and maintenance, and the coordination between the outlook of the arch rib, main bridge, and viaduct. The primary beam of the viaduct is posttensioned concrete box-shaped beam, which has 5 open webs. The width of each open web is about 5.1 m. And the four longitudinal webs are the major weight-bearing components. To apply posttensioning prestress in the longitudinal direction could provide a reverse bending moment for the primary beam, which would counterbalance the deadweight of primary beam and that

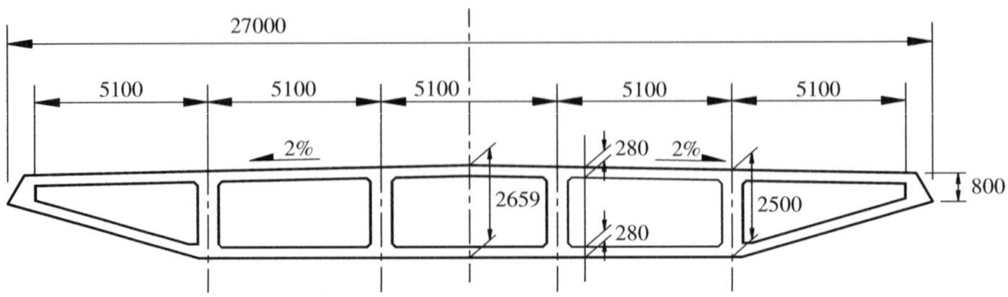

Fig. 6.29 Typical cross section of prestressed concrete primary beam of a viaduct

of the deck. It is also a practical application of the balance design of prestress that has been introduced in Sect. 5.3.

8. **Bridge construction**

The construction of bridge employs a series of temporary supports and suspenders to hoist and install prefabricated arch ribs and segments of the primary beam. After the completion of foundation and abutment, temporary supports are usually erected to raise suspenders of the viaduct that had been fixed to both banks of the river. Afterward, the suspenders of the viaduct and other hoisting devices are employed to place the arch ribs and the primary beam segment by segment upon the temporary supports. After all segments are hoisted and installed in place, tension cables of the arch ribs and connecting components located on the surface of the deck are to be installed and prestressed. This method of construction has some practical successes in construction.

9. **Summary**

The design of Nanning Bridge synthetically took a lot factors into consideration, including functions, aesthetics, economics, and the convenience of construction. The structural system of this bridge is composed of one horizontally winding deck and two pieces of supporting arch ribs. From a structural point of view, the inclined arch ribs of the bridge and the primary beam of the deck provide balance and support for each other, which rationalized the force bearing of the structure. The concrete abutment of the arch was posttensioned so as to suit to the complicated stress state at the ends of the arch. The curved outlook of the approach bridge also expresses a sense of aesthetics and continuity in its display.

This plan of design has gone through substantial and sufficient analysis and research, including the checking calculation of the strength, the stiffness, the stability, the construction sequence, the seismic response and the aerodynamics of the entire structure, and the phenomenon of self-excited oscillation included, which proved the feasibility of this plan.

10. **Discussions**

(1) What are the big idea and the features of Nanning Bridge that were emphasized in the design? Why were they reasonable and effective?

(2) How do you take the mutual balancing and mutual support between the two inclined arch ribs and the primary beam of the deck? What is the mutual restriction between the angles and the heights of these two arch ribs? What is the connection between the curvature of the inclined arch ribs and that of the horizontally winding deck?
(3) Could you work out a simplified procedure of calculation with reference to the concept of design of Nanning Bridge and the calculation theories of the arch structure?
(4) The inspiration of this bridge came from Professor T.Y. Lin earlier design of Ruck-A-Chucky Bridge [2], and Nanning Bridge could be regarded as an extension of the conception of Ruck-A-Chucky Bridge (as shown in Fig. 6.30). Try to discuss the similarities and the dissimilarities between these two bridges and figure out the construction feasibility of Nanning Bridge. A simple introduction of Professor T.Y. Lin earlier design of Ruck-A-Chucky Bridge is as followed.

Ruck-A-Chucky Bridge is located in the upstream of a dam in California, USA. It spans over a distance of about 400 m, and the valley it overspanned has a depth of about 137 m. The bridge has a horizontally winding deck, the supporting system of which is comprised of inclined tension cables that were anchored to the slopes of the surrounding hills. The reason behind the adoption of a winding bridge is that a tunnel had to be dug out in the surrounding hills for the approach bridge if the original plan of a straight bridge was taken. To avoid onerous excavation of the hills, the engineers investigated the terrains of the embracing hills and decided that a horizontally winding bridge could be suspended with steel tension cables that were anchored to the sides of the hills. The central arc angle of the bridge is about 50° so as to connect the roads on both sides of the valley. Due to the novelty of the structural form, graceful presentation of the architectural aesthetics, and an excellent combination of these two, this design of Ruck-A-Chucky Bridge stood out in 923 proposals and was awarded first prize in American Progressive Architecture in 1979.

Fig. 6.30 The design of Ruck-A-Chucky Bridge [22]

However, due to a series of reasons, Ruck-A-Chucky Bridge was not put to the construction in the end. Put aside the economic causes, the technical issues involved are as follows [2] (Structural concepts and systems for architects and engineers, TY Lin; Sidney D. Stotesbury, 1988 2nd Ed. New York: Van Nostrand Reinhold Co. ISBN: 0442259034 9780442259037):

1. The points and directions of the anchorage of the suspension cables were to be determined by scrutinized optimization analysis, while the rock mass and other geological conditions of the valley might not be able to provide corresponding support for those determined points of anchorage and suspension cables in their required directions.
2. The force bearing of suspension cable is a matter of high statical indeterminates, and an optimization is needed when the redistribution of the internal force is concerned. The adjustment of the distribution of the internal force needs to be carried out real time. And the construction and maintenance of a winding suspension bridge and its cables had practical difficulties in store.
3. Ruck-A-Chucky Bridge had to span over a 400-m-wide, 137-m-deep valley, in which case a simultaneous section-by-section hoisting and installation of this bridge and the anchorage of suspension cables to the cliffs around were *devoid of matured practical* construction experiences.

6.5.3 *Jiangyin Yangtze River Bridge*

1. **Project Background [44]**

Jiangyin Yangtze River Bridge (as shown in Fig. 6.31) has a 1385-m-long main span and a 193-m-high main tower. It provides service for all six lanes with a width of 37.7 m (1.5-m-wide sidewalks on both sides included). The construction began in 1994 and completed in October, 1999.

Fig. 6.31 Jiangyin Yangtze River Bridge [44]

2. Structural system and construction characteristics

Jiangyin Yangtze River Bridge is the first bridge with a span over 1000 m in China, and the total length of this bridge is about 3 km. The north of the bridge is multi-span prestressed continuous rigid frame bridge, and the northern and southern approach bridges are prestressed concrete beam bridges. The primary beam of the deck took the form of an oblate box-sectioned steel beam (as shown in Fig. 6.32) so as to reduce unfavorable effects induced by the lateral wind load. The sag/span ratio of the pair of cables is 1/10.5, and the tower is a portal reinforced concrete structure. Figure 6.33 shows the elevation of Jiangyin Yangtze River Bridge.

There have been many successes in the construction of large-span suspension cable bridges in the world. However, due to the unique conditions of geology, hydrology, and climate of Jiangyin and Jingjiang, many issues had to be worked on throughout the process of construction. Innovative designs and technologies employed are as follows:

(1) Northern anchorage and open caisson. The anchorage of suspension cables is the key in the design and construction of suspension cables bridges. The northern and southern anchors have to uphold and stretch the main cables of the

Fig. 6.32 Box-sectioned steel primary beam [44]

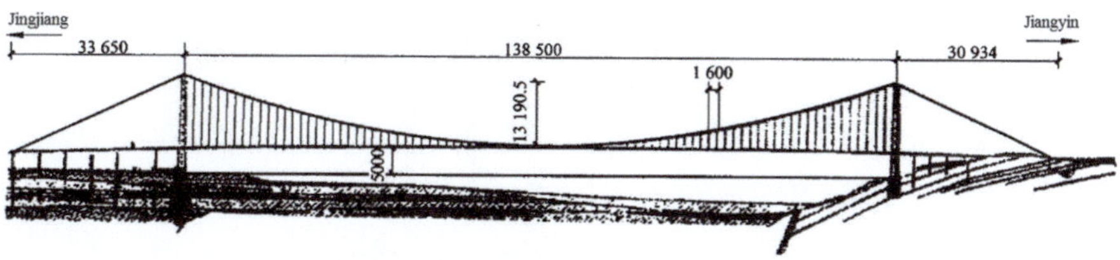

Fig. 6.33 Elevation of Jiangyin Yangtze River Bridge [44]

bridge which weighs 64,000 tons. And the northern anchor has its place on an alluvial plain. As for the underground open caisson, it is 69 m long, 51 m wide, and 10 m deep. The sinking of the caisson has to go through 4 different layers of soil, which easily leads to serious problems such as deflection and torsion. And the sinking process lasts for 20 months. The northern anchor chock is a gravitational friction anchorage structure (as shown in Fig. 6.34), and the southern anchor chock is a gravitational rock-embedding anchorage structure.

(2) Main cable setup. The main cables are the major weight-bearing components of Jiangyin Yangtze River Bridge, which hoist 18,000 tons of steel deck, 5000 tons of asphalt pavement and live loads from vehicles, etc. There are two main cables in Jiangyin Yangtze River Bridge, and together, they weigh 8450 tons. There are 169 minor stands of cables, and each strand weighs 50 tons. The erection of the main cable employed the prefabricated parallel wire strand (PPWS) method. To erect the strands of cables that weigh 50 tons each and are 2200 m long proposed numerous unpredictable technical difficulties throughout the processes of traction, tension, taking the form, and the adjustment of cables. Besides, since the construction was all about overhead working, the weather had quite an impact, which posed great difficulty to the setup (as shown in Fig. 6.35).

(3) Deck pavement of the bridge. On the surface of box-sectioned steel bridge, there had to be a 5-cm-thick layer of asphalt concrete. Since the box-sectioned steel beam has a high thermal conductivity, in which case it would be very hot in the summer and very cold in the winter, the merits of the asphalt could hardly be fully utilized. The ground surface temperature of Jiangyin varies from −15 to 70 °C. Without the help of any previous examples, relevant research and experiments lasted for more than a year and eventually derived a plan for the deck pavement of the bridge that could specifically meet the requirements of Jiangyin Yangtze River Bridge.

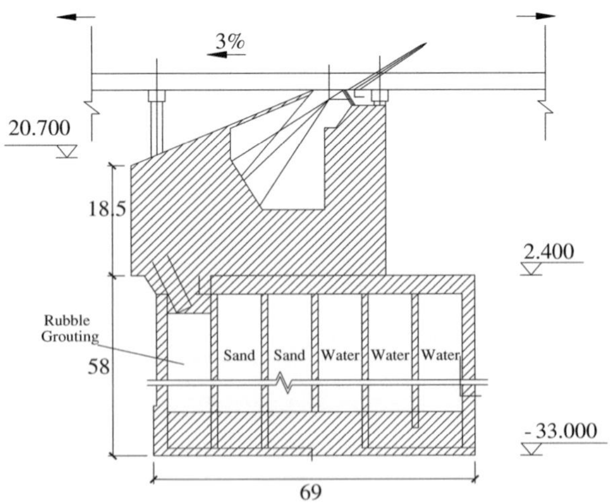

Fig. 6.34 Northern anchor and the open caisson

Fig. 6.35 Jiangyin Bridge in construction [44]

(4) Northern tower construction. It was the first time to apply the advanced technology of climbing framework and the technique of "second casting" of horizontal beams to the construction of the tower, which greatly expedited the speed of construction and ensured the quality of the structure.

3. **Summary**

Among international bridge construction community, it is quite common to tell the bridge technology level of a country from the suspension bridges built. As for the indicator of the level of advancement, it is the "span." Jiangyin Yangtze River Bridge won first prize in Eugene-Figo award issued by the International Bridge Congress. This award was named after the renowned American bridge designer Eugene-Figo to honor notable landmark bridges worldwide that were recently completed with a sense of imagination and a spirit of originality. The winning of this award indicates that our technology of bridge construction is getting mature nowadays.

4. **Discussion**

(1) Jiangyin Yangtze River Bridge is a typical suspension cable bridge. Could you calculate the tensile force of the major suspension cable of this bridge with the method explained in Chap. 7 in Sect. 7.3?
(2) If Jiangyin Yangtze River Bridge switched the structural system to a cable-stayed system, could you estimate the height of the tower?
(3) Could you propose some other mixed structural systems so as to solve the issue entailed in the large span of this bridge?

6.5.4 Gibraltar Strait Bridge

1. **Project background [45]**

Gibraltar Strait Bridge is meant to connect continental Europe and continental Africa. The design of this bridge was originally made by a British–French joint construction company, and they did explore quite a number of strait-crossing options, such as the subsea tunnel, floating bridge, and cross-sea bridge. Professor T.Y. Lin reconsidered the original plan of 30 km Route B and proposed a 14 km Route A (as shown in Fig. 6.36) together with a 5000 m plan that involved a new type of bridge that combined cable-stayed bridge and suspension cable [39] (as shown in Fig. 6.37). Anterior to multiple demonstrations from the experts, this last

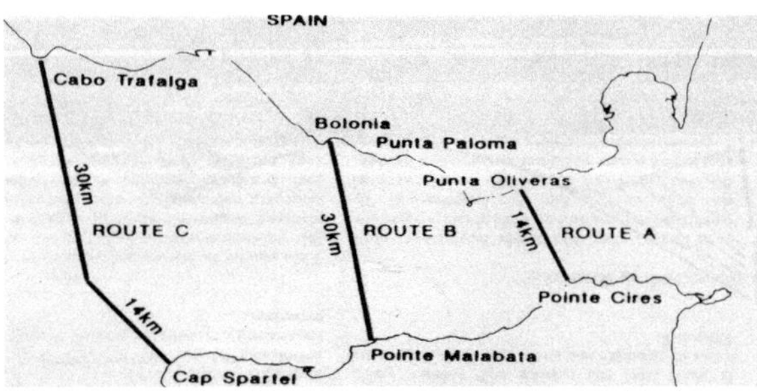

Fig. 6.36 Routes of Gibraltar Strait Bridge [45]

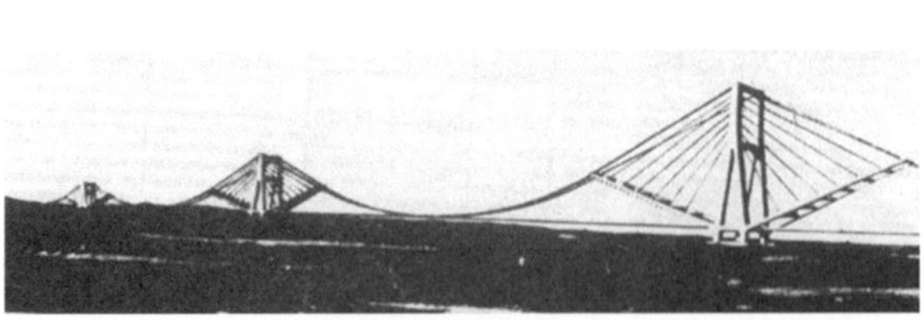

Fig. 6.37 Cable-stayed and suspension hybrid bridge proposed by Lin [45]

plan was awarded top prize in the contest held by the International Association of Bridge and Structural Engineering (IABSE) in 1994.

The construction of a cross-sea bridge in the Gibraltar Strait posed two challenges for civil engineers. One is the selection of the type of bridge due to the deep water and the large span of the strait. And the other is the challenge from the conditions of navigation as in the guarantee of safe passage of vessels under the bridge since more than 50,000 ships navigate across the strait every year. Theoretically, the longest span made of conventional steel that a suspension bridge could have is about 3000 m with the pier submerged in the water for about 200–300 m. However, the span and the depth of water of the Gibraltar Strait are way over the numbers mentioned above. Thereby, the core issue of this plan lies in whether it is possible to go beyond the restrictions of bridge construction with the help of modern technology? Is it possible to build a 5000-m span and a pier that could survive 450 m deep water?

2. **Design plan of Gibraltar Strait Bridge**

Originally, 30 km Route B was taken (as shown in Fig. 6.36) for that the depth of water of 14 km Route A was more than 900 m. However, the fact that the water depth halfway of Route A was merely 450 m was overlooked at first. Therefore, the relatively shallow middle section could be utilized in the construction of a pier along with another two piers, respectively, built in 360 and 80 m deep water. Afterward, a dual-arched bridge with a 5000-m span could be employed to span over the distance, which greatly raised the feasibility of construction of this bridge.

The design plan of Gibraltar Strait Bridge proposed by T.Y. Lin has another two features:

(1) The plan of Route A consists of three piers altogether with a spacing of 5000 m, while that of Route B includes 11 piers with a spacing of 2000 m, in which case the possibility of collision between vessels and piers could be largely reduced. Meanwhile, Route A has two 5000-m-wide waterways that provide simultaneous to-and-from service for the ships, which has more safety reserve in comparison with ten 2000-m-wide waterways or seven 3000-m-wide waterways.

(2) Route A has a length of 14 km, which is much shorter than the 30 km of Route B and therefore considerably cuts the cost of construction, which includes the cost of construction of the bridge itself, that of the construction period, that of the maintenance and management after the completion of construction, and that of the operation and management of navigation.

3. **Design for the upper structure of this bridge**

The world's longest suspension bridge, the Akashi Kaikyo Bridge in Japan with a span of 1990 m, has been completed in 1991. Experts believe that with current material and technology, the span of a suspension bridge could be up to 3000 m in theory; with the help of even higher strength steel cables and towers, the span could be up to 5000 m. In the design for the upper structure of this bridge, if a 3000-m span of suspension bridge could be realized and afterward be integrated with cable-stayed frames so as to add 1000-m side span to each side, then a span of 5000 m could be completed as shown in Fig. 6.37. Such a design in the essence was to maintain the theoretically possible 3000-m span of suspension bridge and make cable-stayed supports and suspenders function as side suspension bridge.

When it comes to the design of this ultra-long bridge, optimization has to be conducted to the upper structure, in which process the influence of the natural environment of this strait and the action of wind load are to be considered; besides, the effects and the stability of aerodynamics are to be referred to in the design of the main structure. Connecting members are to be employed to enhance the connection between major components so that the integrity of the structure is assured. Meanwhile, in regard to vibration control, sufficient damping should be guaranteed for the structure.

In this stage of design, geometrical topology of the structure has to be determined first and then goes the ratio between the sag and the span along with the vertical elevations of all supporting components; at last, there is the sectional design of all components of structure.

1. **Design for the upper structure**

Geometric dimensions of the upper structure bears close relation with the stress state of the structure. And the key is to determine the ratio between the sag and the span as well as the vertical elevations of all supporting components. To determine the sag of the main cables, stress analysis of the main cables with various ratios between the sag and the span has to be conducted.

Assume that the ultimate tensile strength of high-strength steel wire within the suspension cable is 1900 MPa and the allowable stress of high-strength steel cable is taken as 45 % of its ultimate strength, 855 MPa. With the help of Eqs. (6.5) and (6.6), we could derive the span/sag ratio $S = L/y$, the deadweight of suspension cable q_{self} (t/m), and the relation between the unit length effective load q_{eff} and the horizontal internal force H of the suspension cable (Table 6.1).

From Table 6.1, we could see that with the same span/sag ratio, the internal force of suspension cable H is proportional to the span L. The span/sag ratio S has a significant impact upon on the internal force of suspension cable H. Specifically, H decreases with the reduction of S. As shown in the table, when the span is 3000 m

Table 6.1 Main cable stress comparison of different span–sag ratios

No.	L (m)	Sag of cable y_o (m)	$S = L/y_o$	Force in cable H (kN)	Self weight of cable q_{self} (kN/m)	Effective loads in cable q_{eff} (kN/m)	Effective coefficient of cable a	a_i/a_1
1	3000	375	8	30,000	2738	26.5	2.65	1.0
2	5000	625	8	50,000	4565	11.93	1.193	0.45
3	5000	909	5.5	34,370	3138	21.86	2.186	0.825

long and S is 8, the internal force of suspension cable is 30,000 kN. And when the span is 5000 m long and S is modified to 5.5, the increment of the internal force is somehow limited, being 34,370 kN, and is almost at the same level with the internal force of suspension cable when the span is 3000 m. And this makes it possible to control the internal force of suspension cable through the reduction of S when the bridge has a large span. In the design for this bridge, high-strength steel cables that could span over a distance of 3000 m when S is 8 were employed to span over a distance of 5000 m with a reduced S of 5.5 and the sag of this suspension bridge y_o was 909 m.

2. **Comparison between the designs for the upper structure**

Multiple plans are available for the upper structure of this bridge. Plan a and plan b as shown in Fig. 6.38 are conventional suspension cable structures. They have relatively small span and are quite suitable for shallow water cases. Plan c and plan d as shown in Fig. 6.38 have comparatively large span, and they are applicable to deepwater cases. And plan d was selected through comparative analysis [46, 47]. The analysis is briefed as follows.

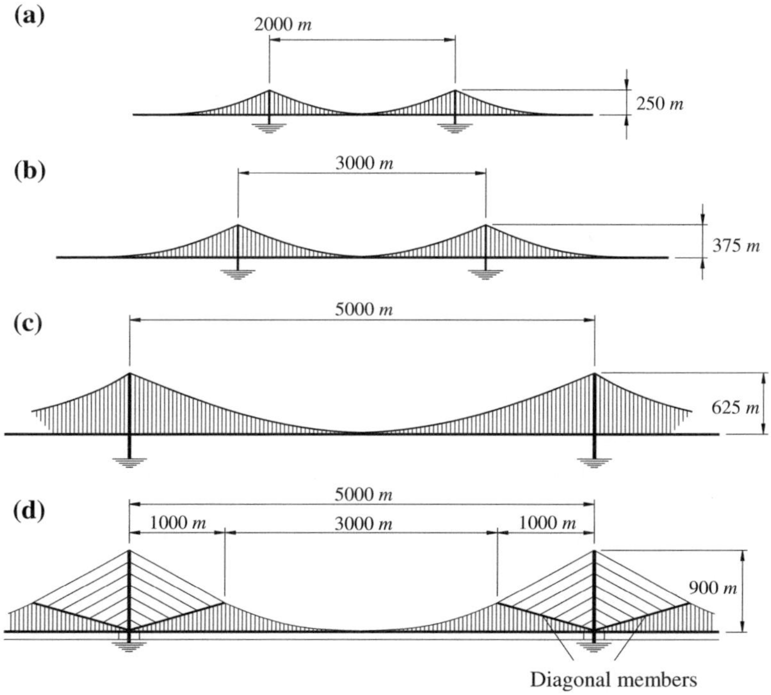

Fig. 6.38 Comparison between the designs for the upper structure of bridge [45]

Table 6.2 Key dimensions of the major components in this design for the bridge

No.	Members	Size (m)
1	The total span of bridge	5000
2	The suspension span of bridge	3000
3	The sag in mid-span	909
4	The height of tower	909
5	The length of inclined tension members	1139

From Fig. 6.38d, when S is taken as 5.5, the sags of the starting and ending point y_C and y_D for the mid-span 3000 m are as follows:

$$y_C = y_D = 3000/5.5 = 545 \text{ m}$$

Coincide the midpoint of suspension cable with the deck, and then, the height of the left and the right tower are as follows:

$$H = y_o = 5000/5.5 = 909 \text{ m}$$

The length of the inclined tension member is $[1000^2 + 545^2]^{1/2} = 1139$ m.

Thereby, the overall dimensions of the major components of the upper structure as shown in Table 6.2.

The overall span of the bridge; the mid-span of the suspension bridge that is hoisted by inclined supporting arm members; the mid-span sag of the suspension cable; the height of the tower; the length of the inclined tension members; the distance between the inclined supporting arm member and the deck; and the lateral distance between the inclined supporting arm member and the pier.

3. **Design for the pier structure**

Design for the lower pier structure is an integral part of that for the whole structure, and we will confine the discussion to the conceptual design alone. Since it is design for the pier base that is settled in 450 m deep water, numerous factors are to be considered, especially the high frequency of shipping across Gibraltar Strait, in which case there is high possibility for collision between the piers and the vessels and serious consideration is called for. Other plans for construction are as follows:

(1) Multi-column pier
Spatial truss tower that is comprised of a number of columns is to be constructed under the water. Nowadays, when it comes to the construction of drilling platform of offshore engineering, there have been actual examples of construction of multi-column pier in the depth of more than 400 m. The advantage of pier of this kind lies in its relatively high stiffness, in which case the deformation at the top of the deck shall be small if collision with vessels does occur.

(2) The construction of pier within the cofferdam
A cofferdam is to be built first, and then, pier is to be constructed within the cofferdam. This approach is applicable in shallow waters. Whereas in deep ocean areas like this, the construction of cofferdam calls for enormous amount

of work and is faced with great challenges from the surrounding environment. Thus, it is a risky and costly choice which shall not be taken.

(3) Prefabricated concrete caisson pier
Concrete caisson is prefabricated on earth and then shipped to the position of the pier. This approach has accumulated plenty of practical experiences, and when it is in deep waters, properly sized prefabricated concrete caissons could be piled up and be less affected by the restriction of the depth of water.

Specific conditions of this bridge include the following: a large span, deep water, and complicated environment for construction. These factors are to be considered, and the method of prefabricated concrete caisson is suggested for the construction of solid pier for it has matured experiences to refer to.

4. **Discussion**

(1) What did readers learn from this example? After going through detailed analysis, we know that a bridge with a 5000-m span could be constructed in the depth of 450 m with materials available (1900 MPa high-strength steel wire and C100 high-strength concrete). Will readers be able to summarize the innovative ideas of the designer as well as the technology involved?

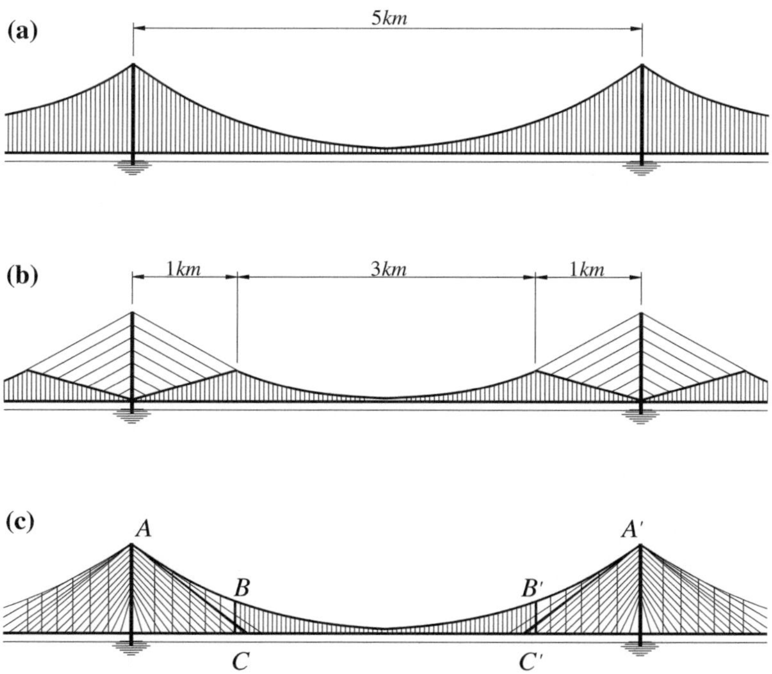

Fig. 6.39 Structural designs for Gibraltar bridge. **a** Traditional suspension bridge. **b** Cable-stayed suspension bridge. **c** Cable-stayed suspension bridge

(2) When it comes to the design of this bridge, practitioners considered conventional suspension cable structure, cable-stayed suspension cable structure, and cable-stayed suspension bridge as shown in Fig. 6.39. Could you point out the advantages and disadvantages of these three designs? In particular, as for the two plans as shown in 6.39b, c, what are the commonalities and fundamental difference between them?
(3) How to conduct the analysis and calculation of the internal force of the suspension cable, that of the main tower, that of the inclined tension members, and that of the inclined tension cables of the cable-stayed suspension cable structure as shown in Fig. 6.39b?
(4) How to conduct the analysis and calculation of the internal force of the suspension cable, that of the main tower, and that of the inclined tension cables of cable-stayed suspension bridge as shown in Fig. 6.39c?

6.6 Conclusions

In this chapter, we explored horizontal large-span structures, the structural form of which has drawn the attention of engineers ever since high-rise structures. In regard to the features of horizontal large-span structures, three large-span structural systems were introduced in this chapter: the arch structural system, suspension cable structural system, and cable-stayed structural system. Simplified design and calculation of these systems were discussed, which could be well applied to the early stage of structural design.

This chapter provided and emphasized the discussion and the analysis of practical examples of construction, which include Shanghai Lupu Bridge, Guangxi Nanning dual-saddled inclined bridge, Jiangyin Yangtze River suspension cable bridge, and Gibraltar Strait Bridge. They are all large-span structures that have been known for the challenges they posed. Through the discussion of these classic examples of construction, readers could have a deeper understanding of the application of mechanical concepts in various types of horizontal large-span structures. Meanwhile, readers could widen the horizon of their engineering background, which does not require only the knowledge of design and calculation but also relevant technology of construction.

With the example of Shanghai Lupu Bridge, the structural system and method of construction of half-through arch bridge with 550-m span were introduced. The design and construction of this bridge integrated that of three different types of bridges: cable-stayed bridge, the arch bridge, and suspension cable bridge. This structural system was especially engineered to adapt to the soft soil foundation of Shanghai. And this project leaves room for the readers to ponder over, specifically which the aspects of technology do you think that engineers shall pay attention to in their efforts to overcome the challenge posed by soft soil foundation of Shanghai through the construction of "No. 1 arch in the world" so that the city could provide a better view?

Guangxi Nanning Bridge is a dual-saddled inclined bridge that was comprised of two asymmetrically inclined arches, within the two planes of which the tension members were installed. From the plane layout of the bridge, we could see that the deck is also displayed as a curve within the plane. Designers ingeniously took advantage of three arches that functioned in three different planes to support the deadweight of the bridge and the horizontal force from the deck. In this case, the functions of the arch structural system have been fully utilized. Moreover, the bridge has a graceful figure and blends in perfectly with the surrounding environment, which does not only embody the union of architectural aesthetics and mechanics but also represent the integration and mutual influence of conceptual thinking from structural engineers and architects.

At the completion of construction, Jiangyin Yangtze River Suspension Bridge had the largest span in China. The feature of this case was that the geological conditions of the construction site were way below the standard and the anchorage of suspenders was an outstanding issue. The northern anchor was located in an alluvial plain, and therefore, the approach of open caisson was adopted. As for the southern anchor, the method of gravitational rock-embedding anchorage was employed. Since the anchor was oversized and had to go through four layers of soil, difficulties were posed to both design and construction. Besides, due to the ultra-large span of the structure, the installation of the main cables was faced with unprecedented difficulties. Apart from wind tunnel test, air-opening-installed oblate box-sectioned steel beam was adopted as the primary beam of the deck, which clearly indicates that designers had sufficiently considered the resistance to wind vibration of this large-span suspension bridge. And such thinking is quite noteworthy.

A grand conception to connect continental Europe and continental Africa was embodied in the design of Gibraltar Strait Bridge. With available material and technology, the designer overcame the ultra-long limitation of 5000 m through an innovation of the type of structure. Although the plan was not carried out due to non-technical reasons, readers could learn the spirit of daring innovation. The designer disintegrated issues encountered and solved them, respectively, in the selection of the type of structure (as in how to turn the calculation of 5000-m span to that of 3000-m span and in the adoption of cable-stayed suspension cable bridge) and in the selection of the pier base (to select among the choices of multi-column pier, pier within the cofferdam, and prefabricated concrete open caisson pier). From this perspective, it is not only a case of bridge design, but also an example of how systematic engineering is applied. It revealed how the designer managed to turn an engineering issue so complicated that there was barely any solution into a broken-down operable design issue of structural engineering.

Chapter 7
Comparative Design and Analyses of Spatial Structures

7.1 Characteristics of Spatial Structures

Horizontal large-span structural system regards structure as a serial of independent planar components. The insufficiency of out-of-plane stiffness, strength, and stability put limits on the application of horizontal large-span structures. However, when the restraining action of spatial components is taken into consideration and mutual support and enhancement could be well played, the bearing capacity or the distance spanned over could be greatly increased. Spatial structural system could be regarded as an extension of horizontal large-span structural system with consideration of spatial effects. On the one hand, it is necessary for the spatial structural system to control vertical strength and stiffness of the structure. On the other hand, since the height of spatial structures has reached certain level, horizontal strength and stiffness have to be regulated, too. As indicated in Sect. 3.1, spatial structural system could be classified in accordance with the form of composition and mechanical characteristics as follows:

1. Spatial truss structure or spatial grid structure

Spatial truss structure or spatial grid structure is developed from planar truss structure. By adding out-of-plane connections to a number of planar trusses, the spatial truss structure is formed and a planar force transfer is changed into a spatial force transfer. When there is no primary or secondary truss in regard of spatial force transference, spatial truss structure of such kind is also named spatial grid structure. With the integration of dual-direction or multi-direction planar truss (parallel-stringed truss and arc-/arch-shaped truss included), spatial force transfer could be well conducted and members support each other in both out-of-the-plane and within-the-plane fashion. Therefore, the overall strength, stiffness, and stability of such structures are much improved.

S. Lin and Z. Huang, *Comparative Design of Structures*,
DOI 10.1007/978-3-662-48044-1_7

2. Spatial latticed shell structure

Latticed shell structure is composed of curved boards or bar members, and its curved surface could be single-curved surface or double-curved surface. With the constraint of rigid framework (or support), various spatial structural forms with distinct features of configuration could be realized to span over a large distance. To counterbalance the bending moment within the span, spanning structures evolve from planar one-way stressed beam to one-way curved arch and then to spatial latticed shell. Along the evolution, the materials are utilized with more and more efficiency. Shell structures are formed by rotating or horizontally moving the axis of the arch with reference to a certain law. With single- or double-curved arch, the action of a two-way arch could take action. Similar to the action of the arch, latticed shell structures are compressed in all directions under the action of the bending moment, and thus, the materials are used more efficiently. Latticed shell structure is up for large-scale industrial production, and it has the advantages of being befitting on-site ground assembly, integral mechanized hoisting, etc.

Latticed shell structure is widely applied. Its composition requires a thin and curved surface along with relatively rigid side support or constraint. On the basis of meeting these two prerequisites, the architectural configurations of thin shell structures are quite varied.

Suspension cable structures could also be made into spatial latticed shell structures by rotating or moving the suspension line horizontally, and these new derivatives are named cable net structures. The membranes or steel cables are tensioned in all directions and are able to conduct spatial force transference.

3. Membrane structures

Membrane structures are made by stretching tight the membranes over planar or distorted brackets or frames. They have smooth curved surface and minimum surface area under the given boundary conditions. When steel cables are hung down, they could bear the loads through the axial force. Similarly, membranes, due to their capacity of two-way tensioning, could undertake normal loads that are perpendicular to their surfaces the way crossed suspension cables do. Therefore, membrane structure is equipped with the action of two-way suspension cable, thanks to its two-way curvature. Although membranes are not able to resist compression and have potential instability when they are slack, a mature method of applying prestress to the membrane to stabilize the structural formation enables membrane structures to undertake local perpendicular loads. Thanks to the lightweight of membrane structures, and easy and fast construction, the application could also be extended to various kinds of temporary structures.

4. Other spatial structural systems

With constant development of construction materials and technologies, numerous novel spatial structures spring up in recent years, such as beam string structures and spatial deployable structures. Beam string structure is a special form of parallel truss structure. It is to replace the lower chord of parallel truss structure with prestressed steel cable, which further lessens the deadweight of the structure, and to apply prestress simultaneously, so the structure takes its formation as a beam string structure. The terminal of Shanghai Pudong Airport (Shanghai, China) adopted one-way beam string structure and expressed the architectural aesthetics in a distinctive way. Beijing National Stadium (Beijing, China) employed two-way beam string structure, so the structure could span over a greater distance in both directions and better meet the requirement of spatial stability.

Classified by the way of deployment, spatial deployable structural system could be divided as spatial retractable structure and spatial deployable/foldable structure. The former is usually applied to large-span roof structures and could open and close the roof in a push–pull or rotary fashion. As for the latter, it has been widely applied for that such structure could fold or unfold like an umbrella and has a significantly small folded volume which facilitates the transportation. The design and construction of spatial deployable structure need the cooperation from all professions such as architects, civil engineers, mechanical engineers, and electronic engineering (automated control).

7.2 Design for Spatial Truss and Grid Structural System

Spatial truss or grid structure is an ideal form of structure for large-span and mid-sized-span structures. Planar trusses from varied directions are integrated and provide support for each other, in which case components could well play the role of spatial force transfer. The structures have relatively large overall stiffness and high stability. Thus, structure of such kind has frequently applied in large-scale exhibition halls or stadiums.

Spatial truss or grid structure is an improvement on planar truss system, and it is by enhancing lateral connections that the structure is able to transfer forces over the space. Grid structure usually refers to the spatial truss structure that has no distinction of primary/secondary truss and is narrowly spaced. The spatial truss or grid structure as shown in Fig. 7.1 is comprised of two orthogonally inclined trusses that share web members; as for the upper and lower chords in this structure, they are placed in an alternating fashion in two directions. Spatial truss or grid structure is of multiple static indeterminates, and related precise calculation has to be done with computer software. Nonetheless, in the stage of conceptual design, spatial truss could in practice be regarded as a two-way slab structure, the calculation of which could be carried out in accordance with simplified calculation of slab system introduced in Sect. 4.4. Calculation formula (4.1)–(4.8) of slab system could be employed in the conceptual design of spatial grid structure. As for slab structures, checking calculation have to be conducted to the shear resistance of slabs at the supports of columns. Similarly, spatial truss or grid structure has to conduct checking calculation of the shear resistance at the supports of columns and when necessary installs column caps or enlarges the sectional dimensions of the web member at the supports of spatial truss or grid structure so as to counterbalance the

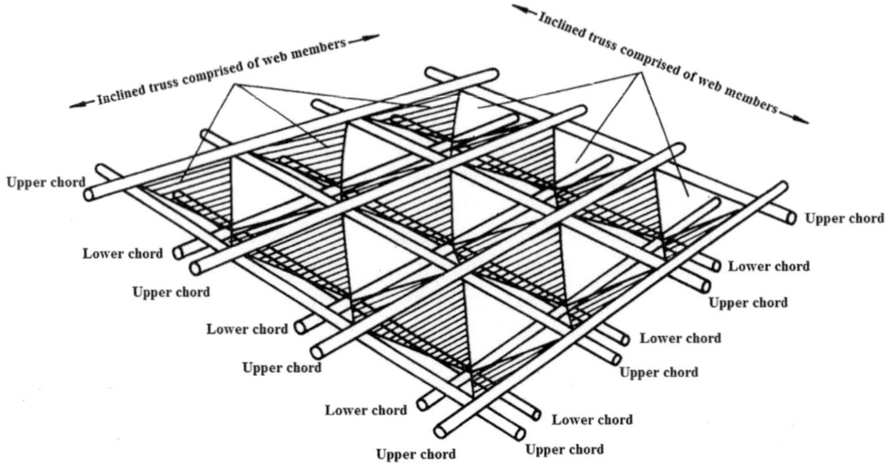

Fig. 7.1 Function of spatial truss or grid structure

enormous shear force in those positions. Figure 7.2a illustrates the situation when the column cap is installed within the planar boundary of spatial truss or grid structure and local shear resistance of the structure is enhanced by the enlargement of sectional dimensions of the web members. Figure 7.2b is the situation when the column cap is installed on the plane created by all upper chords and the shear resistance is provided by locally heightened spatial truss or grid structure. Figure 7.2c illustrates the situation when the column cap is installed between the plane created by the lower chords and the top of the column and local shear resistance is enhanced in this way.

Let us take two-way orthogonal truss system as an example to explain mechanical characteristics and simplified calculation of this system. As shown in Fig. 7.3, spatial truss could be simplified as corresponding intersectant planar truss and then further simplified as independent planar truss, in which case the calculation of deflection, bending moment, and shear force could be conducted so as to

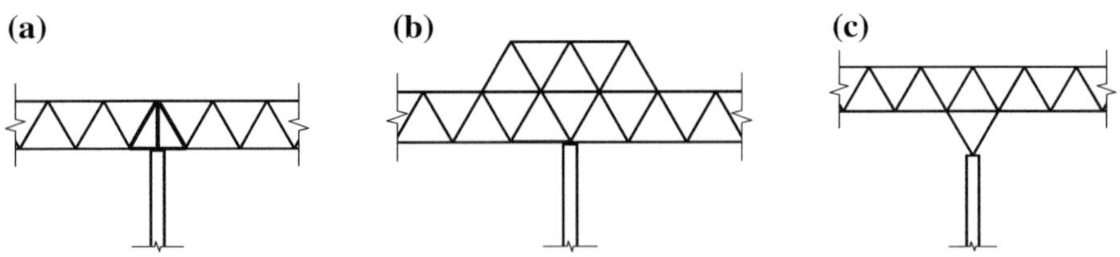

Fig. 7.2 Three ways to install column caps to spatial truss or grid structure. **a** Within the plane. **b** Upon the plane. **c** Under the plane

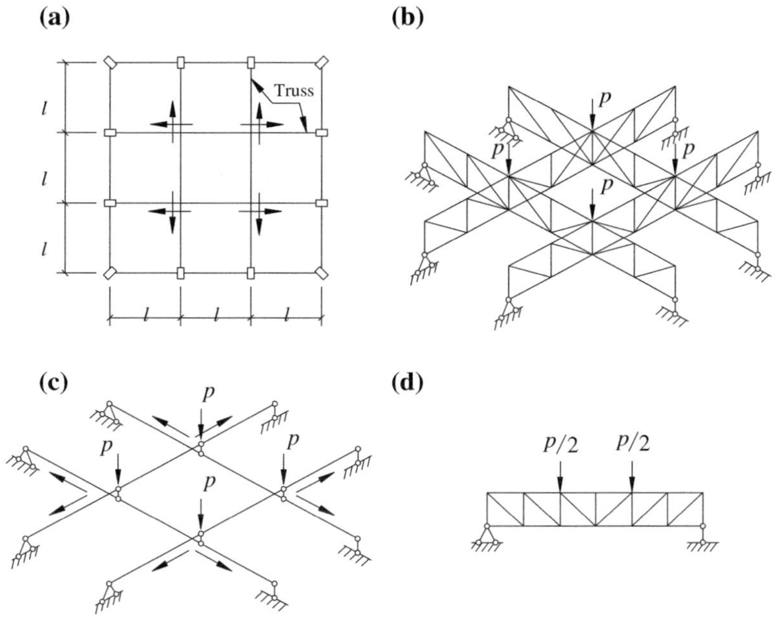

Fig. 7.3 Stress analysis and simplified calculation of spatial truss. **a** Layout. **b** Loading. **c** Force flow. **d** Truss analysis

determine the internal force of all members of the truss. Such simplified calculation assumes that (1) the displacement of the intersectant planar truss at the intersection is identical and (2) horizontal displacement is ignored. When it comes to the calculation of the internal force of all members of the truss, assumption is made that the bending moment is undertaken by the upper and lower chords together and the web members take only the shear force.

In accordance with distance spanned over and stress state of the structure, spatial truss or grid structure could be distinguished as double-layered or three-layered. Generally speaking, the more layers the spatial truss or grid structure has, the greater the force-bearing capacity is. It goes along with a longer spanned-over distance and an increment of construction difficulties and the steel consumption. Thereby, double-layered spatial truss or grid structure is a frequently applied form of structure, due to its efficiency. Conventional types as shown in Fig. 7.4 include intersectant planar truss, cone-shaped grid structure, pyramid-shaped grid structure, and hexagonal-pyramid-shaped grid structure. With constant development of modern design, calculation, and construction technology, spatial truss has more flexible forms. An example would be the fish-bellied spatial truss beam of Shanghai New International Expo Center that will be introduced in Sect. 7.6 of this book.

As shown in Fig. 7.4, spatial truss or grid structure usually takes geometric invariants as the smallest structural units including cone, pyramid, and hexagonal

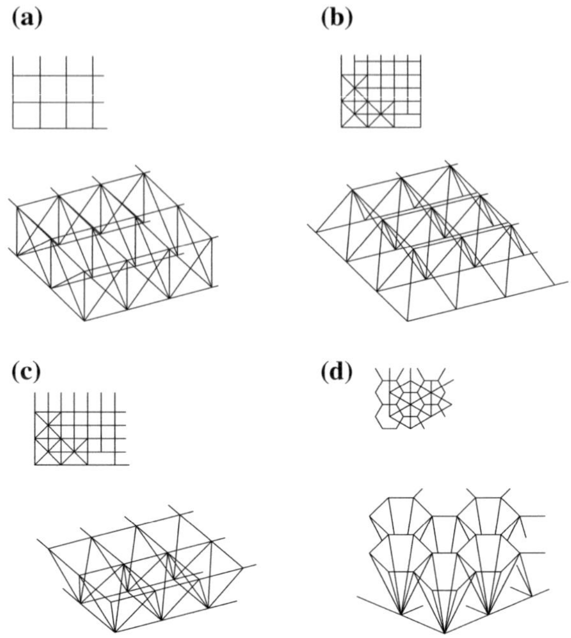

Fig. 7.4 Classification of spatial truss or grid structures. **a** Grid truss. **b** Triangle pyramid. **c** Rectangle pyramid. **d** Hexagonal pyramid

pyramid. As illustrated in Sect. 2.4, these geometric invariants serve as the basis for geometric stability of the entire structure. Due to the large number of members and joints, and the complexity of the stress state and supports, geometric stability of the entire structure has to be analyzed in the early stage of conceptual design. The geometric invariance of the structure under the action of all potential external loads needs to be checked. The analysis of geometric invariance of spatial truss or grid structure has to meet the condition as follows [40]:

$$N = B + S - 3J \geq 0 \tag{7.1}$$

In this formula above, all members are assumed to be hinged and each joint has three degrees of freedom. And B is the total number of members; S the degrees of freedom of the constraints at the support; J the total number of joints; and N the number of indeterminates. When N is less than zero, the structure is geometrically variable; when N is equal to zero, the structure is statically determinate; and when N is greater than zero, the structure is statically indeterminate. Apart from the conduction of geometric analysis with the help of the formula above, the analysis of optimized arrangement has to be conducted with practical construction experiences [41, 42]. Besides, the matrix of overall stiffness of the structure has to be checked so as to guarantee that the spatial truss or grid structure as a whole meets the requirements of strength, stiffness, and stability.

Since the internal force varies to different parts of the spatial truss or grid structure, equally spaced differentially sectioned or differentially spaced equally sectioned grid structure could be employed to reduce the material consumption along with the construction cost. The selection of spatial truss or grid structure has to consider not only the requirements from architecture and structure but also the connecting joints and the types of support.

7.3 Design for Latticed Shell Structural System

As a bionic structure, the earliest inspiration of shell structures came from sea shells, egg shells, etc. General features of shell structures include a thin curved surface and relatively rigid supports.

Hyperbolic arch shell structure comes from rotating the planar axis of the arch to the central line, moving the planar axis of the arch horizontally along or rotating it to a straight line or a curved line and the corresponding curved surfaces are called surface of revolution, surface of translation and ruled surface (as shown in Fig. 7.5). Thereby, hyperbolic arch shell structure largely undertakes compression-featured action of the arch. When the shape of the arch ideally coincides with the intrinsic pressure line, the force flow could be transferred from the top of the arch via the radial rib down to the support, leaving only compression and no bending moment within the shell of the arch. However, similar to a stressed planar arch structure, it is quite difficult to guarantee that the shape of hyperbolic arch shell structure is always kept coincided with the intrinsic pressure line, and therefore, local buckling shall appear within the shell of hyperbolic arch. To avoid local buckling instability, hoops shall be installed to resist the bending moment. As long as there exists horizontal hooping action, hyperbolic arch shell could undertake only tension or compression and maintain the action of stressed membrane. As a result, the thickness of hyperbolic arch shell generally is smaller than that of planar arch structure. As for the majority of hyperbolic arch shells, hooping action is of significance. First of all, the restriction of hoops keeps hyperbolic arch shell non-deployable. As long as there is sufficient restriction of hoops, the shape of

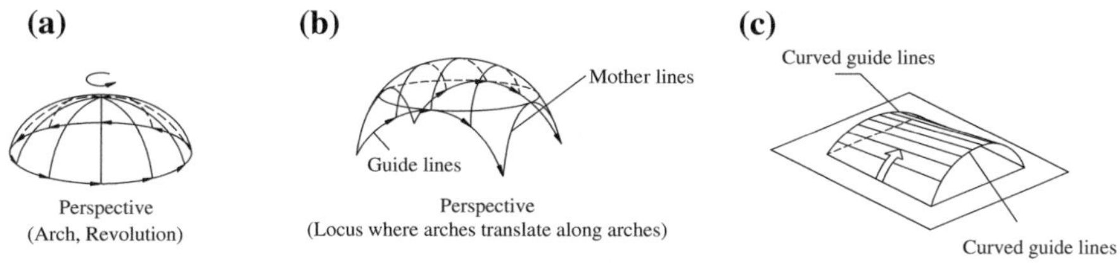

Fig. 7.5 Formation of hyperbolic arch shell. **a** Surface of revolution. **b** Surface of translation. **c** Ruled surface

hyperbolic arch shell could manage to stay away from the intrinsic pressure line and guarantee that there is no buckling instability occurring to the curved surface. When the intrinsic pressure line is above the surface of the arch shell, hoops act to undertake compression; when the intrinsic pressure line is below the surface of the arch shell, hoops act to counterbalance tension (as shown in Fig. 7.6).

Edged ring beam plays an important role in the stabilization of hyperbolic arch shell for that hyperbolic arch shell is spatially non-deployable. To unfold a hyperbolic arch shell, edged ring beam has to be destroyed and the curved surface torn apart. As long as the edged ring beam provides sufficient restriction, hyperbolic arch shell could steadily take on the loads. The calculation of the internal force of edged ring beam could be conducted with the following simplified calculation method of the equilibrium of isolated body. As for the simplified calculation of other spatial structures, similar equilibrium of the free body could be adopted.

Figure 7.7 displays the sectional drawing and plane layout of hyperbolic arch shell. The diameter of its circular bottom is D, and the height is h. The structure undertakes vertical uniform load with the resultant force being W. Calculate tension T within the edged ring beam at the bottom.

Assume that there is no bending moment and only membrane action within hyperbolic arch shell and the resultant supporting force R is tangent to the curved surface of the arch shell. The supporting force R breaks down to vertical supporting

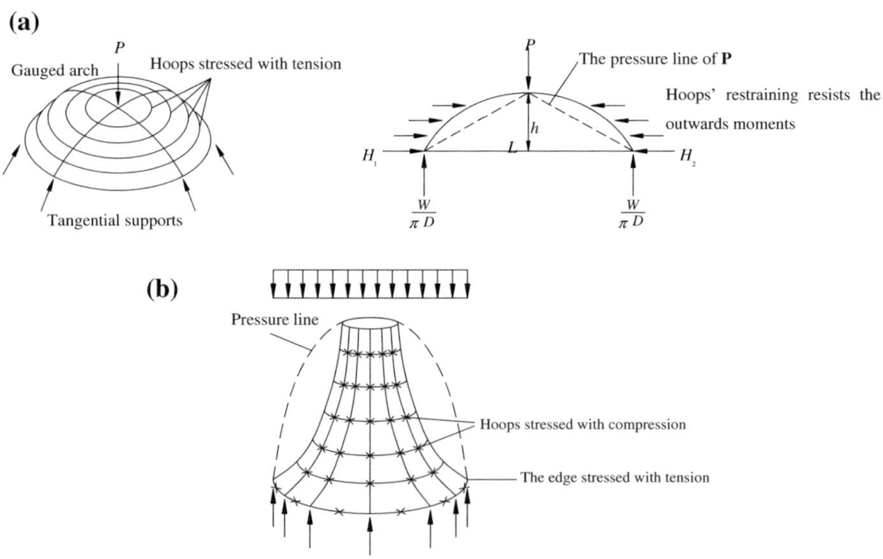

Fig. 7.6 Hooping actions. **a** Hoops resist tension when the intrinsic pressure line is below the surface of the arch shell. **b** Hoops resist compression when the intrinsic pressure line is upon the surface of the arch shell

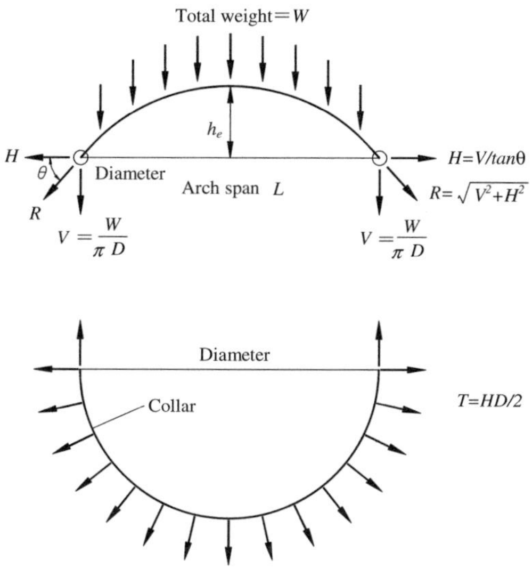

Fig. 7.7 Simplified calculation of edged ring beam of the shell

force V and horizontal thrust H. From the equilibrium of forces, we could derive the following:

$$V = \frac{W}{\pi D} \tag{7.2}$$

$$H = \frac{W}{\pi D \tan\theta} \tag{7.3}$$

$$R = \sqrt{H^2 + V^2} \tag{7.4}$$

$$T = \frac{HD}{2} = \frac{W}{2\pi \tan\theta} \tag{7.5}$$

From Eqs. (7.2) to (7.5), the greater the θ results, the smaller the horizontal thrust H and tension T. When θ is close to 90°, horizontal thrust H and tension T are close to zero.

Suspension cable structure could also be formed through spatially rotating or horizontally moving catenary, and the spatial latticed shell structure derived is simply called cable net structure. Shells in this case undertake tension from both directions and they come out in dish-shaped cable net and saddle-shaped cable net. Figure 7.8 illustrates frequently applied hyperbolic single-layered dish-shaped cable net structure and hyperbolic single-layered saddle-shaped cable net structure.

As shown in Fig. 7.9, spatial latticed shell structure could be classified as shell structure with positive Gaussian curvature and that with negative Gaussian curvature in accordance with the type of surface of the shell and the characteristic of the curvature. When the curvatures in both directions concave or convex simultaneously,

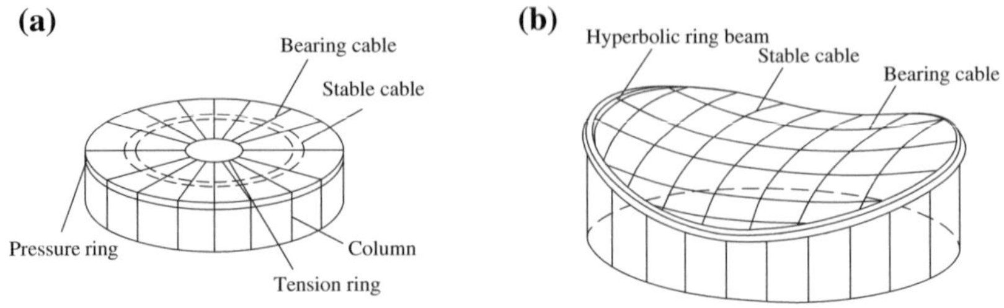

Fig. 7.8 Cable net structures. **a** Hyperbolic single-layered dish-shaped cable net structure. **b** Hyperbolic single-layered saddle-shaped cable net structure

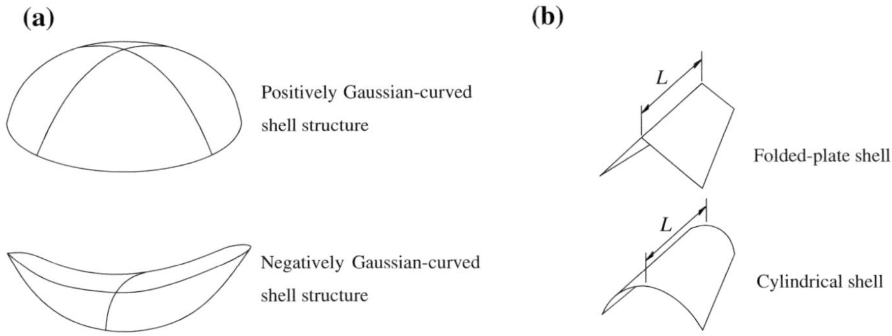

Fig. 7.9 Classification and characteristic of spatial latticed shell structure. **a** Double-curved shell structure (non-deployable). **b** Single-curvature cylindrical shell or zero-curvature folded plate (deployable)

then the structure is positively Gaussian-curved shell structure; otherwise, the structure is negatively Gaussian-curved shell structure. Positively or negatively Gaussian-curved shell structure could not be unfolded on a planar surface without being torn apart and therefore is designated as spatial non-deployable body. As for single-curvature cylindrical shell or zero-curvature folded plate, since they are deployable on a planar surface, they are generally named spatial deployable body.

7.4 Design for Membrane Structural System

The feature of membrane structure lies in its outstanding lighting performance. In the day time, natural light could be fully utilized; at night, when the internal lighting is on, the structure turns into a luminous body and serves as a luminescent resort at night. Compared with traditional daylighting through glass, membrane structure has better photo-permeability. Even in the case of a relatively small light angle, natural light could be sufficiently utilized by membrane structure so as to reduce the expenses on heating and lighting of structure. The glass fiber employed by membrane structure is produced by melting the glass into filaments with extremely small radius and knitting them together. Glass fiber cloth, after baking-coated with PTFE or PVC, is equipped with high strength and durability, and its maximum tensile strength could be up to 3400 MPa, which means it could undertake quite an enormous amount of tension with its small cross-sectional area. PTFE or PVC has stable chemical properties and sound uvio resistant performance and its normal service temperature ranges from −73 to +232 °C. A notable disadvantage of membrane structure lies in insufficient sound insulation, and therefore, unfavorable influences such as the internal echo and the interference from outdoor noises shall be avoided in the design of sound insulation for membrane structure.

According to the stress state and the type of load, membrane structural system could be classified bearing into three categories: air-supported structure, tensioned membrane structure, and dome structure. As for the air-supported structure, it is commonly applied in the construction of gymnasiums and stadiums. The making of its shape comes from the overpressure of the air (0.1 overpressure, supplied by air ventilator), and the structure itself functions as the support for the roof. An example would be the air-supported structure that was employed in American Pavilion in Osaka World Expo, 1970. And the tensioned membrane structure takes its form by integrating the tensile stress of the membrane itself, supporting members and tensile cables as one structural system, commonly known as tent structure. The terminal building of Denver International Airport in the USA is a typical project of this kind. Dome structure comes in the third and is made through the integration of cable network and membrane. And the classiest project of this kind goes to the Georgia Dome, Atlanta, as the roof structure of the main stadium in Atlanta Olympics, 1996, which is also the project that will be discussed in Sect. 7.6.

Air-supported structure rests the entire structure on the air that is filled into the interior of the membrane and takes advantage of wind-induced air lift to counterbalance some of the deadweight and external loads. Constant load accounts for an insignificantly small proportion in the design for air-supported structure, while the wind and dynamic action from the live loads are actually the major concern. Besides, since the membrane in a slack state has no actual stiffness, pretensioning has to be applied so that the membrane is equipped with certain amount of prestress and is able to undertake live loads that are perpendicular to the surface of the membrane without buckling. Another issue goes to the avoidance of vibration under the action of wind load. Practical projects of construction more often than not employ model structures and wind tunnel tests to determine dynamic performance of the membrane structure and to avoid dynamic issues like the vibration at the edge.

Tensioned membrane structure and cable net structure are actually keel-typed membrane structures. As for structures of this kind, the steel keels or steel cables support the entire structure, while the external membrane functions only as the closure of the structure rather than the major load-bearing structure. The design and calculation of steel keels or steel cables are the same as those of spatial latticed shell structure. Beijing National Aquatics Center called the "Water Cube" is a typical keel-typed membrane structure (as shown in Fig. 7.10). The exterior of the membrane employed ETFE (polyethylene-PTFE copolymer), and the air cushion of membrane structure is supported by the keel of double-layered plate-typed grid structure. Plate-typed grid structure is of its own kind, and it is constructed according to the "bubble theory" proposed by Kelvin. Specifically, it is to scale up the microscopic molecular of water to the proportion of a structure. Geometrically speaking, plate-typed grid structure is formed by the repetition of an integration of one tetrakaidecahedron with fourteen dodecahedrons. Spatial grid is completed after the evacuation of air from the polyhedrons mentioned above, and plate-typed grid could then be cut out according to the requirements of the structure. From Fig. 7.10, we could see that the steel keel of the "Water Cube" is double-layered plate-typed grid structure. As for Fig. 7.11, it is a keel-typed membrane stadium and the steel keel is single-layered latticed shell structure (as shown in Fig. 7.11).

Fig. 7.10 Double-layered grid structure

Fig. 7.11 Single-layered latticed shell structure

The design of membrane structure shall be conducted from an overall perspective of the structure and carries out an overall analysis of the stress state of the structure. Most membrane structures are hyperbolic paraboloids, and therefore, stress analysis in two radial directions has to be conducted. Specific procedures of analysis include spatial modeling, shape determination, stress analysis, and construction drawing. These aforementioned steps are usually completed with the help of computer.

7.5 Design for Spatial Deployable Structural System

Classified by the way it unfolds/folds, spatial deployable structural system could be divided as spatial retractable structure and spatial deployable/foldable structure. The former is usually applied to large-span roof structure and opens or folds the roof according to the operational requirements, making sure of a normal function of the structure with an open or a closed roof. As for the latter, it has been widely applied to not only the structures but also the articles for daily use such as umbrellas, tents, mosquito curtains and even aircrafts and pharmaceutical supplies (as shown in Fig. 7.12). We will focus our discussion on spatial retractable structure that has been frequently applied to large-span roof structures.

Spatial retractable structure in the early stage mainly took the form of fold-typed membrane structure, the majority of which was medium-sized or small-sized. The way it opened had intrinsic defaults, and therefore, the malfunction of fold/unfold operation or a torn membrane was commonplace under the action of wind and rain. Modern spatial retractable structure employs rigid structures such as arch-shaped steel structure, arch shells, spherical shells, spatial truss, or grid structure as the supports; as for the roof, lightweight high-strength materials such as membrane and metal plate are employed; modern mechanical control technology has been integrated in the making of a safer and more convenient fold/unfold operation of the

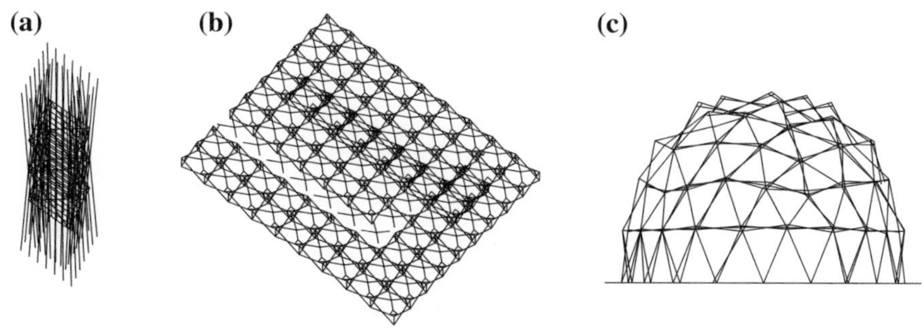

Fig. 7.12 Spatial deployable/foldable structures. **a** Folded state. **b** Unfolded state (*flat*). **c** Unfolded state (*curve*)

structure. Till now, dozens of large-scale spatial retractable structures have been constructed around the globe.

The way that the roof opens and folds is of significance in the design for spatial retractable structure, and it has direct impact on the functions, the form, and the mechanical control system of the structure. Retractable roof system usually consists of a number of monolithic components, through the movement, rotation, overlapping, and stacking of which the opening and closing of the roof are realized. In line with different modes of movement of these opening/closing units of roof, spatial retractable structure could be classified as parallel-shifting style, pivoting style, unfold-opening style, and composite-opening style. As for parallel-shifting style, it could be subdivided into horizontal-shifting type and spatial-moving type. And the pivoting style could be subdivided into vertical pivoting type and horizontal pivoting type. Figure 7.13 illustrates typical ways of opening and closing for spatial retractable structures. Figure 7.13a show parallel-shifting style, and one representative structure would be the Ocean Dome in Japan, covering a building area of over 30,000 m^2. Figure 7.13b is vertical pivoting style, and a representative is Pittsburg Civil Arena in the USA, the roof of which covers an area of 12,000 m^2. Figure 7.13c displays unfold pivoting, and a representative goes to Sendai Dome in Japan, the building area of which is almost up to 20,000 m^2. Figure 7.13d displays a composite-opening style.

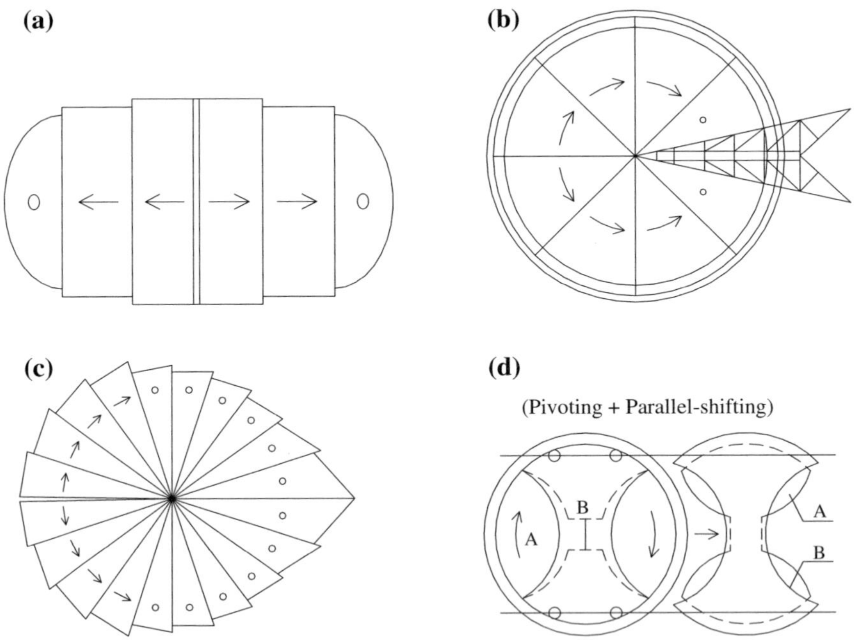

Fig. 7.13 Different opening and closing for spatial retractable structures [46]. **a** Parallel-shifting style. **b** Pivoting style. **c** Unfold-opening style. **d** Hybrid pivoting and parallel-shifting style

Specifically, the two pieces of A would be vertically pivoted toward the position of B and partially opened, and then, this partially opened B would be horizontally moved as a whole. A representative is the Ball Dome in Japan, covering a building area of 1000 m^2 or so.

The design of spatial retractable structure has to take into consideration the factors such as the load transfer in opened, semi-opened, or closed state, and the dynamic effects induced by the movement of the roof [47]. On the other hand, to assure a normal function of mechanical control, temporary and long-term deformation of the roof in the aforementioned three states has to be strictly controlled. Since the roof system is comprised of a number of monolithic components, the issue of waterproofing when the roof is closed calls for special attention when retractable structure is employed.

7.6 Comparative Analyses of Sophisticated Spatial Structures

7.6.1 Shanghai New International Expo Center

1. **Project Background [48]**

Shanghai New International Expo Center is located in Pudong New District (as shown in Fig. 7.14). Each standard exhibition hall has a steel-roof-projected area dimension of 192 m × 80 m, and the internal space after the closure of the external curtain wall is up to 164 m × 70 m × 11 m. The steel roof installed in each standard exhibition hall is typical large-span spatial truss structure, as shown in Fig. 7.15. Many an innovation was developed through the process of design and construction of steel-structured Shanghai New International Expo Center, and we will focus our discussion on the conceptions and controversies of this design.

Fig. 7.14 Shanghai New International Expo Center [48]

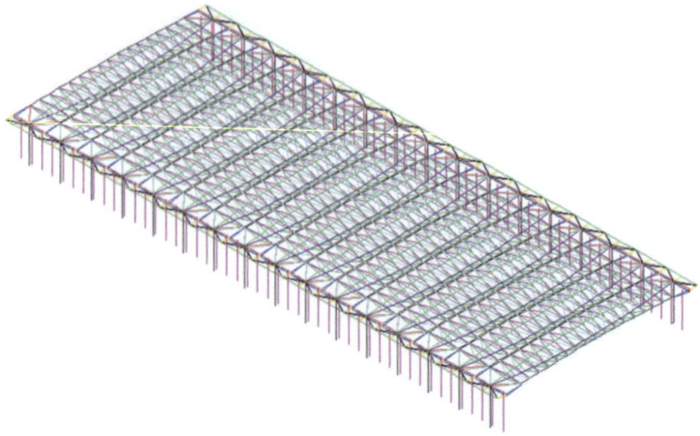

Fig. 7.15 The steel roof system of the exhibition hall [48]

2. **The concepts of design**

The upper main structure of the exhibition hall of Shanghai New International Expo Center is a steel spatial truss system, in which system the roof loads are largely transferred from the roof truss to the side column and then to the foundation. Horizontal loads undertaken by the external maintenance system along with horizontal seismic loads are transferred by large-lateral-stiffness steel roof system to the grouped tension columns of the side walls and to the wind-resistant columns of the end walls. Grouped tension column system is comprised of steel circular tube columns and prestressed tension members (as shown in Fig. 7.16) and functions as shear walls. Since large-span spatial truss system has relatively small vertical stiffness, "flexible" structural system is adopted in the conception of such structures, which allows a relatively large range of elastic deformation for the roof system in its normal working condition. Mechanical characteristics of such structure are as follows:

(1) Spatial truss employed specialty cast steel frame joints that are made from GS-20Mn5 V tempered heat-treated steel. They are fabricated with casting techniques and guarantee a smooth transfer of the bending moment at the joints.

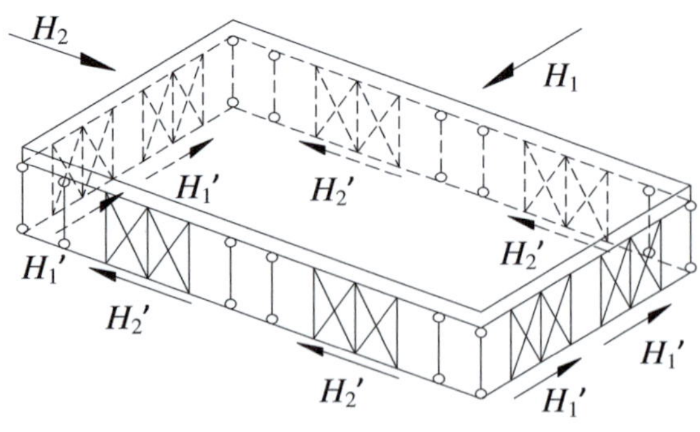

Fig. 7.16 The force analysis of the structural system

(2) "Flexibility" was adopted in the conception of structure. The roof system of standard exhibition hall is comprised of 16 fish-bellied diamond-sectioned truss beams (as shown in Fig. 7.17). Fifteen diamond-shaped frames are welded together with the upper-, the lower-, and the middle-positioned chords on the two sides so as to produce one fish-bellied curved truss beam. Two truss beams are bell-and-spigot-jointed via the fifteen sets of connecting plates installed with the middle-positioned chords and then fastened with high-strength bolts. Meanwhile, middle-positioned chords on the two sides of the roof truss are connected with adjacent truss beams via horizontal tension members and are applied with prestress. The roof truss and columns are hinged together, and so it is between the column and its footing with anti-shear pins installed within. The steel tube-shaped column is about 14 m high and has a 325 mm × 14–16 mm section with a large slenderness ratio. Column footings and reinforced concrete foundation are connected via previously embedded high-strength foundation bolts.

(3) Single truss beam of the roof has a large span, an enormous amount of weight, and a high level of difficulty in construction. A single roof truss provides support for a span of 72 m with 4 m overhang on both ends, and the maximum dead-weight of a single roof truss has exceeded 1000 kN. The actual construction employed the full-shoring temporary supporting system. Certain degree of weight-induced settlement deformation is allowed for the roof truss after an orderly removal of temporary supports at the completion of construction.

3. **Summary**

Figure 7.16 is a simplified structural system and load transfer analysis of the New International Expo Center. The roof could be simplified as a plate with certain stiffness; prestressed grouped tension columns could be regarded as equivalent shear walls; as for other external columns, since the top and the foot of the column are hinged with other structural members, they could be taken as swinging columns. Figure 7.16 illustrates the path of force transference of this structural system: Vertical loads are transferred from the spatial truss of the roof to the swinging columns and grouped tension columns on the four sides and then onto the foundation. Horizontal wind and seismic loads are transferred to the foundation through the equivalent shear walls on the four sides.

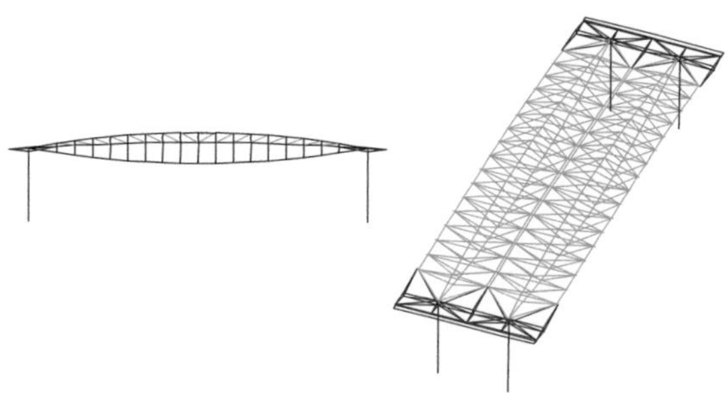

Fig. 7.17 A fish-bellied truss beam

The conceptual design for the structural system of New International Expo Center was highly original and sparked quite a debate over the schematic design. Especially, since the spatial truss has a large span within which a prestressed system is installed, considerable difficulties were posed to the construction. Prestressed grouped tension columns could only be applied with prestress after a complete installation of the entire structure, and the highly effective integral hoisting technique was not an option with the support of swinging columns. Instead, time-and-labor-consuming full-shoring support was adopted, which resulted in a considerable increase in the cost and time span of the construction.

It took several periods of construction for the engineers of New International Expo Center to overcome the aforementioned difficulties and make it a success. The experiences gained in the construction of this project were a profound comprehension of the conceptions and intensions of design, a reasonable arrangement of techniques and procedures, and a thorough process control of the construction.

4. **Discussion**

(1) Since a large number of swinging columns are contained in this large-span four-sided supporting system, which does not contribute to the lateral stiffness and the stability of the entire structure at all, how could several sets of grouped tension columns meet the requirement of lateral stiffness and the stability of the structure?
(2) The conception of swinging columns requires that the spatial truss system of the roof has a relatively large integral stiffness, which inevitably increases the amount of steel employed and the cost of construction of the entire structure. Casting this conception of swinging columns aside, could readers propose a better vertical substructural system?
(3) The conversion of structural system was the problem to solve in the actual construction. How to safely remove the full-shoring support for the construction of the roof?

7.6.2 *Beijing National Center for the Performing Arts*

1. **Project Background [49]**

The project of National Center for the Performing Arts is carried out in the west of the Great Hall of the People and in the south of western Chang-an Avenue. It covers a building area of about 165,000 m^2 in total. The main structure takes up about 105,000 m^2 and underground affiliated facilities 60,000 m^2 [49] (as shown in Fig. 7.18).

The main structure of National Center for the Performing Arts is comprised of external retaining structure, internal opera house, music hall, theater, public hall, and other affiliated sections. Major internal load-bearing structure is reinforced concrete frame. External retaining structure is steel latticed shell. The latticed shell

Fig. 7.18 National opera [49]

is displayed as semi-ellipsoidal. Its west-to-east major axis is 212.20 m, and its south-to-north minor axis 143.64 m. The height of this structure is 46.3 m. An enormous steel spatial latticed shell serves as the major load-bearing structure for the entire ellipsoid. As for the exterior of the structure, titanium plates were employed. The central part of the structure adopted a gradually widening glass curtain wall. The main structure is surrounded by an artificial lake; the northern main entrance is an 80-m-long underwater corridor; as for the southern entrance and other passageways, they were built underwater as well.

2. **Structural system**

The external structural system of National Center for the Performing Arts employed an ellipsoidal hyperbolic arch shell. The framework of the shell is comprised of ring beams at the top and ribbed truss beams. Those ribbed truss beams were connected with each other by connecting members and diagonal braces (as shown in Fig. 7.19). Figure 7.20 is the analysis of the conceptual design of the main steel shell. Ring beams at the top undertake compression and have two purposes to serve: On the one hand, a reliable connection of radial ribbed beams at the top could be assured of, which facilitates the construction, and on the other hand, the side thrust from local shells at the top could be undertaken and support N′ is provided, as shown in Fig. 7.29.

3. **Summary**

(1) The external structure of National Center for the Performing Arts is the spatial shell structure as shown in Fig. 7.20, and readers could conduct simple design and calculation over this.
(2) Controversies and questions still exist over the design of National Center for the Performing Arts, the very conception of which is faced with restrictions

Fig. 7.19 Main steel shell structure [49]

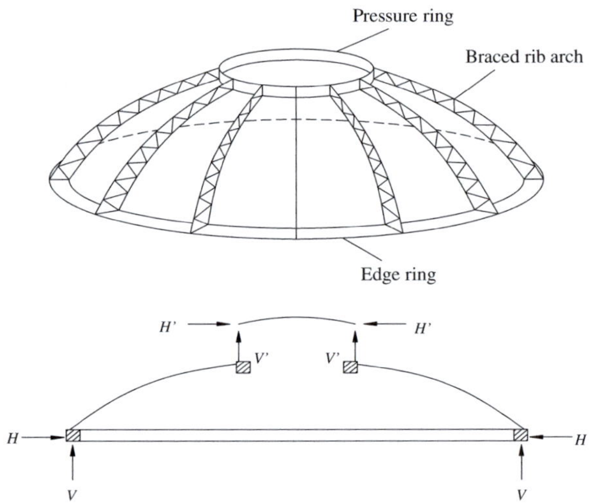

Fig. 7.20 Conceptual design

and challenges from structure, architecture, environment, safety, energy conservation, sustainable development, daily maintenance, and sanitation. If readers are so inclined to know about issues that must be considered in the programming and design of significant structural projects, the following discussion over this aforementioned project could be carefully studied as the reference.

4. **Discussion**

Being a landmark of Beijing, the symbolic meaning of aesthetics of National Center for the Performing Arts was the key issue in the selection of design. However, the architectural design of National Center for the Performing Arts proposed by French architect Paul Andrew has been disputed all the way. Relevant controversies have been summarized as follows:

(1) The conflict of architectural design between "traditionalism" and "modernity." Some deem that the conception of this structure was way too modern and could hardly agree with the architectural styles of the surrounding environment of Tiananmen Square. Andrew believes that the location of the structure had decided the symbolic significance of this structure. Being next to the Great Hall of the People, which symbolizes the highest power of the nation, National Center for the Performing Arts shall be the representative of the culture and architecture of the new century, radiating a sense of culture and history. Architecture of every age has distinguishing features and style of its own, which has never been stagnant but represented the continuation of an era. What is your opinion?
(2) The safety issue of the structure. Since Charles de Gaulle Airport designed by Paul. Andrew collapsed, the safety of National Center for the Performing Arts was once questioned. Although Andrew explained that the collapse of Charles de Gaulle Airport was not the fault of the architectural design itself, people remained suspicious about the safety of the structure. Nonetheless, since the whole design and calculation have gone through the audition administered by authorities, the reliability of the structure should have been guaranteed.
(3) The safety of the entrance. National Center for the Performing Arts is surrounded by an artificial lake; the northern entrance is an 80-m corridor underwater; as for the southern entrance and other channels, they are set in the same fashion. Such an arrangement has potential safety hazard. Once the water permeates, it shall lead to the same catastrophe as water permeation does to the channel of a coal mine, the consequences of which are way too obvious. Precaution against such disaster calls for our attention. Please analyze the threats to the people from the permeation of water at the entrances of the structure and contemplate upon the measures of precaution and improvement.
(4) Light pollution. When the sunlight hits the external wall and the roof of the structure, the light reflected by bright titanium alloy panels of walls and the roof gives rise to light pollution to the surrounding environment, which was not taken into account in the process of design and is not easy to solve. Readers could try to propose some solutions.
(5) Energy conservation and sustainable development. The enormous public space (212.2 m × 143.6 m × 46.3 m) created by the ellipsoidal shell roof of the structure consumes a staggering amount of energy in summer and winter due to the air-conditioning. Even if there is only one performance to put on in one theater, the structure as a whole consumes a considerable amount of energy. Waste of energy of such kind does not agree with the principle of energy conservation and sustainable development.
(6) The cleaning and environmental hygiene of titanium alloy plates on top of the ellipsoidal roof. Since sandstorm is not avoidable in short terms in Beijing, the cleaning of titanium alloy plates on top of the ellipsoidal roof becomes an issue of prominence. The glittering effects that the architect expected from the titanium alloy ellipsoidal roof might not be met with. A safe and highly effective cleaning still remains as an issue to solve. Readers could try to propose some solutions.

7.6.3 *The Georgia Dome*

1. **Project introduction [50]**

Georgia Dome has an installation of 70,500 seats and adopted the largest tensioned cable-supported membrane roof in the history of modern architecture. It was designed by Matthys Levy of New York Weidlinger Associates. The entire stadium except for the roof is made of reinforced concrete. The roof covers an area of about 37,000 m^2 and takes its form of saddle-shaped hyperbolic paraboloid with the help of tensioned cable-supported membrane (as shown in Fig. 7.21). The membrane is supported by tensioned cables, and these cables are connected to two ellipsoidal tensioned rings on the upper part of the roof, undertaking the tension transferred from the tensioned cables. In this way, all the roof loads are bore by the tensioned cables, leaving only tensile stress for the membrane. Such is the design helps with the lightweight, low cost, and expedited construction.

2. **The conception of design and the characteristics of structure**

Figure 7.22 illustrates the internal structure of Georgia Dome. On top of the roof, two layers of box-sectioned ring beams are installed; on the periphery of the ring beams, short steel columns are set; the tops of these short columns are fastened to the tensioned cables and the bottoms to the box-sectioned ring beams. In this case, after prestress is applied through the tensioned cables to the short columns, these columns, being compressed, spring up the box-sectioned ring beams (Fig. 7.23 illustrates the details of box-sectioned ring beams and the supporting columns from the internal side of the roof). As for the tensioned cables, they are fastened to the side walls and columns on the periphery of reinforced concrete main hall.

Fig. 7.21 Georgia Dome [50]

Fig. 7.22 Internal structure of Georgia Dome [50]

Moreover, from Figs. 7.22 and 7.23, we could see that none of these short columns are grounded, and therefore, the views of the audience are completely interference proof. Meanwhile, the balance of the roof itself is guaranteed and that is the highlight of this structural system.

Figure 7.24 displays the distribution of membrane and the arrangement of tensioned cables of Georgia Dome. The pictures clearly show the connections between

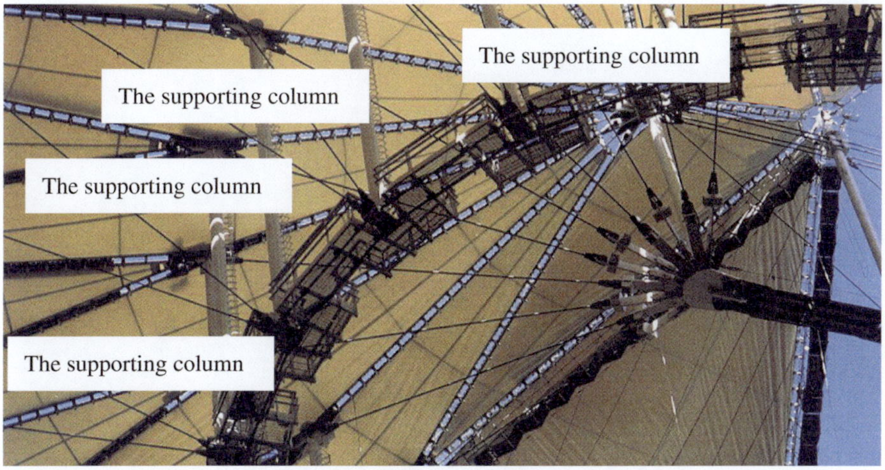

Fig. 7.23 Circular beams and supporting columns of Georgia Dome [50]

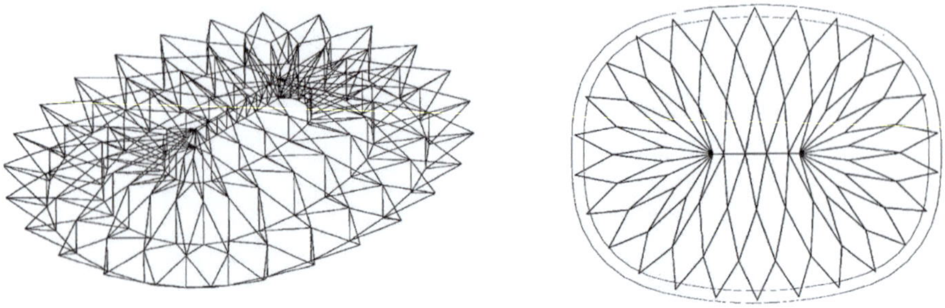

Fig. 7.24 Distribution of membrane and the arrangement of tensioned cables of the Dome [50]

box-sectioned ring beams, tensioned cables, and short columns on the periphery of the ring beams. Meanwhile, we could also see the longitudinal truss connection between the centers of two ellipsoids. Figure 7.25 is the actual construction site of Georgia Dome. In the picture, box-sectioned ring beams along with 52 supporting columns were simultaneously hoisted so as to form the central hoop structural system of the roof. This central hoop and central truss along with other cable systems were hoisted together (as shown in Fig. 7.25).

3. **Summary**

The roof system of this example was constructed with simultaneous hoisting of two box-sectioned ring beams and 52 short supporting columns. Being fastened to the upper tensioned cables, these short columns have a self-balancing system of their

Fig. 7.25 Construction site of Georgia Dome [50]

own. None of these short columns are grounded, and therefore, no view of the audience is interfered and that is the highlight of this project. Since all the roof loads are undertaken by tensioned cables and the tensile stress is the only force that the membrane has to take on, the structure is featured with a lightweight, a low cost, and an expedited construction.

4. **Discussion**

(1) Georgia Dome employed tensioned cables to bear the loads and made a self-balancing structural system out of two box-sectioned ring beams and short supporting columns. Readers could analyze the mechanical characteristics of major components and the path of force transfer.
(2) How did engineers carry out the construction of major load-bearing structural system of the roof system? Readers are suggested to think about and discuss the procedure of construction.
(3) In light of the arrangement of tensioned cables in the main stadium as shown in Fig. 7.24, could readers analyze the formation of major load-bearing structural system of the roof system? How to conduct the design of tensioned cables and the prestress of short columns? Could this structural system retain its balance without the application of prestress?

7.6.4 Kunming Peacock Art Square of World Garden Expo

1. **Project introduction [51]**

Peacock Art Square, completed in the year of 2001, is located in Kunming World Expo Garden. The architecture of Peacock Art Square is retractable in nature (as shown in Fig. 7.26). The membrane roof over the area of performance and that over the area of the audience could be gradually opened from a closed state according to the need of the performance. The membrane roof over the section of the audience could gradually open with the lift-up of the jack as if a peacock is gracefully spreading its tail. Meanwhile, the membrane roof over the section of performance is pulled to the two sides with the spurting of a musical spring. A design of such fashion does not only satisfy the need of view from the audience for entertaining performances, such as the scenery in major scenic area, musical spring, water-screen movies, and laser show, but also place the audience in an atmosphere of nature, the strong sensation of which is the artistic influence on people that is brought forward by retractable architecture.

2. **The conception and features of the structure**

The structure was conceived as a retractable structure, the opening of which was designed the way of a peacock spreading its tail as people could see in Yunnan, China, as shown in Fig. 7.27. The membrane roof over the section of the audience was designed to be the tail of the tail-spreading peacock (as shown in Fig. 7.28b and c),

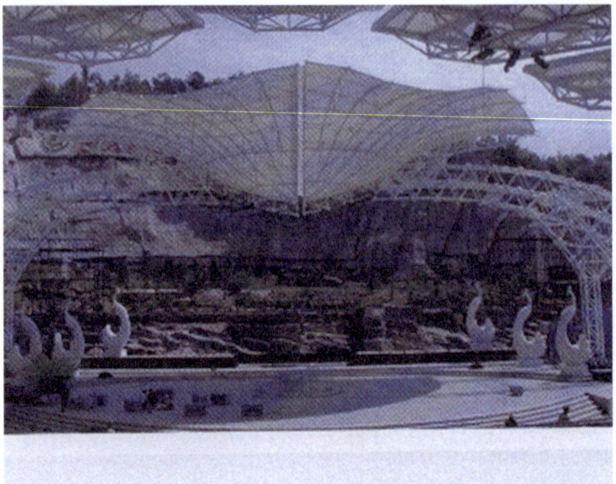

Fig. 7.26 Kunming Peacock Art Square (roof closed) [51]

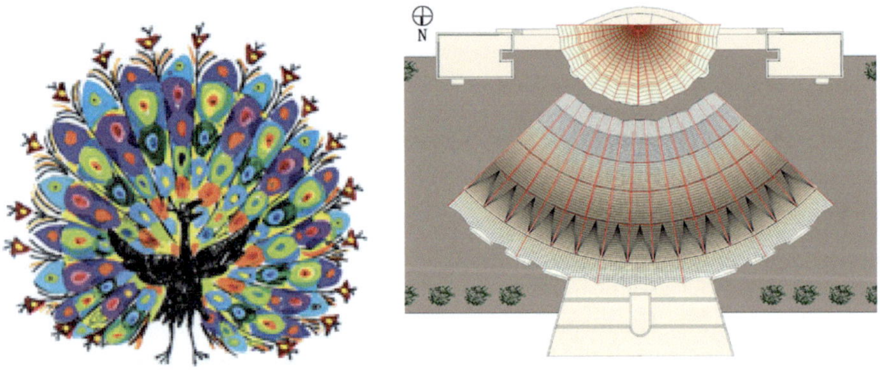

Fig. 7.27 Conception of the tail-spreading peacock [51]

and the membrane roof over the performing stage could be pulled to the two sides as if there were two lovely little peacocks dancing in the air (as shown in Fig. 7.28a). Figure 7.29 is the Peacock Art Square anterior to the opening of the membrane roof.

Structural design and the construction of retractable structure require full cooperation from architecture, mechanics, and automatic control, and the integration of relevant technologies is faced with considerable difficulties. The settlement in the field of construction has to be strictly controlled so as to guarantee that the operational requirements of mechanical equipment are met with. Meanwhile, the deviation of production and installation of steel structure has to be restricted to millimeters, and the construction demands high accuracy. The membrane employed in membrane structure has features such as lightweight, high strength, diffuse

Fig. 7.28 Tail-spreading opening process of the membrane roof [51]. **a** Membrane roof over the performing stage is being pulled to the two sides. **b** The membrane roof over the audience opened from the middle. **c** The membrane roof over the audience completely opened

Fig. 7.29 Opened Peacock Art Square

reflection, and self-cleaning function, which is quite advanced. To secure the reliability of the structure and its mechanical system, Peacock Art Square employed lockup devices from the stage of design and fixed the membrane roof over the performing stage to predesigned rotating position, enabling the spatial movement of the roof when it is opened (as shown in Fig. 7.29). A reasonable setting of multiple fronts of safety and strict adherence to operating instructions makes sure that the opening of the roof is secured and reliable. All the operations are controlled by intelligentized central automatic control system.

The opening of the membrane roof over the audience employed a leverage action and the main structural design and calculation adopted the model displayed in Fig. 7.30.

3. **Summary**

Thanks to the novel architectural configuration and function-based retractable design, Kunming Expo Garden Peacock Art Square remarkably combined the functions of the structure with surrounding environment and therefore won an international architectural design prize awarded by IStruct (International Structural Engineer Association) in 2005 and two respective second prizes issued by the Chinese Ministry of Construction and the Chinese Ministry of Education.

4. **Discussion**

(1) How did Kunming Expo Garden Peacock Art Square manage to provide shade and a view for performances while blending in the environment?
(2) Which field of knowledge was concerned in this project? Do civil engineers and architects still need to learn new things?
(3) This is the first major retractable structure completed by a team of young designers. Which abilities do the readers believe that engineers need to be

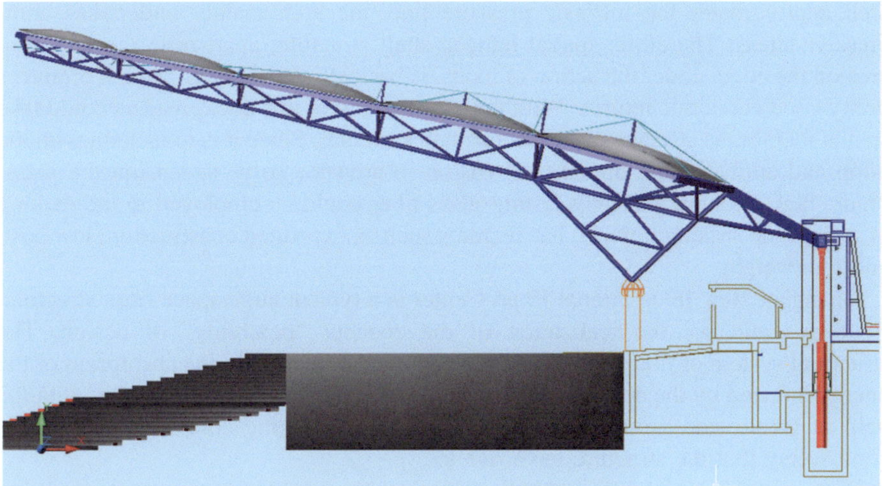

Fig. 7.30 Model of an opening roof

equipped with? What is the reader's opinion about the ability of adaptation, cooperation, creation, presentation, and organization in successfully completing the task?

Conclusion

In order to span over a long distance and create a large space, the roof system usually employs the form of a large-scale spatial structure. The design of structure of such kind requires a sufficient knowledge of material properties as well as mechanical performances of structural forms of all kinds so as to create uniquely configured highly efficient spatial structure. In this chapter, we discussed spatial structural systems of all kinds. Starting from the features of spatial structures, we, respectively, discussed spatial truss structure, spatial grid structure, spatial latticed shell structure, membrane structure, spatial developable structure, etc. In addition, we analyzed four representative projects: Shanghai New International Expo Center, Beijing National Center for the Performing Arts, Georgia Dome, and Kunming Expo Garden Peacock Art Square.

Spatial truss structure is an ideal structural form for large-span and medium-sized-span structures. And it takes trusses from different directions to support each other and form the path of spatial force transference. In the stage of conceptual design, spatial truss could be regarded as an equivalent plate; the bending moment upon the truss could be counterbalanced by the bending moment generated by the tensile and compressive axial force of the upper and lower chords; as for the web members, the shear force is the only force to resist. And this is an effective way of approximate analysis.

Spatial latticed shell structure is in nature a spatial curved surface created by rotating or moving horizontally the planar axis of the arch. When the axis of the arch approximates the intrinsic pressure line, the arch mainly undertakes compressive stress. Therefore, spatial latticed shell structure, apart from the restricted area on the edge, under the action of loads, is basically compressed, which produces pretty good economic returns. With the introduction of this book, readers could take spatial membrane structure as a special shell structure. Since it is to restrain with the hoop and apply certain amount of tension in advance so as to tension the membrane, high-strength and long-lasting glass fiber could be employed in the making of membrane structure, which has features such as expedited construction, low cost, and lightweight.

Shanghai New International Expo Center is a typical large space truss structure. Readers could see the realization of the concept “flexibility” of design. The structural system of this structure makes use of deformation in the absorption of the energy induced by the constant loads and forms a stable supporting system with the help of prestressed grouped columns along with swinging columns. What is the controversy that the structure gave rise to?

The external spatial latticed shell of Beijing National Center for the Performing Arts has a 212.2-m-long axis, a 143.6-m-short axis, and a height of 46.3 m, which could be accounted as an enormous spatial latticed shell structure. Readers could see from this landmark architecture the restraints, conflicts and challenges of all kinds in the aspects of structural system, architectural aesthetics, safety of operation, and environmental protection.

The roof system of Georgia Dome has features such as lightweight, low cost, and expedited construction. The self-balancing structural system was formed by hoisting the two box-sectioned ring beams and supporting columns simultaneously and then fixing short supporting columns to the tensioned cables overhead. None of those short supporting columns are grounded, and therefore, the view of the audience is totally interference free, which is the highlight of this project.

The retractable structure installed in Kunming Expo Garden Peacock Art Square structure is an interdisciplinary success that has integrated knowledge from fields of structure, architecture, mechanics, automatic control, environmental aesthetics, etc. Structural design shall synthetically consider the requirements and limitations of all the aforementioned fields of study. Readers could refer to Figs. 7.27 and 7.30 in the comprehension of actual operation and the conception of deployable structure.

The explanation of theories and the analysis of actual projects of construction provide the readers with an opportunity to think. Through the analysis of aforementioned structural systems and mechanical characteristics, readers are provided with the access to all typical spatial structural systems and their respective approach of analysis and therefore be able to extend the analysis to other newly emerged spatial structures.

Chapter 8
Intelligent Design Based on Knowledge Engineering

8.1 Introduction of Intelligent Design

The design process is an integration of human intelligence through a series of design decision by the designer based on their experience in the past by means of comparison with its functions. Entering the computer era, computer-aided design (CAD) helps designers to liberate them from routine design works. Moreover, by incorporating an expert system and artificial intelligent technologies into CAD system, one may partially replace human intelligence in design by artificial intelligence (AI) technology in computer science; the intelligent CAD system (or ICAD) then emerges. This is an important tendency in modern development of CAD technique, which represents the state of the-art in the 1980–1990s [52, 53].

In structural design, both numerical computation and symbolic inference are used, especially in the preliminary design stage, where a qualitative design decision is adopted for the structural type selection and a quantitative design decision is made for approximate sizing of structural members. Fuzzy model is the most appropriate of these jobs since the preliminary design is aimed to solve global arrangement of layout of structural components and its interrelationships of sizing with many design uncertainties. This is the most important stage in the whole design process and it heavily influences the final consequence of a design outcome and the rationality of the preliminary design will relatively even dominate the quality of the design project.

The design process is characterized by a series of design decisions. Whether it is satisfactory or not will mainly depend on the heuristic judgment of the designer, in other words, on his (her) experiences. This is a process of subjective mental manipulation of the design expert, which forms the kernel of his mental activities in design, so the design expert system is a program of human knowledge and expertise for solving problems. Nevertheless, for a professional expert system, it depends heavily on domain knowledge. To acquire knowledge from domain experts is really a tough, tedious, and troublesome task since human beings are poor knowledge

S. Lin and Z. Huang, *Comparative Design of Structures*,
DOI 10.1007/978-3-662-48044-1_8

transferors, due to the fact that they can express far less than they understand. It forms the main obstacle and bottleneck in building an expert system. Therefore, to incorporate machine learning capability into the system can greatly facilitate and improve the capabilities of the system by learning the facts or samples [54, 55], especially in the knowledge acquisition aspect.

The advantages of machine learning capability for an intelligent design system can be concluded as follows:

- It automates the learning process and can make the knowledge acquisition process faster and less error prone.
- It provides more qualified knowledge than extracted from domain experts manually.
- It accumulates knowledge as "memory" and stores for future applications, which is in such an amount that is impossible to be carried by any individual human expert through memory.
- It is based on the recent progress of AI technology for improving its system behavior, including the improvement of its performance by external domain knowledge; filtering out abnormal solutions through an evaluation module and refining the solution, for instance, through a redesign module etc.

In the intelligent design system, the requirements for machine learning capability can be specified as follows [56]:

- Adaptability of providing a learning capability with explicit classification and sorting functions.
- Flexibility in searching the design parameters in available data base with reasonable selection strategy.
- Effectiveness in evaluating the design alternative in proper way.

The mathematical modeling formulation that we are using in the intelligent design is fuzzy logic and AI technology [57]. For verifying this, we need to explain the geneses through the fundamental statement of indeterministic mathematics. It is well known that the world is full of uncertainty and the engineering design does so. The designer must study the uncertainty characteristics in design environment in order to select the proper mathematical tool accommodating to the characteristics of design problem. Nowadays, there are two main branches of uncertainty studies in engineering [58]: the probability-based theory and the fuzzy mathematics-based theory. Though there is commonality between these two, however, their difference is essential. Each of them deals with events with different natures.

The probability theory deals with those random events which have definite attributes and meaning, however, due to lack of sufficient conditions, there might not be a definite and absolute causal relationship between the occurrence of the event with the corresponding conditions, for which whether the event will be occurred or not is still uncertain. This kind of uncertainty is called the randomness. However, the fuzzy theory deals with those events, which has no definite conceptual meaning by itself and whether or not the examined event belongs to a subscribed concept is still uncertain. The uncertainty from exactly ranking of event, caused the vagueness of extrapolating of conception, called fuzziness.

The objective of the probability theory is to grasp generalized causal relationship from randomness, in other words, the rule of occurring probability or possibility percentage of its occurrence. The goal of fuzzy approach is to grasp generalized relationship of belonging from fuzziness, which is quantitatively defined by membership degree.

During the preliminary design stage of structure, the uncertain information includes the selection of reasonable structural type, the evaluation of seismic zone and zone coefficient Z, and the consideration for structural safety degree, etc. Obviously, since the designers have not yet mastered the details of its interrelationship, this information cannot be processed as a random event, but an event uncertain in degree. Taking the advantages of fuzzy approach into consideration, one could conclude that it seems reasonable to adopt fuzzy reasoning approach in the preliminary design of structures. According to above analyses, the conclusion can be drawn that a qualified machine learning model for structural design can be based on fuzzy theory.

8.2 Fuzzy-AI Model Applied to Comparative Design

Since digitized ICAD technology is a process of a series of intelligent design decision. The latter one is essentially a series of quantitative decision making, in other words, same as the quantitative management in academic point of view. Therefore, the principles in quantitative management can be naturally suitable in the design procedure. We use the technique of "Fuzzy-AI Model" in managerial affairs as to express the management issue quantitatively in degree magnitude, which can be extended to be effective in certain cases in management for decision making [58]. This is why "Fuzzy-AI Model" is not only appropriate to managerial issues but also can be regarded as the foundation of methodology in intelligent design or in ICAD.

8.2.1 *The Fuzzy Expression of P_d*

Refer to Sect. 1.2, the approach of determining design solution P_d through fuzzy comparative design philosophy can be shown in Fig. 1.2: The fuzzy expression of P_d is represented by its fuzzy attributes located in a fuzzy zone of corresponding attribute axis. Whereas, the related fuzzy zone in the attribute axis is determined by fuzzy reasoning of machine learning among attributes of similar samples, associated in n-dimensional design space.

8.2.2 Fuzzy Reasoning of the Intelligent Design

Fuzzy reasoning of the intelligent design process is based on the fuzzy distance of the designed attribute with its corresponding sample attribute as presented in Eq. (1.3)–(1.7) of Sect. 1.2, where Eq. (1.7) can be expressed by a design solution matrix D_{ds} as below:

$$D_{\mathrm{ds}} = \{D_{\mathrm{di}}\} = \begin{vmatrix} D_{d11} & D_{d12} & D_{d13} & \cdots & \cdots & D_{d1m} \\ D_{d21} & D_{d22} & D_{d23} & \cdots & \cdots & D_{d2m} \\ \cdot & \cdot & \cdot & & & \cdot \\ \cdot & \cdot & \cdot & \cdot & \cdot & \cdot \\ \cdot & \cdot & \cdot & \cdot & \cdot & \cdot \\ D_{dn1} & D_{dn2} & D_{dn3} & \cdots & \cdots & D_{dnm} \end{vmatrix} \quad \begin{matrix} (i = 1, 2, 3, \ldots, n) \\ (j = 1, 2, 3, \ldots, m) \end{matrix} \tag{8.1}$$

Denote there are n attributes for the design compared by m samples; D_{dij} in (8.1) is obtained by (1.1) and (1.2), represents the fuzzy distance of jth sample to ith attribute of the design solution. It is noticed that in (1.1), w_i represents the weight of the ith attribute. Each row of (8.1) represents a fuzzy distance vector of a specific sample with the design solution; each column of (8.1) represents a fuzzy distance of a specific attribute vector drawn by each samples. Through (8.1), the design solution expressed by fuzzy quantity of each attribute can be obtained by means of (1.3) and (1.4).

8.2.3 How Fuzzy-AI Model Works in Intelligent Design?

The Fuzzy-AI Model works for intelligent design by integrated procedures of case-based reasoning and machine learning inference of fuzzy distance and by using corresponding fuzzy reasoning through machine learning of selected sample cases.

Based on the concept of state in design space, the difference between two design solution A and B in the same (sub)space can be measured by fuzzy mathematical distance D, which is defined either by membership function difference as shown in (1.1)–(1.3) and is meaningful in view of manipulating design uncertainties. We may use (1.3)–(1.7) for identifying how the fuzzy distance of two attributes could work as to determining each fuzzy attributes of the design solution; in other words, a design solution matrix D_{ds} as presented in (1.7).

8.3 Mathematical Modeling of Design Space

Since design goal can be achieved step by step determininination of design parameters in different design stages, these parameters serve as the attributes of specific design object (or product). The relationship between design parameters can be depicted either by a network or by a tree structure to form a global picture of design states associated with the concept of generalized design space (see Fig. 8.1). Evidently, each node of the network (or tree structure) represents a subspace of certain design parameters (or attributes), and the generalized design space is implied by topological linkage of those subspaces.

A complicated design issue can be expressed by the state in its design space (generalized space) with a series of design parameters (attributes). However, its individual attributes can also be defined in corresponding subspaces. This statement holds true to any level of subspaces.

For simplification reasons, the generalized design space model is illustrated by two-layered hierarchical spaces shown in Fig. 8.1. The state of the design T described in top-level space Π has its attributes $t_i (i = 1, 2, \ldots, n)$, which are correspondingly described in its own subspace Π_i in the lower level of Fig. 8.1 by its subattributes $l_{ik} (k = 1, 2, \ldots, k_i)$. It shows:

For top-level generalized design space $\prod = \{\prod_i\} \quad (i = 1, 2, \ldots \text{n})$

For low-level design subspace $\prod_i = \{l_{ik}\} \quad (k = 1, 2, \ldots, k_i)$

The global design state T is defined by the states of its attributes t_i

$$T = F(t_1, t_2, \ldots, t_i, \ldots, t_n) \tag{8.2}$$

The state of design attribute $t_i (i = 1, 2, \ldots, n)$ can also defined by subattributes

$$t_i = F_i(l_{i1}, l_{i2}, \ldots, l_{ik}, \ldots, l_{iki}) \tag{8.3}$$

It is mentioned that Fig. 8.1 is just the simplest form of tree structure of generalized design space concept. For specific design issue, the design state could be expressed in its design space Π presented by its subspace Π_i in tree structure (Fig. 8.1). If I_i and G_i represent the initial and goal states of top-level design space, $I_l^{(i)}$ and $G_l^{(i)}$ represent the initial and goal states of ith lower level design subspace, then the state expression of design issue can be illustrated by:

For top-level design space [I_t, O_t, G_t], the initial design state of top-level design space is transformed to its goal state G_t through operation $O_t, I_t \xrightarrow{O_t} G_t$.

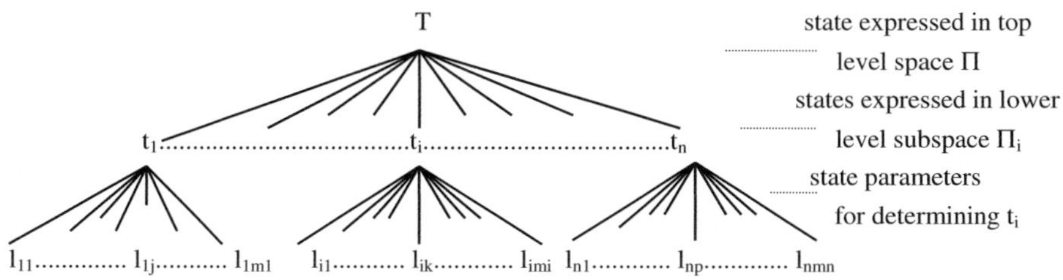

Fig. 8.1 Simplest form of tree structure of generalized design space concept

For lower level design subspace $[I_l^{(i)}, O_l^{(i)}, G_l^{(i)}]$, the initial design state of lower design subspace $I_l^{(i)}$ is transformed to its goal state $G_l^{(i)}$ through operation $O_1^{(i)}, I_l^{(i)} \xrightarrow{O_l} G_l^{(i)}$. Here, $i = 1, 2, \ldots, n$.

As an example of generalized design space concept, assume:

T Design state of preliminary design of aseismic building, which is determined by its main attributes t_1, t_2, …; where

t_1 Structural type of building, which depends upon its attributes

l_{11} height of the building

l_{12} plane configuration of the building

……

t_2 Basic shear subjected by the building(or by unit framework)

l_{21} seismic zone factor

l_{22} total seismic load of the structure (or of the unit framework)

……

t3 ……

……

Thus, we obtain the physical meanings of the elements (nodes) in the hierarchical tree of design space in Fig. 8.1.

8.4 Fuzzy Mathematical Distance in Design Space

8.4.1 *Fuzzy Mathematical Distance by Membership Difference*

We have defined in Eq. (1.1) $D(A,B) = \sum_{i=1}^{m} w_i(d_i) * d_i(A,B)$
where

m the dimension of the (sub)space

w_i influence coefficient of ith factor (or attribute)

$d_i(A, B)$ distance between design solution A and B in ith attribute

obviously, $D(A,B) = D(B,A)$; $D(A,A) = D(B,B) = 0$.
Doubtlessly, the measurement of difference could be extended to hierarchical subspace with uncertain environment of design solutions through Eq. (1.2). $d_i\left(\underset{\sim}{A}, \underset{\sim}{B}\right) = \mu_{\underset{\sim}{A}}(t_i) - \mu_{\underset{\sim}{B}}(t_i)$ where $\mu_{\underset{\sim}{A}}(t_i)$ and $\mu_{\underset{\sim}{B}}(t_i)$ represent the membership functions of design solutions $\underset{\sim}{A}$ and $\underset{\sim}{B}$ with respect to its ith attribute t_i in its Π_i subspace.

The influence coefficient w_i might not be constant for different distance, it should be changeable with certain penalty factors. Further, define

$$w_i(d_i) = a_i * p_i(d_i) \tag{8.4}$$

where: a_i, is the weight of the ith attribute to the design solution. If there are k attributes, then

$$\sum_{i=1}^{k} a_i = 1.0$$

$p_i(d_i)$, the amplified coefficient of w_i (d_i), is suggested to be

$$p_i = 1 + |d_i(A, B)| \tag{8.5}$$

Equation (8.5) shows that the larger the absolute value of distance between A and B, the larger the amplified coefficient p_i, consequently, the larger the influence coefficient $w_i(d_i)$ with penalty amplification, then the distance suffers the same penalty of amplification.

As an example, we assume that the design solution $\tilde{A}$ is compared with another design solution $\tilde{B}$. The state expression of both are determined by

$$T(\tilde{A}) = F(t_{1A}, t_{2A}, \ldots, t_{kA})$$
$$T(\tilde{B}) = F(t_{1B}, t_{2B}, \ldots, t_{kB})$$

State attributes

t_1 represents number of stories
t_2 represents height of the building
t_3 represents total basic shear
……
t_k represents total lateral stiffness of unit frame etc.

The fuzzy expression of state is comprised of a membership function with respect to corresponding attributers:

$$T(\tilde{A}) = \frac{u_{\tilde{A}1}}{t_1} + \frac{u_{\tilde{A}2}}{t_2} + \cdots + \frac{u_{\tilde{A}k}}{t_k} = \frac{0.9}{t_1} + \frac{0.95}{t_2} + \cdots \tag{8.6}$$

$$T(\tilde{b}) = \frac{u_{\tilde{B}1}}{t_1} + \frac{u_{\tilde{B}2}}{t_2} + \cdots + \frac{u_{\tilde{B}k}}{t_k} = \frac{0.85}{t_1} + \frac{0.90}{t_2} + \cdots \tag{8.7}$$

Suppose weight vector $(a_1, a_2, \ldots, a_k) = (0.1, 0.15, \ldots)$.
The fuzzy distance between attributes of design solution $\tilde{A}$ and $\tilde{B}$

$$d_1\left(\tilde{A}, \tilde{B}\right) = u_{\tilde{A}1} - u_{\tilde{B}1} = 0.9 - 0.85 = 0.05$$
$$d_2\left(\tilde{A}, \tilde{B}\right) = u_{\tilde{A}2} - u_{\tilde{B}2} = 0.95 - 0.90 = 0.05$$

……

The amplified coefficient $p_i(d_i)$, by (8.5)

$$p_1 = 1 + |d_1(A, B)| = 1 + 0.05 = 1.05$$
$$p_2 = 1 + |d_2(A, B)| = 1 + 0.05 = 1.05$$

……

The fuzzy distance between design solution $\tilde{A}$ and $\tilde{B}$ by Eq. (1.1):

$$\begin{aligned} D(A, B) &= w_1 \cdot d_1 + w_2 \cdot d_2 + \cdots \\ &= 0.105 \times 0.05 + 0.158 \times 0.05 + \cdots \\ &= 0.00525 + 0.0079 + \cdots \end{aligned}$$

8.4.2 *Fuzzy Mathematical Distance by Nearness Degree*

The mathematical distance can be described by the nearness degree of two events. This is especially efficient when a new design is expected to generate from a set of available similar design solutions. The nearness degree is accounted by fuzzy theory. Suppose there are n engineering project samples, namely: $A_1, A_2, \ldots, A_n$; and the attribute of which is defined by g_j (j = 1,2, …, m) representing the main parameters of the design object to be considered

$$G = \{g_1, g_2, \ldots, g_m\} = \{g_j\} \tag{8.8}$$

As an example of an earthquake-resistant building, G could be expressed by a series of frame parameters.
g_1 number of stories
g_2 height of the building
g_3 total base shear
…
g_m total lateral stiffness of unit framework.

If $\underset{\sim}{G}_{ij}$, the fuzzy subset of ith building A_j to the attribute g_j

$$\underset{\sim}{G}_{ij} = \frac{\mu_{i1}^{j}}{g_{1j}} + \frac{\mu_{i2}^{j}}{g_{2j}} + \cdots + \frac{\mu_{ip}^{j}}{g_{pj}} + \cdots + \frac{\mu_{ik}^{j}}{g_{kj}} = \left\{ \frac{\mu_{ip}^{j}}{g_{pj}} \right\} \quad \begin{matrix} (i = 1, 2, \ldots, n) \\ (j = 1, 2, \ldots, m) \\ (p = 1, 2, \ldots, k) \end{matrix} \tag{8.9}$$

where μ_{iP}^{j} is the membership function of ith building A_j to pth grade g_{pj} of jth attribute g_j of the building.

The fuzzy subset of building to be designed for the grade of jth attribute g_j

$$\underset{\sim}{G}^{\cdot} = \frac{\mu_1^{\cdot}}{g_{1j}} + \frac{\mu_2^{\cdot}}{g_{2j}} + \cdots + \frac{\mu_q^{\cdot}}{g_{qj}} + \cdots + \frac{\mu_k^{\cdot}}{g_{kj}} \tag{8.10}$$

where $\mu_P^{\cdot}$ is the membership function of design alternative B to pth grade of jth attribute g_j of the building.

The nearness degree of the expected design solution with the sampling solution can be obtained by following algorithms: If $\underset{\sim}{A}$ and $\underset{\sim}{B}$ are two fuzzy subsets on universe X, then the interior and exterior products of $\underset{\sim}{A}$ and $\underset{\sim}{B}$ can be expressed by

$$\underset{\sim}{A} \otimes \underset{\sim}{B} = \bigvee_{x \in X} \left[\mu_{\underset{\sim}{A}}(x) \wedge \mu_{\underset{\sim}{B}}(x) \right] \tag{8.11}$$

$$\underset{\sim}{A} \Theta \underset{\sim}{B} = \bigwedge_{x \in X} \left[\mu_{\underset{\sim}{A}}(x) \vee \mu_{\underset{\sim}{B}}(x) \right] \tag{8.12}$$

Then the nearness degree of A and B is defined by

$$N\left(\underset{\sim}{A}, \underset{\sim}{B}\right) = \frac{1}{2} \left[\underset{\sim}{A} \otimes \underset{\sim}{B} + \left(1 - \underset{\sim}{A} \Theta \underset{\sim}{B}\right) \right] \tag{8.13}$$

Substituting Eqs. (8.8) and (8.9) to Eq. (8.11) and (8.12) and then to Eq. (8.13), one may determine the set of nearness degree of expected design solution with each of the sampling solutions for the attribute g_i. For the previous example, k_{ij}, the nearness degree of sample A_i and B in jth attribute is expressed by

$$k_{ij} = (A_i, B)_j \quad (j = 1, 2, \ldots, m;\ i = 1, 2, \ldots, n) \tag{8.14}$$

As concern to specific attribute $g_{i,}$ the top three nearness degree to the expected design solution within the samples can be calculated by Eq. (8.13), which are denoted by k_{1j}, k_{2j}, k_{3j}.

To infer the reasonable information for a new design from the available samples, one should fully utilize the advantages of fuzzy nearness degree between expected design parameters and the available samples by means of estimation vector ε

$$\varepsilon_j = \lambda_j \bigg[k_{1j} * E_{1j} + k_{2j} E_{2j} (1 - k_{1j}) + k_{3j} E_{3j} (1 - k_{1j})(1 - k_{2j}) + \frac{1}{3} (E_{1j} + E_{2j} + E_{3j})(1 - k_{1j})(1 - k_{2j})(1 - k_{3j}) \bigg] \tag{8.15}$$

$$\varepsilon = \{\varepsilon_j\} \quad (j = 1, 2, \ldots m) \tag{8.16}$$

where

ε_j is the *j*th component of estimation vector ε corresponding to *j*th attribute g_j

λ_j is the empirical coefficient of rectification, which influences the estimation and can be chosen through test

k_{1j}, k_{2j}, k_{3j} are the top three nearness degrees in Eq. 8.17 from *n* calculated values by $A_i (i = 1, 2, \ldots, n)$ corresponding to *j*th attribute g_j

E_{1j}, E_{2j}, E_{3j} are the attribute g_j values of the samples corresponding to k_{1j}, k_{2j}, k_{3j}.

During practical calculation, λ_j is suggested as 0.9–1.1. Actually, λ_j is a mutual factor and fluctuated by the conditions of many design factors related to this specific attribute g_j. Usually, λ_j is determined by extensive statistical data processing and comprehensive studies based on the data of existing buildings.

It is clarified that, the nearness degree approach is based on fuzzy theory, which can accommodate the design uncertain environment. The fuzzy subset of design alternative and the samples can be formulated once the scale of classification of the particular attribute is established. For the height of the building (g_j) as an example, the scale of classification could be defined as

$$g_{21} = (<50\,\text{m}) \quad g_{22} = (50\text{–}75\,\text{m}) \quad g_{23} = (75\text{–}90\,\text{m})$$
$$g_{24} = (90\text{–}120\,\text{m}) \quad g_{25} = (120\text{–}150\,\text{m})\ldots$$

If the designer's intention is to design a 20-story building, the fuzzy set of the building height can be expressed as

$$G_2 = \frac{0}{g_{21}} + \frac{0.2}{g_{22}} + \frac{0.9}{g_{23}} + \frac{0.8}{g_{24}} + \frac{0.25}{g_{25}} + \cdots \tag{8.17}$$

Similarly, the fuzzy subset for samples can be constructed as same as Eq. (8.17).

The estimation formula by nearness degree Eq. (8.15) is reasonable. It represents the processes of searching a design state which is likely most near-by what the fuzzy comprehensive state of the samples is. Each term in the parenthesis of Eq. (8.15) represents a weighted term by nearness degree and the fourth term represents a global rectification of the expected value for design solution. The coefficient λ_j can be determined by self-training; i.e., it can number the sampling value as the design expectation, and result to an output value.

8.5 Fuzzy Reasoning and AI-Machine Learning Model

8.5.1 AI-Machine Learning Model

Since we are looking forward for an applicable machine leaning strategy for design expert system in structural engineering, the machine learning approach which is being adopted should have satisfied the following conditions:

- Accommodating to the processes in design practice which is priority a process based on experience rather than logic inference;
- Utilizing fully the available engineering data as well as examples which can serve as the foundation of future design activities;
- Facilitating the available expert system with more powerful system capabilities and potentials, which are reliable and robust;

Starting from the simplest machine learning model, it can be expressed by Fig. 8.2.

In Fig. 8.2, we understand:

Data input from input module is $(x_1, x_2, x_3, \ldots, x_n)$

Data output from the system is $(y_1, y_2, y_3, \ldots, y_m)$

Information stored in the computer is data pair $[(x_1, x_2, x_3, \ldots, x_n), (y_1, y_2, y_3, \ldots, y_m.)]$.

It means once that data pair has been stored, the computer will not work if a job of calculating $f(x_1, x_2, x_3, \ldots, x_n)$ is being offered. Instead, it will provide $(y_1, y_2, y_3, \ldots, y_m)$ from its storage since the learning system makes the computer to remember and to retrieve the consequent data from storage for answer. However, such relation is absolutely restricted; it will break down once the problem has just a single and unimportant change. This situation is quite disappointing as in design environment, it is difficult to maintain every parameter in every case the same.

Supervised learning is aimed to train the system interactively with the training sample and the interferes of designer (user) and builder (system analyst).

The system architecture in Fig. 8.3 shows that the knowledge base is enriched and improved through interactive usage. When the user submits a consulting design project to test the system, he might judge whether or not the system is in good condition by the system response. If there is a feedback information of incorrectness, the system will trace all the related reasoning processes in order to find out the mistakes and explain to the users about the inference procedures. This strategy,

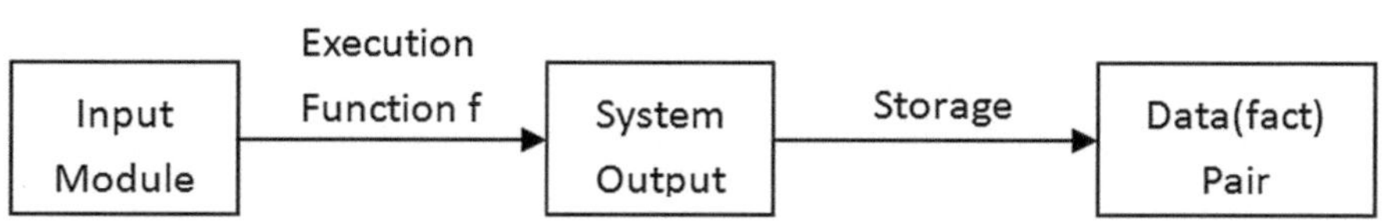

Fig. 8.2 Machine learning model

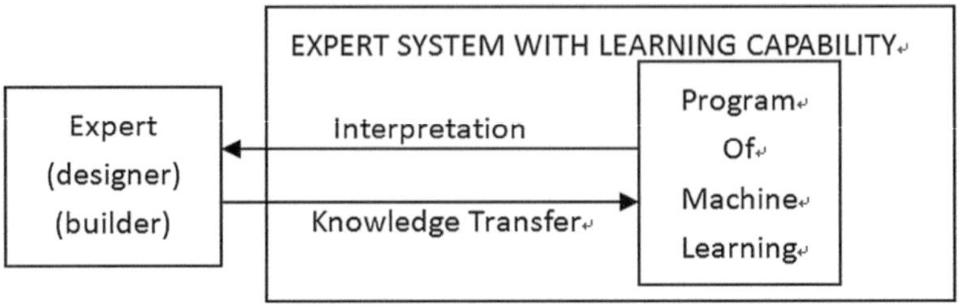

Fig. 8.3 Knowledge base enriched and improved through interactive usage

though it is effective, will rely too much on the system architecture characterized of technical sophistication and cannot be regarded as robust and applicable.

Learning through examples is based on inductive learning. Starting from practical example, the system will conclude some of its abstract rules for guiding system performance in the future. As a matter of fact, the learning procedure of the system through examples can be in different forms. Usually, expert system will be able to induce the laws through the sample and also be able to extrapolate and interpolate design parameters when they do not exactly match with what we have already stored in the database (knowledge base). Therefore, the matching capability of the learning system is strongly expected for its practical use.

According to above-mentioned analyses, we conclude our considerations on the fuzzy-based machine learning system through an example. Let us look at an expert system of earthquake-resistant design of building; since the design environment is full of uncertainties, one cannot achieve a reasonable solution without properly considering of having to deal with uncertainties. On the other hand, as we pointed out previously, in comparison with the probability theory, the fuzzy mathematics can fit much better to our goals and show elasticity in manipulating design issues. Therefore, systematical development of the strategy of fuzzy knowledge representation accommodating to the machine learning is needed for exploring this new field.

Based on the concept of state in design space, the difference between two design solutions A and B in the same (sub)space can be measured by fuzzy mathematical distance D, which is defined either by the membership function difference or by the nearness degree, and is meaningful in view of manipulating design uncertainties.

8.5.2 *Machine Learning Modeling*

As a matter of fact, design engineers found that existing engineering samples are better knowledge carriers than the individual domain experts are. A structural design solution is evolved from the preliminary conceptual layout until detailed structural analysis and design. Among them, the conceptual layout plays a more important role than the latter one. Besides, it is believed that a good design comes from the creativity and imagination of the designer(s); hence, this information is naturally stored in a "best practice" sample, which is available for retrieval through different ways of learning. To learn from good samples is one of the best ways for retrieving information from former domain experts. This is the direct impetus of developing the method of machine learning from samples. For machine learning in design, the problems can be stated as follows:

- How the learning procedures can be formulated? How the sample for supervised learning can be classified by its attributes and become a part of knowledge base for the designed to use?
- How the initial design conditions can be recognized and a design alternative can be identified and processed based on the samples in the knowledge base? After machine learning in structural design, a feasible and reasonable design solution is expected.
- How the assigned design alternative can be synthetically evaluated based on different satisfactory degree with respect to corresponding attributes of the design problem?
- How designer evaluate the design solution according to different satisfaction?

Since designers are dealing with design problems which are characterized of uncertain design environments, each of the processing mentioned above should be capable to incorporate uncertainties. The fuzzy method is used for "soft" design purpose and the machine learning processing of the design is actually based on the technique of fuzzy attributive identification. In order to solve those problems, the following four kinds of machine learning models based on fuzzy theory are proposed:

- Machine Learning Model for Fuzzy Classification
- Machine Learning Model for Fuzzy Identification
- Machine Learning Model for Fuzzy Matching
- Machine Learning Model for Fuzzy Evaluation

8.5.2.1 Machine Learning Model for Fuzzy Classification

Since the essentials of design process are merely identifying the design parameters among the ranked samples in the sample base, which must be exactly categorized quantitatively (including quantitative design parameter and quantized qualitative design parameter) for avoiding the abnormality of a design solution caused by divergence of sample parameters to be used in machine learning.

(1) "Distance" model by membership function

Providing the design sample is $\underset{\sim}{B}$, the samples to be learned from the knowledge base are $\underset{\sim}{A} = \{\underset{\sim}{A}_i\}\ i = (1, 2, \ldots n)$. After examining the distance of *i*th learning sample and design sample *B* in *j*th attribute, the equation is

$$\text{Min}\left[\left|D_j\left(\underset{\sim i}{A}^j, \underset{\sim}{B}^j\right)\right| \middle| \underset{\sim i}{A} \in \underset{\sim}{A}, \quad i = 1, 2, \ldots, n\right] \leq \theta_j \tag{8.18}$$

where

$D_j\left(\underset{\sim i}{A}^j, \underset{\sim}{B}^j\right)$ represents the "distance" of design sample $\underset{\sim}{B}$ and $\underset{\sim i}{A}$ in jth attribute

θj represents certain threshold value

Equation (8.18) is the criteria of sample ranking and classification by fuzzy identification of its attributes. Design sample $\underset{\sim}{B}$ belongs to the same rank of samples $\underset{\sim i}{A}$ $(i = 1, 2, \ldots, n)$ in j-th attribute of the design if Eq. (8.18) is satisfied.

(2) "Distance" model by nearness degree

Similar to Eq. (8.18), the ranking criteria of classification by "distance" based on nearness degree can also be expressed by

$$\text{Max}\left[\left|k_{ij}\left(\underset{\sim i}{A}^j, \underset{\sim}{B}^j\right)\right| \middle| \underset{\sim i}{A} \in \underset{\sim}{A}, \ i = 1, 2, \ldots, n\right] \geq v_j \tag{8.19}$$

where $k_{ij}\left(\underset{\sim i}{A}^j, \underset{\sim}{B}^j\right)$ represents the "distance" of design sample and samples in jth attribute $(0 > k_{ij} > 1.0)$ based on nearness degree.

v_j represents certain threshold value.

Equation (8.19) is also the ranking criteria of the samples by fuzzy nearness identification of its attributes. Design sample $\underset{\sim}{B}$ belongs to the same rank of $\underset{\sim i}{A}$ $(i = 1, 2, \ldots, n)$ of jth attributes of the design solution if Eq. 8.19 is satisfied. The values of θ_j and v_j are decided according to the requirements in ranking. The bigger the θ_j (or v_j), the more the samples in the same type of classification, or vice versa.

8.5.2.2 Machine Learning Model for Fuzzy Identification

For the purpose of generating a new design alternative (or parameter), the fuzzy identification model is used through comparison of initial design parameter with those learned samples in the knowledge base on the basis of similarity, by which the identification process is performed.

Table 8.1 Similarity Degree versus Distance Relationship

Distance scale	Distance location	Similarity degree S
D_1	$\|D\| < 0.05$	0.95
D_2	$0.05 \leq \|D\| < 0.10$	0.90
D_3	$0.10 \leq \|D\| < 0.15$	0.85
D_4	$0.15 \leq \|D\| < 0.20$	0.80
D_5	$0.20 \leq \|D\| < 0.25$	0.75
D_6	$0.25 \leq \|D\| < 0.30$	0.70
D_7	$0.30 \leq \|D\| < 0.40$	0.60
D_8	$0.40 \leq \|D\| < 0.50$	0.30
D_9	$\|D\| \geq 0.50$	0

(1) *Similarity criteria*
The similarity degree is based on "distance" of the design alternatives, where the "distance" can be either measured by membership function or by nearness degree of the alternative with learned samples. Denote a set of "distance"

$$D = [D_1, D_2, \ldots, D_9] \tag{8.20}$$

We also define the similarity of "distance" in Table 8.1.
Substituting the "distance" to Table 8.1, then the designer can evaluate similarity degree s of a pair of design alternative with known mathematical "distance". Based on "pair comparison," the similarity matrix of design alternative to be identified with those learned samples can be established through "distance" concept.

(2) *Fuzzy similarity matrix for identification*
Let us denote $\underset{\sim}{A}_i$ $(i = 1, 2, \ldots, n)$ the learned samples in knowledge base, $\underset{\sim}{B}$ the design alternative to be identified; the "distance" between $\underset{\sim}{B}$ and $\underset{\sim}{A}_i$ can be determined by Eq. (1.1). Taking consideration of combining $\underset{\sim}{B}$ into $\underset{\sim}{A}_i$, it has the pair-comparison distance of a new set $\underset{\sim}{A}_j$ $(j = 1, 2, \ldots, n, n+1)$ as the distance matrix

$$\underset{\sim}{D}\left(\underset{\sim}{A}_p, \underset{\sim}{A}_q\right) = \sum_{i=1}^{m} w_i(d_i) * d_i\left(\underset{\sim}{A}_p, \underset{\sim}{A}_q\right) \tag{8.21}$$

where $p, q = 1, 2, \ldots, n, n + 1$; m represents the number of subspaces, i.e., the number of design attributes. Other signs are similar to those in Eq. (1.1).

As a numerical example of identification, the stiffness of building is examined below. Suppose the learned samples are $\tilde{A}_i$, $(i = 1, 2, 3, 4)$ and design alternative to be identified is $\tilde{B}$, then the new set $\tilde{A}_j$ will be j = 1, 2, 3, 4, 5 with $A_5 = \tilde{B}$ and remain $\tilde{A}_1$ through $\tilde{A}_4$ the same.

Denote the factor set $U = (U_1, U_2, U_3, U_4)$ of certain attribute, say stiffness of the building, then the influence factor set U will influence to A:

U: (U_1 = shape, U_2 = size, U_3 = height, U_4 = thick) We have the membership functions of each sample with respect to each individual influenced factor:

$$\begin{aligned}
\tilde{A}_1 &= (U_1 = 0.8, U_2 = 0.5, U_3 = 0.6, U_4 = 0.6) \\
\tilde{A}_2 &= (U_1 = 0.6, U_2 = 0.6, U_3 = 0.8, U_4 = 0.7) \\
\tilde{A}_3 &= (U_1 = 0.7, U_2 = 0.7, U_3 = 0.5, U_4 = 0.8) \\
\tilde{A}_4 &= (U_1 = 0.5, U_2 = 0.8, U_3 = 0.8, U_4 = 0.7) \\
\tilde{A}_5 &= (U_1 = 0.6, U_2 = 0.7, U_3 = 0.5, U_4 = 0.6) = \tilde{B}
\end{aligned} \tag{8.22}$$

Assume the weighted set of influenced factors is $a_i = (0.25, 0.25, 0.375, 0.125)$, substituting a_i to Eq. (8.4) then Eq. (1.2) to Eq. (8.5) and the both to Eq. (1.1), the fuzzy similarity matrix R can be expressed by either Eq. (1.1) or Eq. (8.21). Numerically, the element of $\tilde{D}$ is determined by Eq. (1.1). We take $\tilde{D}_{23}$ as an example:

$$\begin{aligned}
\tilde{D}_{23} = {} & 0.25 \times (1 + 0.1)(-0.1)^* + 0.25 \times (1 + 0.1)(-0.1) \\
& + 0.375 \times (1 + 0.3)(0.3) + 0.125 \times (1 + 0.1)(-0.1) \\
= {} & [-0.0275 + (-0.0275) + 0.14625 + (-0.1375)] \cong 0.078
\end{aligned} \tag{8.23}$$

where * means $a_1(1 + |d(2,3)|)$ $d(2,3) = 0.25 \times (1 + |0.6 - 0.7|)(0.6 - 0.7) = 0.0275$.

Similar to Eq. (8.23), the fuzzy distance matrix can be formulated

$$\tilde{D} = \begin{bmatrix} 0 & -0.071 & -0.021 & -0.103 & 0.041 \\ 0.071 & 0 & \mathbf{0.078} & -0.032 & 0.132 \\ 0.021 & -0.078 & 0 & -0.10 & 0.058 \\ 0.103 & 0.032 & 0.10 & 0 & 0.160 \\ -0.041 & -0.132 & -0.058 & -0.16 & 0 \end{bmatrix} \tag{8.24}$$

Consequently, the similarity matrix $\tilde{S}$ can be expressed through Eq. (8.25):

$$\tilde{S} = \begin{array}{c} \begin{array}{ccccc} (\tilde{A}_1) & (\tilde{A}_2) & (\tilde{A}_3) & (\tilde{A}_4) & (\tilde{A}_5) \end{array} \\ \begin{bmatrix} 1 & 0.95 & 1 & 0.90 & 1 \\ 0.95 & 1 & 0.95 & 1 & 0.95 \\ 1 & 0.95 & 1 & 0.95 & 0.90 \\ 0.90 & 1 & 0.95 & 1 & 0.85 \\ 1 & 0.90 & 0.95 & 0.85 & 1 \end{bmatrix} \end{array} \begin{array}{c} (\tilde{A}_1) \\ (\tilde{A}_2) \\ (\tilde{A}_3) \\ (\tilde{A}_4) \\ (\tilde{A}_5) \end{array} \tag{8.25}$$

For evaluating the elements of the similarity matrix $\tilde{S}$, two examples are presented:

(i) $\underline{\tilde{S}_{23} = 0.95}$ is determined by $\tilde{D}_{23} = 0.078$, which is located between 0.05 and 0.10 and is categorized by D_2. According to our definition presented in (8.20'), we have $S = 0.95$.

(ii) $\underline{\tilde{S}_{45} = 0.85}$ is determined by $\tilde{D}_{45} = 0.16$, which is located between 0.15 and 0.20 and is categorized by D_4. According to our definition presented in (8.20'), we have also $S = 0.85$.

As shown in the fifth column of (8.24) and (8.25), one can readily identify that the design alternative can be $\tilde{B} = \tilde{A}_5 = \tilde{A}_1$, or we can take certain form of weighted sum of the samples based on its "distance" with the design alternative to be identified.

$$\tilde{B}_r = \tilde{A}_5^r = \tfrac{1}{n}\textstyle\sum_{i=1}^{n} \left(1 - D_{i(1+n)}\right)^k \cdot A_{ir} \qquad (8.26)$$
$$(i = 1, 2, \ldots, n \cdot n + 1; r = 1, \text{ or } = 2, \ldots, \text{or } r = m)$$

when $n = 4$, $r = 4$ and $k = 2$, we have r as the subscript of rth design attribute.

$$\tilde{B}_r = \tilde{A}_5^r = \frac{1}{4}[(1 - D_{15})^2 \cdot A_{14} + (1 - D_{25})^2 \cdot A_{24} + (1 - D_{35})^2 \cdot A_{34} + (1 - D_{45})^2 \cdot A_{44}] \qquad (8.27)$$

where k is weighted exponent, it is suggested $k = 2$.

As a matter of fact, Eq. (8.26) can be also expressed by similarity degree of samples presented in Eq. (8.25).

$$\tilde{B} = \tilde{A}_5 = \tfrac{1}{n}\sum_{i=1}^{n} S_{i(1+n)}^{(k)} \cdot A_{ir} \qquad (8.28)$$
$$(i = 1, 2, \ldots, n \cdot n + 1;\ r = 1, \text{or} = 2, \ldots, \text{or } S = m)$$

when $n = 4$, $r = 4$, and $k = 2$, we have

$$\tilde{B} = \tilde{A}_5 = \frac{1}{4}[S_{15}^2 \cdot A_{14} + S_{25}^2 \cdot A_{24} + S_{35}^2 \cdot A_{34} + S_{45}^2 \cdot A_{44}] \qquad (8.29)$$

Using Eq. (8.21), it can get the fuzzy similarity matrix $S = \{S_{ij}\}$, where the value S_{ij} is decided by D_{ij}. Sequentially, it can decide the value of the rth attribute of the design solution

$$\underset{\sim}{B}_r = \frac{\sum_{j=1}^{n} \left(1 - D_{j(n+1)}\right)^k * A_{jr}}{\sum_{j=1}^{n} \left(1 - D_{j(n+1)}\right)^k} \ (r = 1, 2, \ldots, m) \qquad (8.30)$$

where k, weighted exponent, it is suggested by experience $k = 2$.

(3) *Numerical Example*

As to identify (select) the design parameter U_4 (thickness) of building, one may address the trained sample in the knowledge base

$$A_{14} = 0.6, A_{24} = 0.7, A_{34} = 0.8 \text{ and } A_{44} = 0.7$$

Assume that initially the design parameter (thickness) of building is $B_4 = 0.6$, the identification model (8.26) and (8.28) will give the fuzzy identification results for design consultation as below:

First, by substituting (8.24) to (8.27), we have identified design thickness B_4 by fuzzy distance

$$\begin{aligned} B_4 &= \frac{1}{4}\Big[(1-0.041)^2 \times 0.6 + (1-0.132)^2 \times 0.7 \\ &\quad + (1-0.058)^2 \times 0.8 + (1-0.16)^2 \times 0.7\Big] \\ &= \frac{1}{4}[(0.919 \times 0.6) + (0.753 \times 0.7) + (0.887 \times 0.8) + (0.705 \times 0.7)] \\ &= \frac{1}{4}[0.551 + 0.527 + 0.710 + 0.494] = 0.571 \end{aligned} \tag{8.31}$$

Similarly, substituting (8.25) to (8.29), we have identified design thickness B_4 by fuzzy distance

$$\begin{aligned} B_4 &= \frac{1}{4}\left[1^2 \times 0.6 + 0.9^2 \times 0.7 + 0.95^2 \times 0.8 + 0.85^2 \times 0.7\right] \\ &= \frac{1}{4}[0.6 + 0.567 + 0.722 + 0.506] = 0.598 \end{aligned} \tag{8.32}$$

Compare (8.31) and (8.32) with initial value of thickness in design 0.6, the initial design value should be changed by the identified value. As regard to (8.31) and (8.32), the difference is caused by the similarity criteria. Equation (8.32) is the simplification of (8.31), it takes the advantage of computing simplicity over the (8.32), yet the essential of both are the same.

(4) *Discussion*

It is evident that the fuzzy similarity matrix $\tilde{S}$ is essentially the α-cut set of fuzzy "distance" matrix $\tilde{D}$. The concept of similarity matrix S is meaningful and to be used for sample classification, the classification process with the similarity matrix S is quite easy and straightforward once distinguish it by selecting $S_{ij} = 1.0$. Generally speaking, (8.26) and (8.28) are the same; however, we do prefer (8.28) for simplicity reason.

8.5.2.3 Machine Learning Model for Fuzzy Matching

Although the identification capability can provide reasonable design parameters for the designers based on the trained samples in the knowledge base, nevertheless, in many cases, the designer is aimed to have a reasonable design solution which is not available as exactly coincide with the desired design parameter. Thus the matching capability either by extrapolating or interpolating the existing sample parameter should meet the needs of the ongoing design project.

According to Eq. (8.21), the *r*th attribute value of *B* must be determined by the *r*th attribute value of samples *A* which have the same type as *B*. For simplification, use the attribute value of the closest samples to match it fuzzily. Define:

A the sample set
B the design solution to be decided
m the attribute number of A
u the known attribute number of B
v the unknown attribute number of B

Evidently, $m = u + v$. To the *r*th unknown attribute, the fuzzy subdistance between *A* and *B* can be expressed by *D*. Let us select the three smallest fuzzy subdistances which satisfy Eq. (8.33) and the three samples are:

$$D_r\left(\underset{\sim}{B},\underset{\sim}{A}_a\right) < D_r\left(\underset{\sim}{B},\underset{\sim}{A}_b\right) < D_r\left(\underset{\sim}{B},\underset{\sim}{A}_c\right) < D_r\left(\underset{\sim}{B},\underset{\sim}{A}_h\right)$$
$$(h = 1, 2, \ldots, n;\ r = 1, 2, \ldots v)\ (h \neq a, h \neq b, h \neq c) \tag{8.33}$$

Then the matched *r*th attribute of $\underset{\sim}{B}$ is

$$\underset{\sim}{B}_r = \frac{\left(1 - D_r\left(\underset{\sim}{B},\underset{\sim}{A}_a\right)\right) * A_{ar} + \left(1 - D_r\left(\underset{\sim}{B},\underset{\sim}{A}_b\right)\right) * A_{br} + \left(1 - D_r\left(\underset{\sim}{B},\underset{\sim}{A}_c\right)\right) * A_{cr}}{\left(1 - D_r\left(\underset{\sim}{B},\underset{\sim}{A}_a\right)\right) + \left(1 - D_r\left(\underset{\sim}{B},\underset{\sim}{A}_b\right)\right) + \left(1 - D_r\left(\underset{\sim}{B},\underset{\sim}{A}_c\right)\right)} \tag{8.34}$$

where A_{ar}, A_{br}, A_{cr} are the *r*th attribute value of $\underset{\sim}{A}_a, \underset{\sim}{A}_b, \underset{\sim}{A}_c$.

8.5.2.4 Machine Learning Model for Fuzzy Evaluation

After identifying of the design alternative *B*, it is necessary to further evaluate it by fuzzy synthetic evaluation. Assume that the evaluation attribute set is defined by $U = (u_1, u_2, \ldots, u_m)$, the fuzzy evaluation subset of design alternative *B* with respect to each attribute can be expressed by membership function of satisfaction:

$$\mu(B) = \frac{\mu_B(u_1)}{u_1} + \frac{\mu_B(u_2)}{u_2} + \cdots + \frac{\mu_B(u_m)}{u_m} \tag{8.35}$$

The weight coefficient between $u_i(i = 1, 2, \ldots, m)$ are $p_i(i = 1, 2, \ldots, m)$, then the fuzzy synthetic evaluation indicator η for the design alternative are:

$$\eta = \sum_{i=1}^{m} p_i * \mu_B(u_i) \geq I \tag{8.36}$$

where I represent the satisfactory threshold of the design, the bigger the value of I, the more the fitness with the design. The design alternative can be accepted if Eq. (8.36) is satisfied, or it must be rectified.

8.6 Neural Network Reasoning for Structural Design Parameters

During the preliminary design of structure, design decisions will not merely depend upon calculation rather than the conceptual reasoning of the experts. Mostly, the design alternative comes from the experiences (the design concept of existing buildings) of the design experts. Therefore, a qualified expert system for design should be capable to simulate the perceptional imagination process of an expert for approaching a sound design goal, i.e., the design knowledge representation should first focus on the knowledge of perceptional imagination rather than those of quantitative behavior; for instance, the analysis method or computing algorithms, which have, in some times, disturbed designers' attentions from the more important points of conceptual functions of the design.

As a matter of fact, the "IF–THEN" rules are fairly suitable for describing knowledge based on logic inference; however, it is difficult to simulate the perceptional imagination capabilities of human being during design processes. For a human being is a poor knowledge transmitter, most of the intuitive knowledge can only be the source to drive a decision rather than to be represented for transmission. Therefore, alternative approaches for simulating perceptional imagination capabilities in design are badly needed.

8.6.1 Neural Network (NN)

The knowledge representation of perceptional imagination is difficult and people are halt as engineers extract knowledge from domain experts, which is a bottleneck by knowledge representation and is full of uncertainties. Neural networks constitute a radically different way to the representation. The simplified model of the human neuron is organized into the networks similar to those found in the brain activities. What we will discuss in this section will just provide the principles necessary for understanding the model for preliminary design of structures as a context of the problem, and illustrate that the neural networks could be useful in the structural design through machine learning.

The breakthrough work by Hopfield [59, 82] in 1982 described how the computational algorithm BP (Back Propagation) can be built in neural network architectures capable in performing associate memory and solving optimization problem. The principal strength of back-error propagation is its general pattern of mapping capability, and it does not require any prior mathematical functions for manipulating the processes; yet, the only thing merely needed is the training sample. Eventually, Hopfield's work extended the neural network from 0–1 binary logic to continuous value of its output, which may accommodate a broad spectrum of application, especially for engineering design issues.

We start from a basic physical neural network model of a bio-neuron which was later developed by J.E. Dayhoff in 1990 [60]. The simulation relation is presented in Fig. 8.4. The bio-neuron is modeled by a basic processing unit j in the artificial neural network, where the dendrites are simulated by a series of unit in the input layer and synapse; the output layer. Each processing unit is hierarchically connected with certain connecting strength (weight) to form an artificial neural network. Denote the state of activation of unit j is S_j

$$S_j = \sum_{i \neq j} \left(T_{ji} \cdot a_i - \theta_j\right) \tag{8.37}$$

Fig. 8.4 Simulation relation

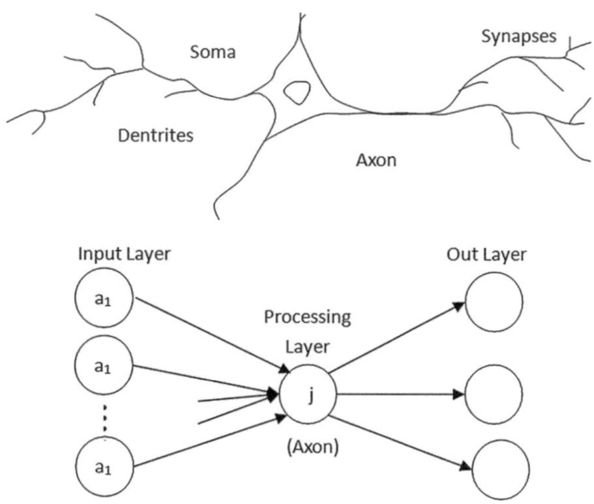

where

T_{ji} is the connecting strength (weight) from unit i of the input layer to unit j;

a_i the output state of input layer unit i;

θ_j the threshold value of unit j;

Then the output of unit j is determined by a sigmoid function

$$a_j = f(S_j) = \frac{1}{1 - e^{-S_j}} \tag{8.38}$$

Obviously, $T_{ii} = T_{jj} = 0$.

Since Hebb's rule presents the learning process of a neural network is essentially a process of successively rectifying T_{ji} for n processing units $(i = 1, 2, \ldots, n)$ that form the network as shown in Fig. 8.5, it follows

$$T_{ji} = \alpha S_j \cdot S_i (\alpha > 0) \tag{8.39}$$

Equation (8.39) represents the connection between unit i and unit j will be reinforced if both S_i and S_j are active.

This situation is accommodated to the "Conditional Refection" effect in biological phenomena of synaptic connections in the bio-science. Figure 8.5 represents a five-layered neural network when back-error propagation method is used for supervised training. Input vector X is applied to the system at input layer units, the system will automatically processing the information until the vector Y is coming out by which we are using later to modify network weights in a manner that the errors of units in the system can be reduced through every subsequent training

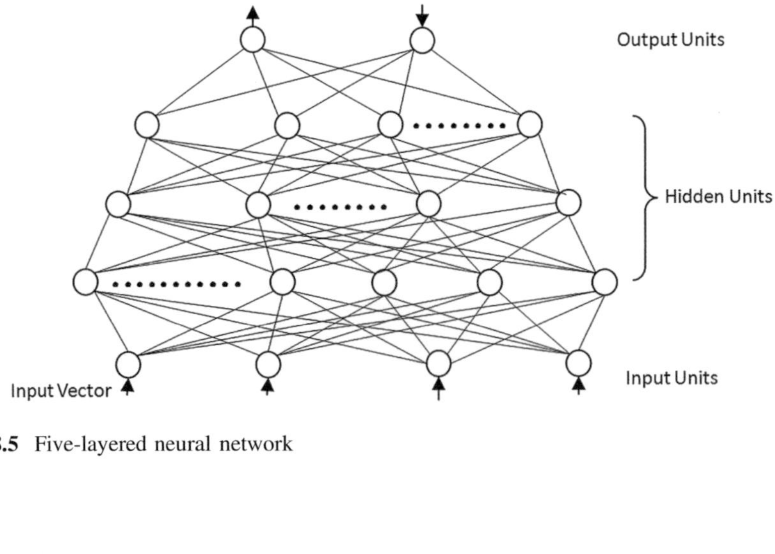

Fig. 8.5 Five-layered neural network

cycle. Denote the back propagating error for jth unit at its output layer is $\delta_j = \Delta f(S_j)$, then we have

$$f'(S_j) = \frac{\Delta f(S_j)}{\Delta a_j} = \frac{\delta_j}{(t_j - a_j)} \tag{8.40}$$

$$\delta_j = (t_j - a_j) \cdot f'(S_j) \tag{8.41}$$

t_j the input value of unit j
$f'(S_j)$ the derivation of a sigmoid function $f(S_j)$
S_j weighted sum of inputs to j unit

Denote the backpropagating error value for j unit at hidden layer

$$\delta_j = \sum_k [\delta_k T_{kj}] \cdot f'(S_j) \tag{8.42}$$

where k, subscript of unit in "front" layer from which the error is back propagated from jth unit.

T_{kj}, weight strength from unit j to unit k.

The adjustment of the connection weights is carried out by using the δ value of the processing unit. Each interconnection weight (from j to i) is adjusted by

$$\Delta T_{ji} = \eta \delta_j a_i \tag{8.43}$$

where

a_i activation level of ith unit
η learning rate, ranging 0.25–0.75
δ_j Error value of unit j.

Thus, a large value of error for unit j results in large adjustment to its incoming weights. On the other hand, a higher activation level a_i of coming unit i will result in a larger adjustment to its outgoing weight T_{ji} as shown in Fig. 8.6. The back propagation error method provides a theoretically sound approach for training the expert system with machine learning capability.

The NN possesses following advantages and could be an appropriate approach for machine learning in various situations:

(1) Natural parallelism: Similar to the brain inference, the system can process information in parallel, similar to the parallel processing in modern computers.
(2) Distributed information storage: The information of the neural net is stored within the links between the nodes of the net. The final node activation (strength) after learning represents the real information distribution of the sample.

(3) Tolerance of system performance: Similar to brain adaptability to errors or ability to conform to uncertainties, this feature represents the immense capability of the human being in processing fuzzy information.
(4) Self-adaptability: The neural network is highly self-adaptable, which can be widely applied in different fields of engineering in view of its powerful capabilities of learning, reasoning, self-organization, and self-rectification.

8.6.2 Fuzzy Neural Network (Fuzzy-NN)

The neural network can store the associated information between input and output by means of weights through learning. It needs a large number of samples and long learning time. Especially, it has drawbacks implicitly; for example, people do not know what are the weights between layers and the reasons during learning. Because, there will be no method to verify the convergence of neural network during its inference. The nonlinear characteristics of the neural network destroy this inference relationship, and no one knows how the input data get the output data through learning while the inference iterates routinely. One of the main advantages of the neural network is its independence on model, but it is difficult for defining the weights. The other problem is how to get the training samples, when to get the samples and how to know whether it is reliable for driving new design?

Nevertheless, fuzzy system encodes directly the associated relations or events, for which is described in vague language and stored in fuzzy matrix. Fuzzy approach combines the advantages of numeric method, symbol model as well as intelligent method. In many cases, it is easy to handle since the dimension of fuzzy matrix is small. While collecting the uncertain knowledge with vague language, and by describing and processing it quantitatively, people will know the inference routine in fuzzy approach even though one cannot directly understand the knowledge stored in the fuzzy system; but it can easily lead the additional heuristic knowledge into the picture. The main drawback of fuzzy system, however, is the accuracy of encoding is low; especially, it is more evident when the inference routine become tedious.

It is known that the fuzzy system and the neural network do not exclude each other since they both process knowledge with numeric method. Therefore, it seems possible to integrate the both for maintaining its advantages. This technology is now promising and has been widely using in fuzzy modeling, system control, pattern recognizing, predication, and diagnosis etc.

8.6.3 *Fuzzy-NN Architecture for Machine Learning*

When examining the roles of machine learning capability in intelligent design of structures, one can identify that there are basically two kinds of approaches, namely: learning capability from available engineering samples and matching capability from available samples' data for design parameters.

8.6.3.1 Classification Learning Capability from Engineering Samples

Learning capability of the system means that using the classification capability, the system is trained by practical engineering samples which are considered as "good" (best practice) experiences to be adopted on for future design. When a sample is learning, its attributes are stored in the engineering fact base of the knowledge base of the system. In the process of learning, the attributes of sample A_i should be trained one after another, as shown in Fig. 8.6, in which

$A_i(i = 1, 2, \ldots, n_i)$	represents the sample set to be trained. In case of when only one sample *B* is being trained (added) to the system, B can be regarded as A_{nj+1} (while n_j is trained sample)
k	represents the subscript $(k = 1, 2, \ldots, n_k)$ for example A_{ik}
m_k	represents the subscript of factors affected to attribute $k\,(m_k = 1, 2, \ldots, n_{mk})$
g_k	represents the subscript of ranking group of $A_{ik}(g_{ik} = 1, 2, \ldots, n_{gk})$
$A_{mk,i,k}$	means the initial input for *k*th attribute, *mk*th factor by sample *i*

The flowchart neural network for classification is show in Fig. 8.7, where four layers of neural nets are scheduled including both input and output layers as well as two hidden layers (Fig. 8.8).

Figure 8.9 is the implementation chart of neural network architecture for machine learning of structural design. The input of the first layer is $d_{i,nk,j} * (1 + d_{i,nk,j})$, where $d_{i,nk,j} = \mu_{i,nk,j} - \mu_{i,nk,1}$, $i = 1, 2, \ldots, n_i$; $j = 1, 2, \ldots, m_{nk}$. The input of second layer is $\sum w_{12}^* d_{i,nk,j}^* (1 + d_{i,nk,j})$, where w_{12} is the weight between the first and second layers. The output of the fourth layer is 1 or 0. If the samples belong to the particular kind, the output is 1, otherwise 0.

After training, the neural network will be able to classify the kind of a new engineering sample, which means getting good membership value for the sample

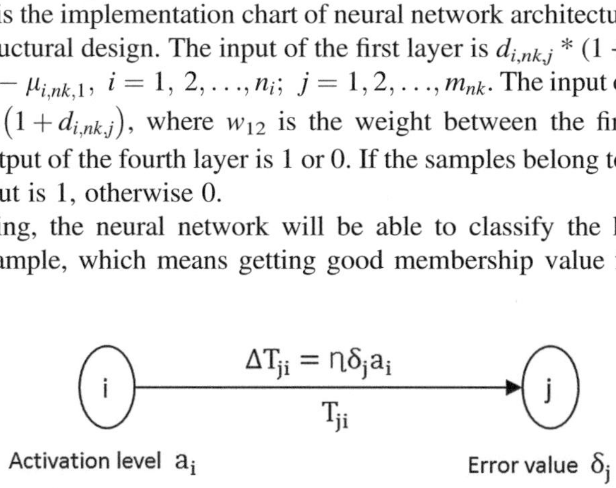

Fig. 8.6 From activation level to error value

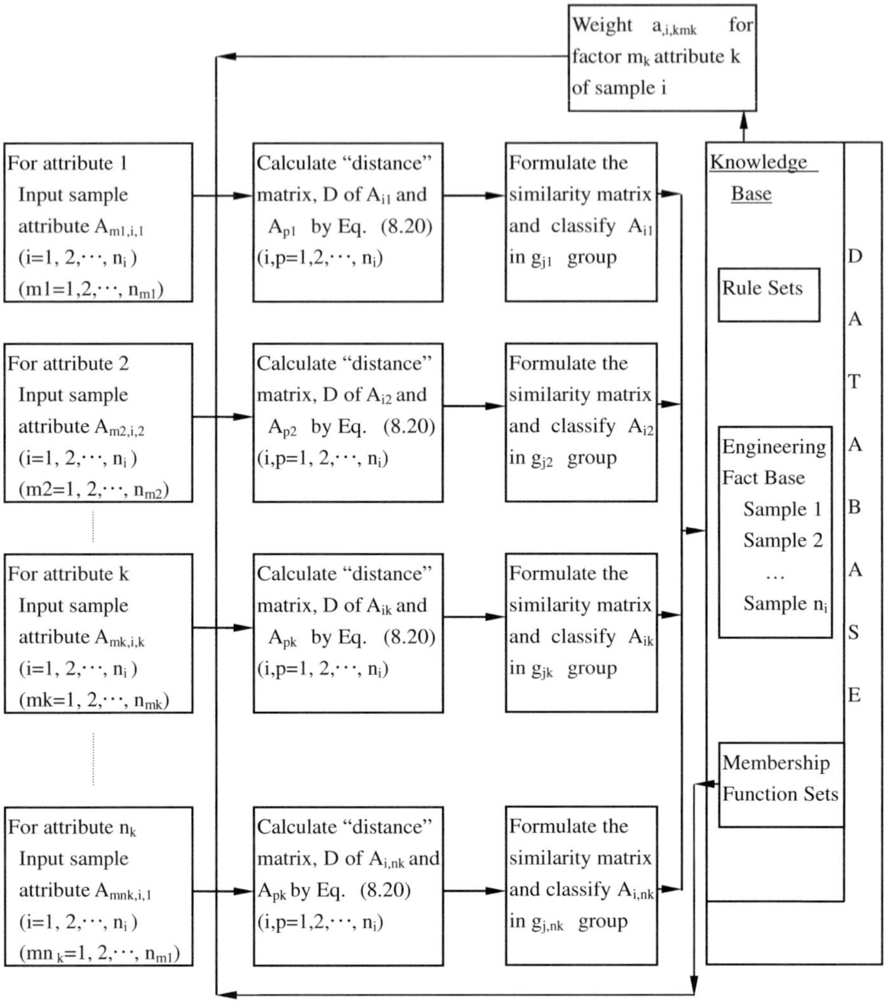

Fig. 8.7 Neural network flowchart for classification

where w_{12} is the weight between first layer and the second layer. In other words, after training, the network gets a good series of weight determined by sample attributes.

8.6.3.2 Matching Capability for Design Parameters

Matching capability of the system means that how a reasonable design attribute is generated by the matching process of the available information of samples stored in the engineering sample fact base of the system. The process starts from identifying

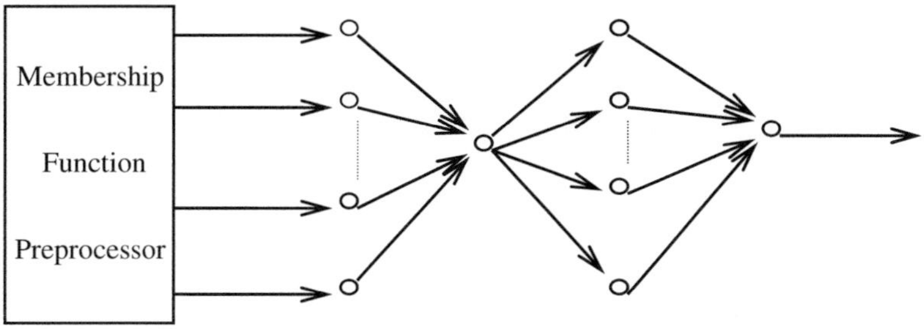

Fig. 8.8 Neural network for classification

three of the nearest known samples (Min. "distance") from the available sample base, then matching these three for optimum fuzzy design attribute.

Figure 8.9 presents the flowchart of neural network processing for matching in a machining learning functional framework, where the neural network architecture can be easily established. In Fig. 8.9, as we have assumed, B_1 is comprised of $B_{11}, B_{12}, \ldots, B_{1,(m1-1)}$; however, $\underset{\sim}{A}_{1,m1}$ is comprised of $\underset{\sim}{A}_{11}, \underset{\sim}{A}_{12}, \ldots, \underset{\sim}{A}_{1,nm1}$, $\underset{\sim}{B}_1$ represents desired design parameter, in which corresponding to $\underset{\sim}{A}_{1,nm1}$ is $\underset{\sim}{B}_{1,nm1}$ of design fuzzy attributive set $\underset{\sim}{B}$.

It is evident that the architecture for learning sample with classification and for identifying the design parameters by matching are almost the same in algorithmic processing. The differences between the two are that the second neural processing

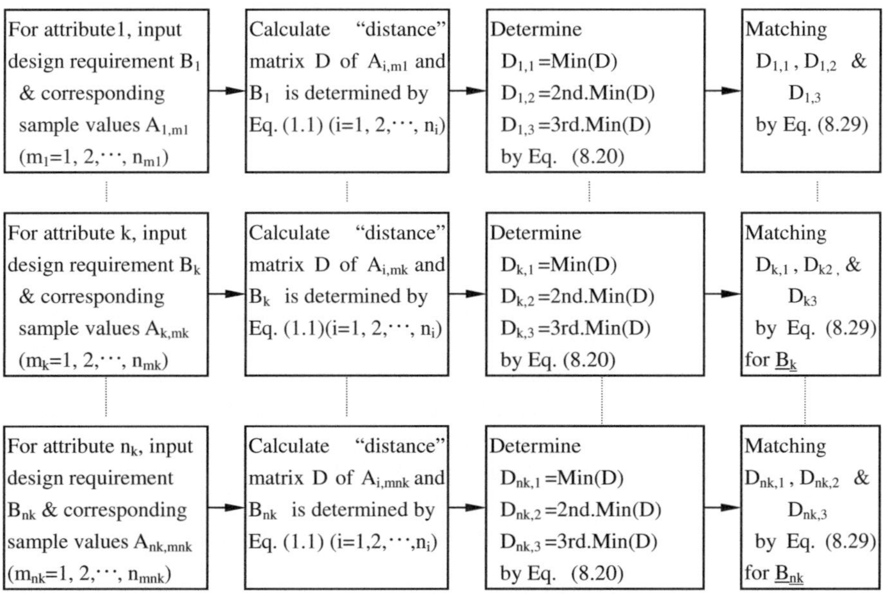

Fig. 8.9 Neural network flowchart for matching

unit in learning case replaces fuzzy "distance" by "nearness degree" for the purpose of providing similarity matrix to third layer neural processing units. This is to execute a process of minimizing selection for the purpose of providing three minimum "distance" to the third layer neural processing units for matching. Similarly, the readers also are able to realize the other two fuzzy identification and evaluation models with fuzzy-NN.

8.6.4 Conclusive Remarks

This chapter discusses the modeling and the adoption of fuzzy machine learning and fuzzy neural network in the intelligent CAD of structures. Four kinds of fuzzy machine learning models are presented, namely, fuzzy classification, fuzzy identification, fuzzy matching, and fuzzy evaluation. The FNN model seems more adoptable since the results of FML (Fuzzy Machine Learning) model will greatly affect by initial selection of membership function. Further efforts must be spent for cardinally improving the membership function. Fortunately, it can be realized theoretically through improvement of assigned membership function by fuzzy machine learning of well-chosen existing samples [62].

8.7 Fuzzy Neural Network Model for Structural Design

Taking the advantages of machine learning through fuzzy neural network, the F-NN procedures for structural design are formulated and discussed in this section. Practical examples of preliminary design of reinforced concrete structure are presented for illustrating the philosophy of the intelligent comparative design [61].

8.7.1 Introduction

Researchers might have experienced that the development of knowledge based expert system (KBES) could be constrained by the effectiveness of knowledge acquisition from the domain experts since the human being is by no means perfect in knowledge transfer. Hence, it is difficult to develop a design system for practical use from the KBES. The expert system for design is characterized by the combination of logic inference and functional evaluation. However, the inference machine of the KBES, expressed by production rule-based system, is just processing the "IF–THEN" rule sets in the knowledge base. It seems impossible that an expert system could cover all necessary information of the design issues.

Efforts have been made by incorporating computational capabilities into the system, such as structural analysis, section design of subsystems etc. Nevertheless, these efforts will not benefit to preliminary design stage, where the concepts and experiences of the designer are dominant. In fact, an intelligent design system should simulate the perceptional imagination processes of the designer; yet, obviously it is far from enough to perform this function by the KBES with "IF–THEN" logic.

Taking into account of the characteristics of neural network mentioned above, it is appropriate to use neural network reasoning model and learning strategies for intelligent comparative design of structures.

8.7.2 Neural Network Back-Error Propagation Learning

See Fig. 8.10.

8.7.3 NN Machine Learning for Sample Classification

Considering machine learning for sample classification in Fig. 8.7, the principle of NN machine learning for sample classification is based on fuzzy similarity for identification as shown in Eq. (8.20). Denote $\underline{A}_i$ $(i = 1, 2, \ldots, n)$ the learned sample in the sample base; $\underline{B}$ the design alternative to be identified. Combining $\underline{A}_i$ and $\underline{B}$ to form a new sample set $\{A_f\}$ (where $f = 1, 2, \ldots, n + 1$), then the "distance matrix" between $\underline{A}_i$ and $\underline{B}$ can be obtained through "pair comparison" in Eq. (8.21), i.e., to compare both $\underline{A}_i$ $(i = 1, 2, \ldots, n)$ and $\underline{B}$ one by one for the importance between each other [56].

It is shown in Sect. 8.5.2.2 that the similarity degree is based on "distance" of the design alternatives, where the "distance" can be either measured by membership function or by nearness degree. Define F_j, the membership function of samples with respect to jth attributes, which is heuristically defined by the domain experts

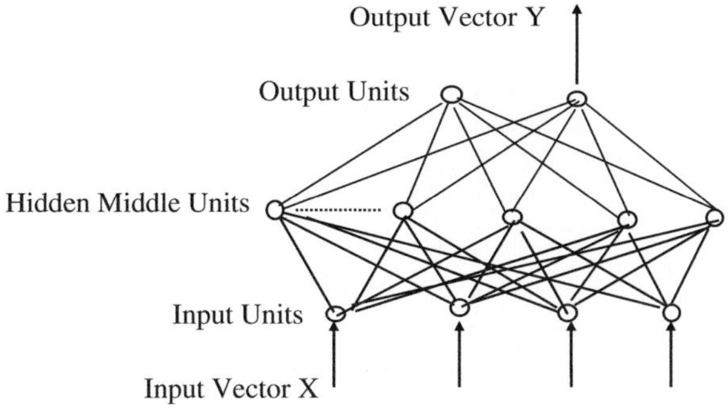

Fig. 8.10 Neural network for learning

beforehand. The similarity degree versus “distance” relationship is determined by membership function values A_{ij} of ith sample and jth attribute. Denote “distance interval” set $\{D\} = \{D_1, D_2 \ldots D_9\}$, then the relationship between the similarity degree $\underline{S}_{ij}$ and the “distance interval” $\{D\}$ between samples is shown in Table 8.1.

The classification of sample is to ranking the attributes of each example. After NN learning, each of m attributes of n samples will be ranked into different attribute groups $\{g\} = \{g_p\}$ ($p = 1, 2, \ldots$). In each group, the attributes are similar in certain degree.

8.7.4 *The Implementation of Fuzzy Machine Learning*

The implementation of fuzzy machine learning processes is started by system analysis of the “sample-attribute-factor” system. For each sample, there are several attributes, and for each attribute, there are still several factor to be affected on it. The notations of which are shown in Table 8.2. Whereas the flowchart of fuzzy machine learning for sample classification is presented in Fig. 8.2 and its NN implementation is presented in Fig. 8.11 in three NN layers with total number of neural processing units 3 nm. Through learning, the sample attributes are classified and stored in the engineering sample fact base.

Table 8.2 “Sample-Attribute-Factor” system chart

Sample Sets	Sample	Attribute	Factor					
		A_{11}	A^1_{11}	A^2_{11}		A^k_{11}		A^{s1}_{11}
		……	……					
	A_1	A_{1j}	A^1_{1j}	A^2_{1j}		A^k_{1j}		A^{sj}_{1j}
		……	……					
		A^1_{1m}	A^1_{1m}	A^2_{1m}		A^k_{1m}		A^{sm}_{1m}
	⋮	⋮	……					
		A_{i1}	A^1_{i1}	A^2_{i1}		A^k_{i1}		A^{s1}_{i1}
Sample set		……	……					
$\{A_i\}$	A_i	A_{ij}	A^1_{ij}	A^2_{ij}		A^k_{ij}		A^{sj}_{ij}
(i=1,2,…,n)		……	……					
		A_{im}	A^1_{im}	A^2_{im}		A^k_{im}		A^{sm}_{im}
	⋮	⋮	……					
		A_{n1}	A^1_{in1}	A^2_{n1}		A^k_{n1}		A^{s1}_{n1}
		……	……					
	A_n	A_{nj}	A^1_{nj}	A^2_{nj}		A^k_{nj}		A^{sj}_{nj}
		……	……					
		A_{nm}	A^1_{nm}	A^2_{nm}		A^k_{nm}		A^{sm}_{nm}

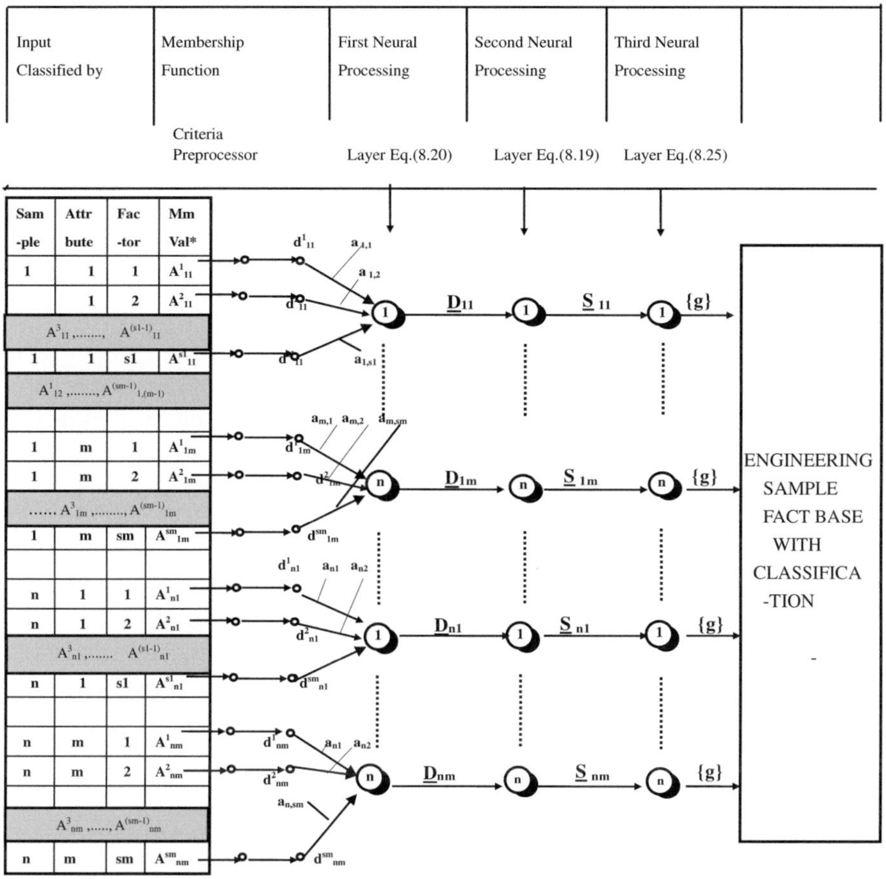

* Mem Val. means membership function value.; $\underline{D}_{ij} = \sum_{k=1}^{sj} a^k_{ij} w^k_{ij} (1 + |a^k_{ij}|)$ (i=1,2,...,n; j=1,2,...,m)

Fig. 8.11 Implementation of fuzzy machine learning for sample classification

(1) The first processing layer is mainly dealing with the fuzzy distance calculation with a pre-processor of manipulating the factorial information as the input data. The factorial "fuzzy distance" is accounted relatively to sample 1.

$$d^k_{ij} = A^k_{1j} - A^k_{ij} \tag{8.44}$$

Therefore, we have the "fuzzy distance" of any attribute j and factor k of sample 1, equals zero

$$d^k_{1j} = A^k_{1j} - A^k_{1j} = 0$$

The distance intensity a_{ij}^k represents the relative deviation of ith sample (jth attribute and kth factor) with of sample 1 (jth attribute and kth factor)

$$a_{ij}^k = d_{1j}^k - d_{ij}^k \tag{8.45}$$

The attributive "fuzzy distance" of attribute j of sample i relative to sample 1 is $\underline{D}_{ij}$, which is a Sigmoid function $f(x)$, where $x = \underline{D}_{ij} + \theta$; and the threshold value takes $\theta \approx 0.5$.

$$\underline{D}_{ij} = \sum_{k=1}^{sj} a_{ij}^k w_{ij}^k \left(1 + \cdots \left|a_{ij}^k\right|\right) \tag{8.46}$$

Input of the 1st layer: the factorial "Fuzzy distance" of samples with respect to specific attribute.

Output of the 1st layer: the attributive "fuzzy distance" $\underline{D}_{ij}$ for jth attribute of ith sample, where symbol k is the superscript of factor order, k represents the number of factors affecting jth attribute.

(2) Processing the attributive distance matrix Θ and similarity matrix $\underline{S}$
Based on Eqs. (8.44), (8.45) and (8.46), $\underline{D}_{ij}$, the fuzzy attributive distance related to sample i, is transformed by Table 8.2 into the similarity degree S_{ij}, the similarity matrix $\underline{S}$ is then determined.

$$\Theta = \left[\underline{D}_{ij}\right] = \left[\sum_{k=1}^{sj} a_{ij}^k w_{ij}^k \left(1 + \left|a_{ij}^k\right|\right)\right] \tag{8.47}$$

$$\underline{S} = \left[\underline{S}_{ij}\right] \tag{8.48}$$

Input of the 2nd layer: the attributive "Fuzzy distance" $\underline{D}_{ij}$ of samples with respect to specific attribute, the matrix $[\underline{D}_{ij}]$ is input to this layer.
Output of the 2nd layer: the "similarity degree" for jth attribute of ith sample S_{ij}, the matrix $[S_{ij}]$ is output from this layer.

(3) Processing of sample attribute classification relative to sample 1, Eq. (8.48) provides the similarity degree S_{ij} for each attribute j of the sample i. The criterion of classification for attributes is

$$\begin{array}{lll} \text{if} & S_{ij} \geq 0.95; & \text{Then group } g_1 \\ \text{if} & 0.95 > S_{ij} \geq 0.90; & \text{Then group } g_2 \\ \text{if} & 0.90 > S_{ij} \geq 0.85; & \text{Then group } g_3 \\ \text{if} & 0.85 > S_{ij} \geq 0.80; & \text{Then group } g_4 \\ & & \cdots \end{array} \tag{8.49}$$

$$\{g\} = \{g_1, g_2, g_3, g_4 \cdots\}$$

If the similarity degree of the attribute satisfies Eq. (8.49), then those sample attributes are in the same rank with the sample 1. Simultaneously, Eq. (8.49) can be expanded to fit different similarity value intervals, and then the attributes of different samples could be accordingly sorted into various groups in the "Engineering Sample Fact Base" as shown in Fig. 8.7.

Input of the 3rd layer: the "similarity degree" matrix $[S_{ij}]$ is input to this layer.

Output of the 3rd layer: the attribute ranking by Table 8.1 with different attribute groups are stored in the "Engineering Sample Fact Base" in form of $\{g\} = \{g_1, g_2, g_3, g_4 \ldots\}$.

8.7.5 Fuzzy-NN Machine Learning for Structural Design

In order to verify the adoptability of above two models, this section presents a practical example of preliminary structural design of RC high-rise building [61].

Engineering samples and original design data (or known design parameters) are presented Table 8.3. There are 37 engineering samples, the first 36 denoting the samples and the 37-th, a testing design alternative, which is actually a real sample but we introduced for checking the results from the two above-mentioned models. The original data of 37-th testing design alternative is listed at the last line of Table 8.3.

Using the two above-mentioned models, one may obtain design parameters from the results of both models presented in Table 8.2. It shows that the results are acceptable; the two models are worthily to be adopted in the structural primary design.

From data given in Table 8.5, the results from two models are well similar to the real data. In fact, the fuzzy machine learning process, its fuzzy reasoning process implies the rectification of heuristic membership function of the uncertainties. Through training by engineering samples, it will obtain reasonable membership values, and then the designer could draw a series of favorable parameters. On the other hand, the fuzzy-NN model and its BP iteration represent the process of rectification of heuristic membership function of uncertain design. The essentials of rectification of membership functions of the design parameters are the same although the expression of modeling is different. The consideration of both results and the real data from engineering sample implies the fact that fuzzy approach is reasonable and the convergence of which, though in different models, is satisfactory. Thus, in theoretical point of view, the intelligent design by fuzzy approach is reliable and convincible for practical application in preliminary design of structures.

The classification of samples is presented below as a numerical example. If the learned sample is $\underset{\sim}{A}_i (i = 1, 2, 3, 4)$ and the new sample to be classified is $\underset{\sim}{B}$, then the new set $\{A_f\} (f = 1, 2, 3, 4, 5)$ with $\underset{\sim}{B} = A_5$ can be setup. Providing the attribute set is $U = (U_1, U_2, U_3, U_4)$, where U_1 represents shape; U_2 represents floor area; U_3, building height and U_4, building width. We have the membership function of each sample with respect to each individual influenced attribute as shown in Table 8.6.

Table 8.3 Engineering samples and original data

Sample no.	Attr. 1	Attr. 2	Attr. 3	Attr.4	Attr. 5	Attr. 6	Attr. 7	Attr. 8	Attr. 9 (tsd. m^2)
1	7.0	3.0	71.04	24	2.90	4.50	7.50	1.40	16.95
2	7.0	2.0	80.50	26	3.00	3.50	5.00	1.50	17.70
3	7.0	3.0	65.90	20	3.00	4.60	7.60	1.03	28.49
4	7.0	2.0	64.07	18	3.30	3.70	4.80	1.60	15.07
5	7.0	2.0	66.89	21	3.00	4.00	6.00	1.30	12.53
6	8.0	2.0	70.00	23	2.95	3.70	9.20	1.23	22.00
7	7.0	1.0	84.10	26	3.10	3.70	5.00	1.70	23.00
8	7.0	2.0	114.5	36	3.00	6.60	8.00	1.85	50.00
9	7.0	3.0	72.60	23	3.00	3.90	9.00	1.40	18.00
10	6.0	2.0	69.75	22	3.00	4.00	5.60	1.17	18.60
11	6.0	2.0	63.60	20	3.00	3.60	7.80	1.80	18.00
12	7.0	1.0	67.80	23	2.80	3.80	4.50	1.30	21.10
13	8.0	2.0	39.50	13	2.90	7.20	9.00	1.33	30.00
14	7.0	2.0	55.00	16	3.30	4.50	4.50	1.80	9.22
15	7.0	3.0	50.40	16	3.00	3.90	6.00	1.22	12.80
16	7.0	1.0	61.60	19	3.00	3.60	6.00	1.60	10.50
17	7.0	2.0	80.90	24	3.40	3.60	6.00	1.70	19.97
18	8.0	2.0	109.0	32	3.30	3.40	4.90	1.40	53.98
19	7.0	2.0	50.30	15	3.20	3.60	6.90	1.40	7.82
20	7.0	2.0	65.90	19	3.00	4.20	4.90	1.60	10.00
21	8.0	2.0	58.80	20	2.90	3.90	5.10	1.25	10.98
22	8.0	2.0	49.80	18	2.70	3.30	5.10	1.25	19.21
23	8.0	2.0	48.60	18	2.70	3.30	4.80	1.25	70.10
24	8.0	2.0	82.30	28	2.70	3.30	5.10	1.46	15.16
25	7.0	1.0	82.90	25	3.00	6.30	7.80	1.24	22.00

(continued)

Table 8.3 (continued)

Sample no.	Attr. 1	Attr. 2	Attr. 3	Attr.4	Attr. 5	Attr. 6	Attr. 7	Attr. 8	Attr. 9 (tsd. m^2)
26	7.0	2.0	86.00	24	3.30	6.00	6.60	1.70	16.00
27	6.0	3.0	95.10	30	2.80	3.00	5.10	1.50	50.80
28	8.0	2.0	39.10	11	3.30	5.00	5.00	1.60	6.80
29	8.0	2.0	67.80	21	2.80	5.00	7.50	1.31	20.00
30	8.0	2.0	76.00	17	4.50	3.30	7.20	1.61	10.20
31	7.0	3.0	69.10	15	4.50	3.30	6.00	1.61	7.40
32	8.0	2.0	55.80	14	3.60	4.20	5.50	1.93	4.62
33	7.0	3.0	55.40	16	3.20	3.60	6.00	1.30	12.82
34	8.0	1.0	56.50	14	3.80	3.30	6.00	1.59	9.57
35	8.0	2.0	35.30	9	3.95	3.60	6.90	1.67	12.23
36	8.0	2.0	50.00	14	3.60	5.40	5.40	1.60	10.60
37	7.0	2.0	65.50	18	3.50	4.20	6.00	1.85	17.9.

Attr. 10	Attr. 11	Attr. 12	Attr. 13	Attr. 14	Attr. 15	Attr. 16	Attr. 17
48.0	2	15.0	60.0	25.0	90.0	90.0	40.0
55.0	2	20.0	55.0	25.0	70.0	70.0	30.0
35.0	2	14.0	50.0	40.0	70.0	70.0	25.0
35.0	2	11.0	60.0	25.0	70.0	60.0	30.0
90.0	2	14.0	50.0	25.0	60.0	60.0	25.0
30.0	4	10.0	60.0	35.0	110.0	110.0	40.0
55.0	4	11.0	50.0	35.0	80.0	60.0	40.0
35.0	3	18.0	0.0	0.0	0.0	0.0	35.0
75.0	5	12.0	60.0	25.0	100.0	60.0	35.0
40.0	2	13.0	50.0	25.0	65.0	65.0	20.0
70.0	2	12.0	70.0	35.0	70.0	70.0	25.0
50.0	2	13.0	60.0	30.0	60.0	60.0	25.0

(continued)

Table 8.3 (continued)

Attr. 10	Attr. 11	Attr. 12	Attr. 13	Attr. 14	Attr. 15	Attr. 16	Attr. 17
30.0	2	18.0	80.0	30.0	90.0	90.0	25.0
60.0	2	22.0	65.0	25.0	70.0	70.0	25.0
50.0	2	11.0	60.0	25.0	60.0	60.0	25.0
55.0	2	16.0	50.0	24.0	80.0	60.0	25.0
45.0	3	20.0	0.0	0.0	0.0	0.0	30.0
35.0	5	13.0	50.0	40.0	75.0	60.	60.0
40.0	2	12.0	60.0	30.0	60.0	60.0	20.0
30.0	4	13.0	55.0	35.0	70.0	70.0	30.0
40.0	3	13.0	0.0	0.0	0.0	0.0	18.0
35.0	2	13.0	60.0	40.0	50.0	50.0	19.0
35.0	3	10.0	0.0	0.0	0.0	0.0	18.0
35.0	3	13.0	0.0	0.0	0.0	0.0	25.0
50.0	2	15.0	60.0	30.0	70.0	70.0	30.0
25.0	2	18.0	60.0	30.0	70.0	60.0	30.0
35.0	2	10.0	80.0	40.0	120.0	100.0	35.0
35.0	3	16.0	0.0	0.0	0.0	0.0	25.0
35.0	2	16.0	65.0	35.0	60.0	60.0	30.0
45.0	2	10.0	60.0	25.0	65.0	60.0	30.0
25.0	2	10.0	55.0	25.0	75.0	45.0	25.0
30.0	2	10.0	80.0	35.0	60.0	60.0	22.0
60.0	3	10.0	0.0	0.0	0.0	0.0	20.0
50.0	2	10.0	60.0	30.0	55.0	55.0	25.0
35.0	1	13.0	65.0	35.0	50.0	50.0	0.0
35.0	2	9.0	60.0	25.0	50.0	50.0	35.0
85	?	?	?	?	?	?	?

Table 8.4 The definition of attributes for sample base in Table 8.3

Attribute no.	Definition	Attribute no.	Definition
Attr. 1	Earthquake density	Attr. 2	Site soil classification
Attr. 3	Total height (m)	Attr. 4	Number of stories
Attr. 5	Height of standard storey (m)	Attr. 6	Typical X-direction span (m)
Attr. 7	Typical Y-direction span (m)	Attr. 8	Standard floor loading (kN/m^2)
Attr. 9	Total architectural area (m^2)	Attr. 10	Basic wind pressure (kN/m^2)
Attr. 11	Type of structural system: 1-frame; 2-frame-shear wall; 3-shear wall; 4-frame/tube; 5-tube in tube	Attr. 12 Attr. 13	Thickness of floor slab (cm) Height of beam cross-section (cm)
Attr. 14	Width of the beam cross-section (cm)	Attr. 15	Height of column cross-section (cm)
Attr. 16	Width of the column cross-section (cm)	Attr. 17	Thickness of the shear wall (cm)

Table 8.5 Data table for a preliminary structural design with two models

Attribute no.	Definition	Real sample	Fuzzy machine learning (FML)	Fuzzy neural network (FNN)
Attribute 11	Structure type	Frame-shear wall	Frame-shear wall	Frame-shear wall
Attribute 12	Thickness of usual slab (m)	0.120	0.141	0.130
Attribute 13	Height of usual beam cross-section (m)	0.500	0.553	0.522
Attribute 14	Width of usual beam cross-section (m)	0.350	0.303	0.358
Attribute 15	Height of usual beam cross-section (m)	0.600	0.671	0.626
Attribute 16	Width of usual column cross-section (m)	0.600	0.650	0.624
Attribute 17	Thickness of usual shear wall (m)	0.250	0.320	0.247

Table 8.6 Membership function A_{ij} of samples

Sample no.	U_1	U_2	U_3	U_4
$\underline{A}_1$	0.8	0.5	0.6	0.6
$\underline{A}_2$	0.6	0.6	0.8	0.7
$\underline{A}_3$	0.7	0.7	0.5	0.8
$\underline{A}_4$	0.5	0.8	0.8	0.7
$\underline{B} = \underline{A}_5$	0.6	0.7	0.5	0.6

Sequentially substituting those data from Table 8.2 to Eq. (8.44), (8.45) and (8.46), the fuzzy distance matrix Θ can be determined by Eq. (8.24). Introducing criteria from Table 8.1, the similarity matrix $\underline{S}$ can be consequently expressed by Eq. (8.25). The notations of "sample-attribute-factor" system are shown in Table 8.6. Whereas the flowchart of fuzzy machine learning for sample classification is presented in Fig. 8.7 and its implementation, in Fig. 8.11.

For evaluating the element in Eq. (8.24), according to Table 8.1, two examples are presented:

(1) $\underline{S}_{23} = 0.95$ is determined by $\underline{D}_{23} = 0.078$ in Eq. (8.25); which is located between 0.05 and 0.10.
(2) $\underline{S}_{45} = 0.85$ is determined by $\underline{D}_{45} = 0.160$ in Eq. (8.25); which is located between 0.15 and 0.20.

From Eq. (8.25) and (8.49), the ranking matrix of the example is

$$R = \begin{bmatrix} 1 & 2 & 1 & 3 & 1 \\ 2 & 1 & 2 & 1 & 3 \\ 1 & 2 & 1 & 2 & 2 \\ 3 & 1 & 2 & 1 & 4 \\ 1 & 3 & 2 & 4 & 1 \end{bmatrix} \quad (8.50)$$

The samples are stored in group at the "Engineering Sample Fact Base" after classification.

For attribute 1: Group 1—Sample 1, 3, 5; Group 2—Sample 2; Group 3—Sample 4.

For attribute 2: Group 1—Sample 2, 4; Group 2—Sample 1, 3; Group 3—Sample 5.

For attribute 3: Group 1—Sample 1, 3; Group 2—Sample 2, 4, 5.

For attribute 4: Group 1—Sample 2, 4; Group 2—Sample 3; Group 3—Sample 1; Group 4—Sample 5.

Based on the sample base presented in Table 8.3, the first 10 attributes of the high-rise building samples are input variables and the seven remaining attributes are the output variables of design solution. The fuzzy-NN machine learning process is to "learn" from these samples, the NN unit strength between each other after learning represents the implicit energy distribution of the system corresponding to the arts of design intelligence.

Consequently, in construction of NN system we consider 10 units for input layer; 7 units for output layer and 15 units for the hidden layer. The parameters in calculation are determined empirically and revised afterward through comparison.

Let us first to train the NN through learning the sample No. 1–37 from Table 8.3, then arbitrarily use samples No. 4, 5, 8, 20 as the testing samples for checking. By inputting first 10 attributes of those samples, we obtain 7 outputs each through NN BP learning. The results from the outputs can be compared with the original 7 outputs for assessing the errors of the system.

Let the learning sample set $\{A_i\} = \{A_1, A_2, \ldots, A_{34}, A_{35}\}$;

Testing sample set $\{A'\} = \{A_4, A_5, A_8, A_{20}\}$,

Output vector of testing sample $V_q = \{a_1, a_2, a_3, a_4, a_5, a_6, a_7\}$; $(q = 1, 2, \ldots)$

Output vector of testing samples through learning $V' = \{a'_1, a'_2, a'_3, a'_4, a'_5, a'_6, a'_7\}$

The error estimation is measured by the mean value of root of the square sum.

$$\varepsilon = \frac{1}{N}\left\{ \left[(a_1 - a_1')/a_1\right]^2 + \left[(a_2 - a_2')/a_2\right]^2 + \cdots + \left[(a_N - a_N')/a_N\right]^2 \right\}^{1/2} \tag{8.51}$$

After NN machine learning by the samples of No. 1–35 from Tables 8.3 and 8.4, we obtain two weight matrices W_{ij} (sizing 10 × 15) and W_{jk} (sizing 15 × 7) as below:

$$W_{ij} = (\text{sizing } 10 \times 15)\begin{bmatrix} -0.081 & -0.374 & 1.187 & \ldots\ldots & -0.396 & 0.253 & 0.030 \\ -0.325 & 0.520 & 0.625 & \ldots\ldots & -0.455 & -0.625 & -0.248 \\ & \ldots & \ldots & \ldots & \ldots & & \\ -0.359 & -0.369 & 0.395 & \ldots\ldots & 0.096 & -0.457 & -0.414 \\ -0.506 & -0.283 & 3.254 & \ldots\ldots & -0.654 & -1.024 & -0.433 \end{bmatrix} \tag{8.52}$$

$$W_{jk} = (\text{sizing } 15 \times 7)\begin{bmatrix} -0.985 & -0.943 & \ldots\ldots & -0.026 & 0.212 \\ 0.009 & -0.408 & \ldots\ldots & 0.012 & -1.054 \\ & & & & \\ -0.696 & -0.072 & \ldots\ldots & 0.144 & -0.599 \\ -0.084 & 0.101 & \ldots\ldots & 0.683 & -0.992 \end{bmatrix} \tag{8.53}$$

Through learning, the NN system is able to perform intelligent capabilities of the design. The results of testing samples manipulated by the system are presented in Table 8.7. It shows that the mean error from NN system for design is less than 6 % and the attributes of high-rise building structure in Table 8.7 is then considered to be acceptable.

The adaptability of NN system means that the design solutions provided by the NN system is rather stable, it will not volatile due to any changing of individual

Table 8.7 Results of Testing Samples through Net

Sample no.	[a] [b]	Attri. 11	Attri. 12	Attri. 13	Attri. 14	Attri. 15	Attri. 16	Attri. 17	Error[c]
4	SB	2.000	11.000	60.000	25.000	70.000	60.000	30.000	0.055
	NN	2.501	13.504	60.843	31.365	73.316	66.933	26.955	
5	SB	2.000	14.000	50.000	25.000	60.000	60.000	25.000	0.036
	NN	1.879	13.521	54.219	29.202	68.979	62.588	25.661	
8	SB	3.000	18.000	60.000	30.000	90.000	80.000	35.000	0.045
	NN	2.589	13.950	59.946	31.193	87.745	73.672	40.367	
20	SB	4.000	13.000	55.00	35.00	70.000	70.000	30.000	0.059
	NN	2.545	13.953	61.020	31.517	74.212	67.825	27.577	

[a]SB—Sample Base value; [b]NN—Neural Network value; [c]—calculated by Eq. (8.51)

Table 8.8 Comparison of testing sample solutions with and without the input of attribute 9

Sample no.	Case	Attri. 11	Attri. 12	Attri. 13	Attri. 14	Attri. 15	Attri. 16	Attri. 17
4	A	2.501	13.504	60.843	31.365	73.316	66.933	26.955
	B	2.336	13.320	61.795	32.271	71.735	66.571	26.179
5	A	1.879	13.521	54.219	29.202	68.979	62.588	25.661
	B	1.913	13.710	51.995	28.240	67.448	59.919	27.197
8	A	2.589	13.950	59.946	31.193	87.745	73.672	40.367
	B	2.605	14.826	57.432	32.266	90.171	77.774	44.007
20	A	2.545	13.953	61.020	31.517	74.212	67.825	27.577
	B	2.342	13.370	62.580	32.694	73.235	67.953	26.876

Case A—Testing sample solutions with the input of attribute 9
Case B—Testing sample solutions without the input of attribute 9

design attributive input. Table 8.8 shows the testing design solutions with and without the input of attribute 9 (total floor area). The comparison shows that both results mentioned above are almost the same.

Conclusion

It is mentioned that the fuzzy machine learning and the neural network learning are the fundamental approaches in the intelligent systems for structural design. The preliminary results presented in this chapter show that the previous one is devoted to the classification of samples, which form the sample base of the system, sorted by its attributes; the latter one aims to "cultivate" the system through NN learning in order to accumulate the design expertise of human being in the system.

It is necessary to point out that since the number of the samples used in this chapter is not adequate, the characteristics of the building are also different from each other. It seems that further efforts must be paid for collecting more samples of various kinds until the sample base can be improved for practical use. Nevertheless, the methodology of intelligent design introduced in this chapter is actually based on comparative design philosophy. It has been much encouraged in this chapter that even by using such an incomplete sample base, the numerical error estimation of test examples show the mean error of the samples inferred by NN system can be limited within 6 %.

Nevertheless, for achieving benefits from intelligent comparative design of structure, the input fuzzy membership function is the key for success, hopefully the fuzzy machine learning method can effectively solve the problem by [62]. However, for achieving our goals of benefit by intelligent comparative design of structure, there will be a lot of common efforts to be spent for researchers and practitioners in the future.

S. Lin and Z. Huang, *Comparative Design of Structures*,
DOI 10.1007/978-3-662-48044-1

Appendix A

A.1 Example of Conceptual Design for Single-Story Multi-span Steel Factory Building

Previous chapters of this book have introduced in detail some indispensable mechanical concepts and methods of simplified calculation in modern structural design. Now we'll come to the application of these concepts and methods in practical construction project, specifically, in the conceptual design for single-story multi-span steel structural factory building.

1. **Project Introduction**

This is a construction project of an industrial factory building, taking up a construction area of about 20,000 m^2. The factory building is comprised of two single-storied multi-span steel structural sections A and B (as shown in Fig. A.1). Each covers an area of 10,000 m^2. In light of the length limitations of this book, we will take factory building A as an example and talk about the conceptual design of this part of the factory only.

2. **The purpose of conceptual design**

    (1) It is to select a better structural system and the way of arrangement, in which case the resultant structure should not only be up to the standard but also minimize the cost and guarantee the safety.

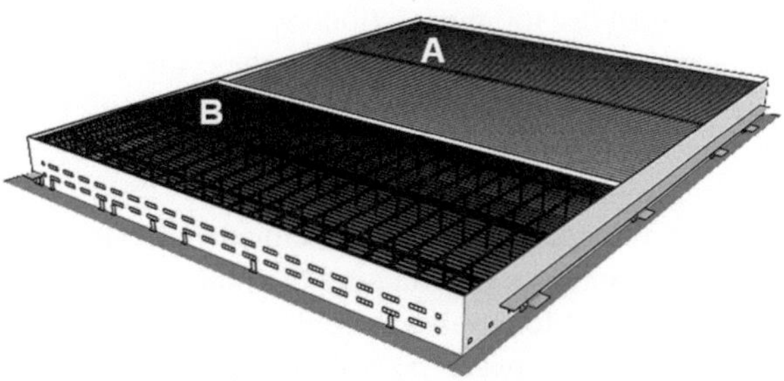

Fig. A.1 Rendering of this prospective factory building

S. Lin and Z. Huang, *Comparative Design of Structures*,
DOI 10.1007/978-3-662-48044-1

(2) Through the estimation of steel consumption of the main structure, the appraisal of economic efficiency and applicability of the structural system could be better made along with the total cost of construction.
(3) Conceptual design could serve as a guide to the following preliminary design and detailed design. The control of the most critical section and the combination of loads ensures the safety and rationality of structural design.

3. **Conditions of design**

(1) Proprietor-provided plane and elevation layouts,
(2) Proprietor-provided live loads,
(3) Preliminary survey report of foundation soil.

4. **Design norms**

(1) Load Code for the Design of Building Structures (GB50009-2001), 2001, Beijing,
(2) Technical Specification for Steel Structure of Lightweight Building with Gabled Frames (CECS102), 2002, Beijing,
(3) Code for Design of Steel Structures (GB50017-2002), 2002, Beijing,
(4) Code for Design of Building Foundation (GB50007-2002), 2002, Beijing,
(5) Code for Design of Concrete Structures (GB50010-2002), 2002, Beijing.

5. **The assumptions**

(1) Assume the structural system works in an elastic state.
(2) Live loads are simplified as equivalent uniformly or linearly distributed loads, conducted with reference to "Load Code for the Design of Building Structures".
(3) All supports or connecting joints are assumed to be rigid connections, in which case the bending moments are transferred.
(4) The slope of the roof is assumed to be 5 %.

6. **The diagram of simplified calculation of the main steel structure**

According to the design principles of optimization of steel structures, multiple spans of single-storied portal frames of steel structure are employed in this project. Since the load of crane shall be imposed on each span, identical-sectioned wide-flanged I-shaped steels are employed in each section of column of the main rigid frame; as for the section of the beams, differential-sectioned wide-flanged I-shaped steels are employed to reduce the consumption of steel. Since the crane goes over each span of the structure, beams and columns are rigidly connected and so it is with the columns and the foundation, which enables the transference of the bending moment.

With reference to the requirements of construction technology and the optimization principles of structure, the main structure of factory building A adopts the portal frame of four spans. The span crosses over a distance of 25 m and there's a spacing of 6 m between main rigid frames. The side column is 12 m high, as shown in Fig. A.2.

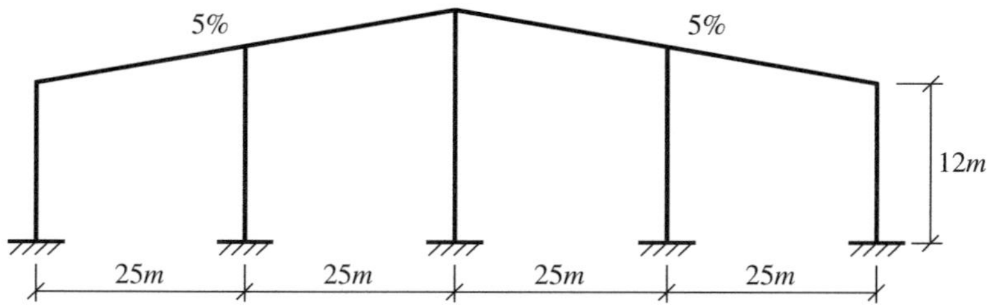

Fig. A.2 Diagram of simplified design and calculation of the main steel structure

7. **Estimation of the loads in structural design**

The estimation principles of loads introduced in Chap. 2 shall be applied. Taking into consideration the requirements of construction technology for industrial project, major loads as listed in Table A.1 have to be considered in the design of this factory building.

8. **The estimation of the internal forces of the cross section under the action of the most unfavorable combination of vertical loads**

According to the Load Code for the Design of Building Structures (GB50009-2001), the most unfavorable vertical uniform load of beam is 1.2 × deadweight + 1.4 × overhanging + 1.4 × load of snow or live load upon the roof

$$w = 1.2 \times 0.45 + 1.4 \times 0.10 + 1.4 \times 0.30 = 1.10\,\mathrm{kN/m^2}$$

The deadweight of the roof structure (beam, purlin, and the panel included) is estimated to be 0.45 kN/m^2.

The uniform load derived from the equation has to be transformed into equivalent uniform linear load q, being borne by each single-span portal frame.

Since the spacing between the frames is 6 m, we have

$$q = 6\,\mathrm{m} \times 1.10 = 6.6\,\mathrm{kN/m}$$

Table A.1 Summary of major loads for factory building A

Type	Load
Uniform roof live load	0.3 kN/m^2
Uniform roof snow load (Shanghai area)	0.2 kN/m^2
Overhanging roof load	0.1 kN/m^2
Wind load (Shanghai area)	0.55 kN/m^2
Type of crane	12.5 t × 24 m
Total weight of the crane	19.316 t
Weight of the crane itself	1.879 t
Uniform floor live load	30 kN/m^2
Load of the forklift	15 t

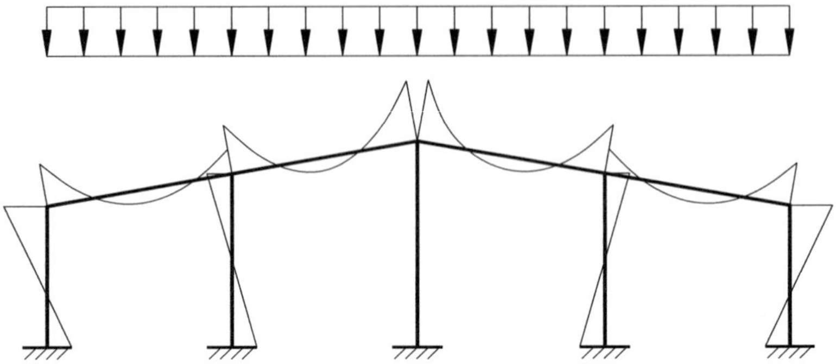

Fig. A.3 The distribution of bending moment of major rigid frame under vertical loads

The maximum bending moment and shear force borne by the primary beam of the portal frame are estimated as (the primary beam is assumed to be continuous beam)

$$M_{\max} = qL^2/12 = 6.6\,\text{kN/m} \times 25\,\text{m} \times 25\,\text{m}/12 = 343.75\,\text{kN m}$$
$$Q_{\max} = qL/2 = 6.6\,\text{kN/m} \times 25\,\text{m}/2 = 82.5\,\text{kN}$$

The maximum axial force of the central column is estimated as (the load of crane along with possible combinations considered; crane weighs 19.316 t)

$$N_{\max} = qL + 4(Qc) = 6.6 \times 25 + 4 \times 1.1 \times 0.8(125 + 193.16/2) = 944.96\,\text{kN}$$

The maximum bending moment and axial force of the side column is estimated as (the load of crane and possible combinations considered)

$$M_{\max} = (2/3)qL^2/12 = (2/3)343.75 = 229.17\,\text{kN m}$$
$$N_{\max} = 0.5(qL + 4(Qc)) = 0.5 \times 944.96 = 472.48\,\text{kN}$$

Figure A.3 is the diagram of bending moment distribution of major rigid frame under the action of most unfavorable vertical loads.

9. **Estimation of internal force of the cross section under the action of most unfavorable combination of horizontal loads**

    (1) The most unfavorable combination of horizontal loads that parallel with the axis of the major rigid frame is 1.4 × wind load + horizontal braking force of the crane. Specifically,

$$\text{Wind load}: hw = 1.4 \times (0.8 + 0.7) \times 0.55 \times 6 = 6.01\,\text{kN/m}$$

0.8 and 0.7 in the above equation are the shape coefficients of the wind load

$$\text{Horizontal braking force of the crane: } hc = 2 \times 0.7 \times 0.1(125 + 18.79) = 20.13\ \text{kN}$$

125 kN in the equation above is the lifting weight of the crane and 18.79 kN the deadweight of the crane itself.

Assume the action point of the crane is at the midpoint of the side column (the most unfavorable position), then the maximum bending moment and shear force of the central column (the height being 14.5 m) are estimated as

$$M_{\max} = 6.01 \times 12 \times (2/8) \times 14.5\ \text{m}/2 + 20.13\,\text{kN} \times 14.5\,\text{m}/8 = 165.95\,\text{kN m}$$

According to the method of simplified calculation introduced in Chap. 4 Sect. 4.7 concerning portal frame and frame structure, we assume that the wind-induced shear force distributed to the central column is twice that to the side column and the inflection point is the midpoint of the column.

$$Q_{\max} = 6.01 \times 12(2/8) + 20.13/2 = 28.10\,\text{kN}$$

The maximum bending moment and shear force of the side column are estimated as

$$M_{\max} = 6.01 \times 12 \times (1/8) \times 12\,\text{m}/2 + 0.5 \times 20.13 \times 12\,\text{m}/8 = 69.19\,\text{kN m}$$
$$Q_{\max} = 6.01 \times 12 \times (1/8) + 0.5 \times 20.13 = 19.08\,\text{kN m}$$

Figure A.4 is the distribution diagram of the bending moment of major rigid frame under the action of the most unfavorable combination of loads that parallel with the axis of the frame.

(2) The most unfavorable combination of horizontal loads that are perpendicular to the axis of major rigid frame

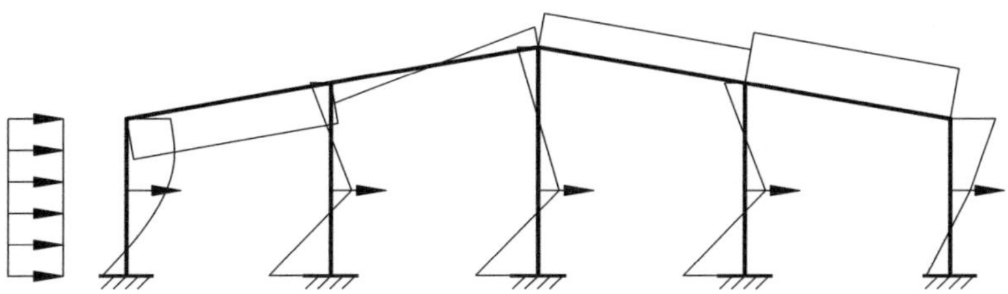

Fig. A.4 Distribution diagram of the bending moment of rigid frame under horizontal loads

Horizontal braking force of the crane:

$$hc' = 2 \times 0.7 \times 0.1(12.5\,\mathrm{t} + 19.316\,\mathrm{t})/2 = 22.27\,\mathrm{kN}$$

And the maximum bending moment and shear force of the central column are estimated as

$$M_{\max} = 22.27\,\mathrm{kN} \times 14.5\,\mathrm{m}/8 = 40.36\,\mathrm{kN\,m}$$
$$Q_{\max} = 22.27/2 = 11.14\,\mathrm{kN}$$

The maximum bending moment and shear force of the side column are estimated as

$$M_{\max} = 0.5 \times 22.27 \times 12/8 = 16.7\,\mathrm{kN\,m}$$
$$Q_{\max} = 0.5 \times 11.14 = 5.57\,\mathrm{kN\,m}$$

10. **Estimation and selection of cross section of beam and side column in major rigid frame**

The control internal forces within the section of beam and side column under the action of load combinations are summarized as

$$Mx_{\max} = 343.75\,\mathrm{kN\ m}$$
$$My_{\max} = 16.7\,\mathrm{kN\ m}$$
$$Q_{\max} = 82.5\,\mathrm{kN}$$
$$N_{\max} = 472.48\,\mathrm{kN}$$

(1) Checking of the bending moment:
Assume the major rigid frame adopts steel Q345, whose design strength is 310 mpa, the sectional modulus of the selected section shall meet the requirement

$$W > M_{\max}/310 = 343{,}750{,}000/310 = 1{,}108{,}871\ \mathrm{mm}^3$$

Symmetrical wide-flanged I-shaped section H400 × 280 × 8 × 16 could be adopted as the maximum cross section for both the side column and the beam.
The differential section of the beam is H500-300 × 280 × 8 × 16 and the calculation of deformation and strength in conceptual design adopts the equivalent identical section H400 × 280 × 8 × 16. The area moment of H400 × 280 × 8 × 16 is

$$Wx = 2 \times 280 \times 16 \times 200 + 8 \times 400 \times 400/6 = 1{,}952{,}000\,\text{mm}^3 > 1{,}108{,}871\,\text{mm}^3 \quad \text{OK!}$$
$$Wy = 2(16 \times 280 \times 280/6) = 418{,}133\,\text{mm}^3 > 16{,}700{,}000/310 = 53{,}871\,\text{m}^3 \quad \text{OK!}$$

(2) Checking of the shear force:
The maximum shear stress of the cross section is

$$\tau_{\max} = 1.5 \times 82.5\,\text{kN}/(8 \times 400) = 38.67\,\text{MPa} < [\tau] = 180\,\text{MPa} \qquad \text{OK!}$$

(3) Checking of the axial compression:
The maximum axial compressive stress within the section of the side column is

$$\sigma_{\max} = 472{,}480/(8 \times 400 + 2 \times 16 \times 280) = 38.86\,\text{MPa} < [\sigma] = 310\,\text{MPa}$$
OK!

(4) Checking of vertical deformation of the beam:
Sectional moment of inertia is first estimated:

$$Ix = 2 \times 280 \times 16 \times 200 \times 200 + 8 \times 400^3/12 = 401{,}066{,}667\,\text{mm}^4$$

The combination of loads selected for the checking of deformation (dissimilar with the one chosen in the checking of strength is 1.0 deadweight + 0.5 overhanging load + 0.5 live load upon the roof

$$w' = 0.45 + 0.5 \times 0.1 + 0.5 \times 0.3 = 0.65\,\text{kN/m}^2$$
$$q' = 6\,\text{m} \times 0.65\,\text{kN/m}^2 = 3.9\,\text{kN/m}$$

The maximum vertical deformation of the beam is

$$\Delta b = (1/384) \times 3.9\,\text{kN/m} \times 25^4\,\text{m}^4/\left(2.06 \times 10^5\,\text{N/mm}^2 \times 401{,}066{,}667\,\text{mm}^4\right)$$
$$= 0.048\,\text{m} < L/400 = 0.0625\,\text{m} \quad \text{OK!}$$

(5) Checking of horizontal deformation of the side column:
Sectional moments of inertia are first estimated:

$$Ix = 2 \times 280 \times 16 \times 200 \times 200 + 8 \times 400^3/12 = 401{,}066{,}667\,\text{mm}^4$$
$$Iy = 2 \times 16 \times 280^3/12 = 58{,}538{,}666\,\text{mm}^4$$

The selected combination of loads is different from the one in the checking of strength.
The combination of horizontal loads that parallel with the axis of the major rigid frame is as follows:

Wind load + horizontal braking force of the crane. Specifically,
Wind load: $hw = (0.8 + 0.7) \times 0.55 \times 6 = 4.95$ kN/m
Horizontal braking force of the crane: $hc = 2 \times 0.7 \times 0.1$ $(125 + 18.79) = 20.13$ kN
Assume the action point of the crane is at the midpoint of the side column (the most unfavorable position).
The shear force imposed upon the side column is estimated as

$$Qx = 4.95 \times 12 \times (1/8) + 0.5 \times 20.13 = 17.49 \text{ kN m}$$

The deformation of the side column under the action of axis-paralleling horizontal loads is estimated as

$$\begin{aligned}\Delta c &= (Qx)h^3/(12EI) \\ &= 17.49 \text{ kN} \times 11^3 \text{ m}^3/\left(12 \times 2.06 \times 10^5 \text{ N/mm}^2 \times 401{,}066{,}667 \text{ mm}^4\right) \\ &= 0.0235 \text{ m} < 0.0275 \text{ m} = h/400 \quad \text{OK!}\end{aligned}$$

The deformation of the side column under the action of axis-perpendicular horizontal loads is estimated as

$$\begin{aligned}\Delta c &= (Qy)h^3/(12EI) \\ &= 5.57 \text{ kN} \times 12^3 \text{ m}^3/\left(12 \times 2.06 \times 10^5 \text{ N/mm}^2 \times 58{,}538{,}666 \text{ mm}^4\right) \\ &= 0.007 < 0.03 = h/400 \qquad \text{OK!}\end{aligned}$$

11. **Estimation and selection of cross section of the central column in major rigid frame**

(1) The control internal forces of the cross section of the central column under the action of the most unfavorable combination of loads, derived from previous calculation, are

$$\begin{aligned}Mx_{\max} &= 165.95 \text{ kN m} \\ My_{\max} &= 40.36 \text{ kN m} \\ Q_{\max} &= 28.1 \text{ kN} \\ N_{\max} &= 944.96 \text{ kN}\end{aligned}$$

(2) Symmetrical wide-flanged I-shaped section H300 × 260 × 8 × 16 is adopted and the according area moments are

$$Wx = 2 \times 260 \times 16 \times 150 + 8 \times 300 \times 300/6 = 1{,}368{,}000 \text{ mm}^3$$
$$Wy = 2(16 \times 260 \times 260/6) = 3{,}605{,}300 \text{ mm}^3$$

(3) Checking of the bending moment, shear, and axial force:

$$Mx_{\max} = (Wx) \times 310 = 1{,}368{,}000 \times 310 = 424.32\,\text{kN m} > 165.95\,\text{kN m} \quad \text{OK!}$$
$$My_{\max} = (Wy) \times 310 = 3{,}605{,}300 \times 310 = 111.76\,\text{kN m} > 40.36\,\text{kN m} \quad \text{OK!}$$
$$\tau_{\max} = 1.5 \times 28.1\,\text{kN} / (8 \times 300) = 17.56\,\text{MPa} < [\tau] = 180\,\text{MPa} \quad \text{OK!}$$
$$\sigma_{\max} = 944{,}960/(8 \times 300 + 2 \times 16 \times 260) = 88.15\,\text{MPa} < [\sigma] = 310\,\text{MPa} \quad \text{OK!}$$

(4) Checking of deformation:
The sectional moments of inertia are

$$Ix = 2 \times 260 \times 16 \times 150 \times 150 + 8 \times 300^3/12 = 205{,}200{,}000\,\text{mm}^4$$
$$Iy = 2 \times 16 \times 260^3/12 = 46{,}869{,}000\,\text{mm}^4$$

The selected combination of loads in the checking of deformation is different from the one in the checking of strength.

The combination of axis-paralleling horizontal loads is wind load + horizontal braking force of the crane.

Wind load $hw = (0.8 + 0.7) \times 0.55 \times 6 = 4.95$ kN/m

Horizontal braking force of the crane $hc = 2 \times 0.7 \times 0.1(125 + 18.79) = 20.13$ kN

Assume the action point of the crane is at the midpoint of the side column (the most unfavorable position).

The shear force imposed upon the central column is estimated as

$$Qx = 4.95 \times 12 \times (1/8) + 0.5 \times 20.13/2 = (7.43 + 5.03)\ \text{kN m}$$

The deformation of the central column under the action of axis-paralleling horizontal loads is estimated as follows:

$$\begin{aligned}\Delta c &= h^3/(12EI) + (Qx'')h^3/(192EI) = (Qx'/12 + Qx''/192)h^3/EI \\ &= (7.43\ \text{kN m}/12 + 5.03\ \text{kN m}/192) \times 14.5^3\ \text{m}^3/\left(2.06 \times 10^5\ \text{N/mm}^2 \times 205{,}200{,}000\ \text{mm}^4\right) \\ &= 0.032\ \text{m} < 0.036\ \text{m} = h/400 \quad \text{OK!}\end{aligned}$$

The deformation of the central column under the action of axis-perpendicular horizontal loads is estimated as follows:

Horizontal braking force of the crane is

$$\begin{aligned}hc' &= 2 \times 0.7 \times 0.1(12.5\ \text{t} + 19.316\ \text{t})/2 = 22.27\ \text{kN} \\ \Delta c &= (Qy)h^3/(192EI) \\ &= 11.14\ \text{kN} \times 12^3\ \text{m}^3/\left(192 \times 2.06 \times 10^5\ \text{N/mm}^2 \times 46{,}869{,}000\ \text{mm}^4\right) \\ &= 0.003 < 0.036 = h/400 \quad \text{OK!}\end{aligned}$$

12. **Estimation of steel consumption of major rigid frame**

The steel consumption of each major rigid frame in factory building A:

$$1.1\big[(8 \times 400 + 2 \times 16 \times 280) \times 10^{-6} \times 7850 \times (25 \times 4 + 12 \times 2) \\ + (8 \times 300 + 2 \times 16 \times 260)10^{-6} \times 7850 \times (14.5 \times 3)\big] = 17.045\ \mathrm{T}$$

The coefficient 1.1 in the equation above has taken into consideration the extra steel consumption entailed in the connecting plates and the indispensable structural supports.

Unit steel consumption is as follows:

$$17{,}045/(4 \times 25 \times 6) = 28.41\ \mathrm{kg/m^2}$$

The total steel consumption of major rigid frames in factory building A is as follows:

$$1.05 \times 28.41 \times 10{,}000\ \mathrm{m^2} = 298\ \mathrm{T}$$

The coefficient 1.05 in the equation above comes from the extra steel employed in wind-resisting columns.

13. **Simplified calculation of the supporting ground and foundation**

According to above static calculation, the loads in side column are

$$\begin{aligned} Mx_{\max} &= 229.17\ \mathrm{kN\,m} \\ My_{\max} &= 26.7\ \mathrm{kN\,m} \\ Q_{\max} &= 19.08\ \mathrm{kN} \\ N_{\max} &= 472.48\ \mathrm{kN} \end{aligned}$$

According to above static calculation, the loads in middle column are

$$\begin{aligned} Mx_{\max} &= 165.95\ \mathrm{kN\,m} \\ My_{\max} &= 40.36\ \mathrm{kN\,m} \\ Q_{\max} &= 19.08\ \mathrm{kN} \\ N_{\max} &= 944.96\ \mathrm{kN} \end{aligned}$$

With reference to the results of preliminary survey of the supporting ground and taking into consideration the supporting capacity of the ground and a rational form of force bearing, independent below-column pile foundation is adopted, as shown in Fig. A.5. The independent below-column pile foundation consists of a triangle-shaped pile cap and three prestressed-prefabricated supporting piles. Diameter of the pile is 400 mm and the spacing between piles is 1200 mm. The

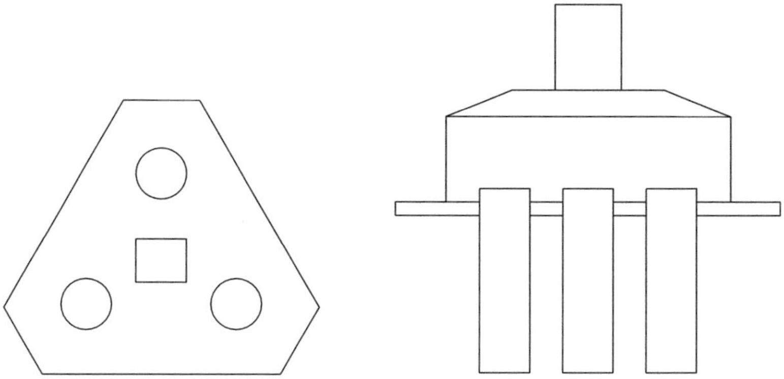

Fig. A.5 Diagram of below-column pile foundation and the pile cap

length of the pile cap is no less than 2 m and the thickness no less than 700 mm. Based on the distribution of soil layers, the length of the pile is determined to be 13 m.

The checking of bending and compressive resistance of the central pile cap:

$$Mx = My = (N_{\max}/3)\left(s - 3^{1/2}c/4\right) = (766\ \text{kN}/3)\left(1200 - 3^{1/2}0.866 \times 400\ \text{mm}/4\right)$$
$$= 268\ \text{kN m} > 229.17\ \text{kN m}\ > 165.95\ \text{kN m}\quad \text{OK!}$$
$$N_{\max} = 3 \times 766\ \text{kN} = 2298\ \text{kN} > 944.96\ \text{kN} > 472.48\ \text{kN}\quad \text{OK!}$$

14. **Simplified design and calculation of floor**

With reference to the condition of soil layers of the supporting ground and a strict restriction upon the settlement of floor, piles are needed in the treatment of the supporting ground under the floor. Prestressed-prefabricated 300-mm-diameter piles, the length of and the spacing between which are respectively 8 m and 2.5 m × 2.5 m, fit the job.

The checking of vertical loads:

$$(30 \times 2.5 \times 2.5) = 188\ \text{kN} < Ra = 259\ \text{kN}\quad \text{OK!}$$

300-mm-thick C30 reinforced concrete floor is preselected, as shown in Fig. A.6. Readers could manually design and calculate the two-way double-layered reinforcements of the floor.

The design of the author goes with two-way double-layered Grade II reinforcements, the diameter of and the spacing between which, respectively, are 16 and 150 mm, for the plate strip above the column and 12-mm-diametered and 200-mm-spaced two-way double-layered reinforcements for the mid-span plate strip. And the average reinforcement consumed is

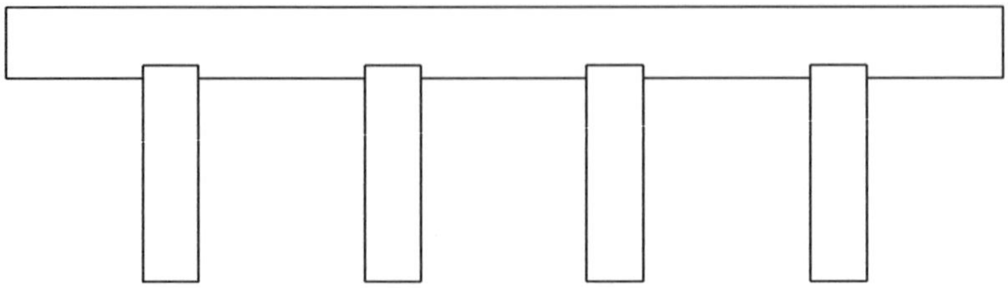

Fig. A.6 Diagram of floor slabs

$$(42.13 + 17.8)/2 = 30\ \mathrm{kg/m^2}$$

15. **Summary of design**

The conceptual design of this project could be summarized as follows:

(1) The design of the major rigid frame is the key in this project. The estimation of loads, the simplification of calculation model and the consideration of the most unfavorable combination of loads are to be conducted with careful attention. Afterward, practical construction experiences and simplified methods of calculation introduced in previous chapters could be referred to in the design and checking calculation. Figure A.7 is the diagram of selected sections for the major rigid frame.

(2) Unit steel consumption of major rigid frame of this factory building is 28.41 $\mathrm{kg/m^2}$, on the basis of which the total steel consumption of major steel structure could be calculated as 298 t along with the probable cost of construction.

(3) The design of the supporting ground and foundation could determine the type of pile foundation as well as the length and type of the pile. And this is quite significant for the estimation of cost of construction concerning this part of the structure.

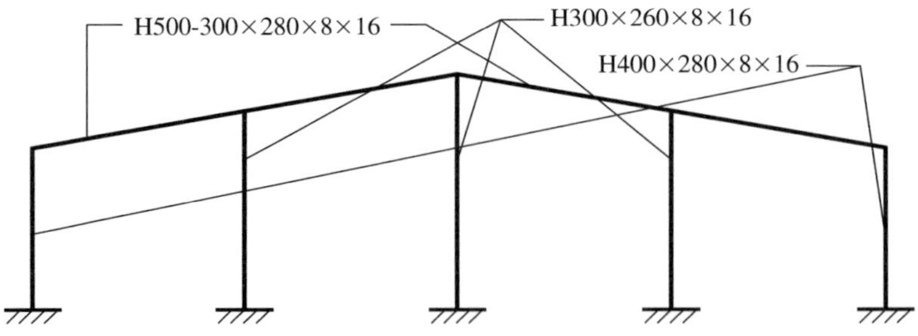

Fig. A.7 Diagram of selected sections for the major rigid frame

A simple design of floor could determine whether or not to adopt pile, technique entailed in the construction of floor and the estimation of reinforcements to be installed within the floor. At the completion of this process, the cost of construction of this part could be estimated, too.

Appendix B

B.1 Example of Conceptual Design of Multi-story Steel Frame Factory Building

In previous chapters of this book, detailed introduction has been made of necessary concepts of mechanics and simplified methods of calculation to be applied in modern structural design. And now we will come to the application of these concepts and methods in construction projects and to the analysis of example of conceptual design of multi-storied multi-span steel frame factory building.

1. **Project Introduction**

The project of building this industrial factory building covers an area of about 30,000 m^2. And factory building C is a four-story steel frame structure (as shown in Fig. B.1), taking up an area of 6400 m^2 with a plane layout of 60 m × 60 m. As for the others, they are all single-storied. Due to the limited length of this book, we will only discuss the conceptual design of the representative four-storied steel frame factory building C in here.

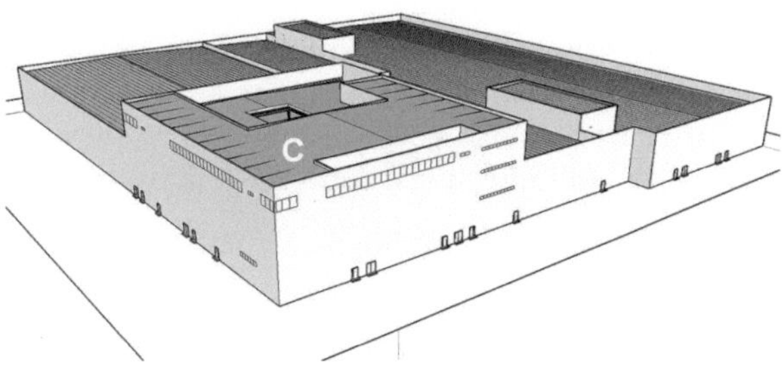

Fig. B.1 Rendering of this prospective factory building

S. Lin and Z. Huang, *Comparative Design of Structures*,
DOI 10.1007/978-3-662-48044-1

2. **The purpose of conceptual design**

   (1) It is to select a better structural system and the way of arrangement, in which case the resultant structure should not only be up to the standard but also minimize the cost and guarantee the safety.
   (2) Through the estimation of steel consumption of the main structure, the appraisal of economic efficiency and applicability of the structural system could be better made along with the total cost of construction.
   (3) Conceptual design could serve as a guide to the following preliminary design and detailed design. The control of the most critical section and the combination of loads ensures the safety and rationality of structural design.

3. **Conditions of design**

   (1) Proprietor-provided plane and elevation layouts,
   (2) Proprietor-provided live loads,
   (3) Preliminary survey report of foundation soil.

4. **Design Norms**

   (1) Load Code for the Design of Building Structures (GB50009-2001), 2001, Beijing,
   (2) Technical Specification for Steel Structure of Lightweight Building with Gabled Frames (CECS102), 2002, Beijing,
   (3) Code for Design of Steel Structures (GB50017-2002), 2002, Beijing,
   (4) Code for Design of Building Foundation (GB50007-2002), 2002, Beijing,
   (5) Code for Design of Concrete Structures (GB50010-2002), 2002, Beijing.

5. **The assumptions**

   (1) Assume the structural system works in an elastic state.
   (2) Live loads are simplified as equivalent uniformly or linearly distributed loads, conducted with reference to "Load Code for the Design of Building Structures".
   (3) All supports or connecting joints are assumed to be rigid connections, in which case the bending moments are transferred.
   (4) Factory building C is assumed to be a frame structure as shown in Fig. B.2. In both directions, there are five twelve-meter spans in the structure.

6. **The diagram of simplified calculation of the main steel structure**

According to the design principles of optimization of steel structures, multi-storied frame structure is adopted in this project. Since all stories are involved in the force bearing, box section is employed for the major columns; as for the primary beams, identical-sectioned wide-flanged I-shaped steel is adopted; all beam–column joints and supports are rigidly connected to make a frame structure.

In light of the construction technology and the principles concerning the optimization of structures, the main structure of factory building C takes form of a

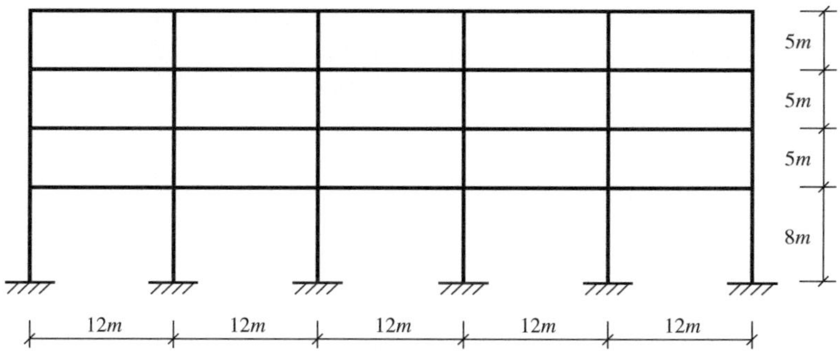

Fig. B.2 Diagram of simplified calculation of the main steel structure

four-storied steel frame structure. The column grid is 12 m × 12 m. The ground floor is 8 m high and the rest are 5 m high, as shown in Fig. B.2.

7. **Estimation of the loads in structural design**

The estimation principles of loads introduced in Chap. 2 shall be applied. Taking into consideration the requirements of construction technology for industrial project, major loads as listed below have to be considered in the design of this factory building (Table B.1):

8. **Estimation of the internal forces of the cross section under the action of the most unfavorable combination of vertical loads**

   (1) The most unfavorable combination of vertical loads on the roof.
   The structural system of the roof is a primary–secondary beam system. The secondary beams are simply supported, spanning over a distance of 12 m with an in-between spacing of 6 m. The primary beams are rigidly supported, spanning over a distance of 12 m with an in-between spacing of 12 m as well.

Table B.1 Summary of major design loads for factory building C

Type	Load
Uniform roof live load	0.3 kN/m^2
Uniform roof snow load (Shanghai area)	0.2 kN/m^2
Overhanging roof load	0.1 kN/m^2
Overhanging floor load	0.3 kN/m^2
Wind load (Shanghai area)	0.55 kN/m^2
Load of crane	0
Uniform live load on the 4th floor	5 kN/m^2
Uniform live load on the 3th floor	7.5 kN/m^2
Uniform live load on the 2nd floor	7.5 kN/m^2
Uniform live load on the ground	30 kN/m^2
Load of the forklift on the ground	5 t

According to the Load Code for the Design of Building Structures (GB50009-2001), the most unfavorable combination of vertical uniform loads is 1.2 × deadweight + 1.4 × overhanging load + 1.4 × load of snow or live load on the roof.

$$w = 1.2 \times 0.45 + 1.4 \times 0.10 + 1.4 \times 0.3 = 1.10\ \text{kN/m}^2$$

The equivalent uniform linear load shall then be (since the spacing between secondary beams is 6 m)

$$q' = 6\,\text{m} \times 1.10 = 6.6\ \text{kN/m}$$

And the maximum bending moment and shear force of the simply supported secondary beam are estimated as

$$M'_{\max} = q'L^2/8 = 6.6\ \text{kN/m} \times 12\ \text{m} \times 12\ \text{m}/8 = 118.8\ \text{kN m}$$
$$Q'_{\max} = q'L/2 = 6.6\ \text{kN/m} \times 12\ \text{m}/2 = 39.6\ \text{kN}$$

Surface load is imposed upon 12-m-spaced, rigidly supported primary beams and

$$q = 12 \times 1.10 = 13.2\ \text{kN/m}$$
$$M_{\max} = qL^2/12 = 13.2\ \text{kN/m} \times 12\ \text{m} \times 12\ \text{m}/12 = 158.4\ \text{kN m}$$
$$Q_{\max} = qL/2 = 13.2\ \text{kN/m} \times 12\ \text{m}/2 = 79.2\ \text{kN}$$

The maximum axial compression of the middle column is estimated as

$$N_{\max} = 1.1 \times 12 \times 12 = 158.4\ \text{kN}$$

The maximum bending moment and axial force of the side column are estimated as

$$M_{\max} = 0.5 \times 158.4 = 79.2\ \text{kN m}$$
$$N_{\max} = 0.5 \times 158.4 = 79.2\ \text{kN}$$

(2) The most unfavorable combination of vertical loads upon the 4th floor
The structural system of the floor is a primary–secondary beam system. The secondary beams are simply supported, spanning over a distance of 12 m with an in-between spacing of 2 m. The primary beams are rigidly supported, spanning over a distance of 12 m with an in-between spacing of 12 m as well.
The most unfavorable combination of vertical loads is 1.2 × deadweight + 1.4 × overhanging load + 1.3 × live load on the roof.

$$w = 1.2 \times 3.5 + 1.4 \times 0.30 + 1.3 \times 5 = 11.12\ \text{kN/m}^2$$

In the equation above, we assume the deadweight of the 4th floor to be 3.5 kN/m^2 and the equivalent uniform linear load (since the spacing between secondary beams is 2 m) would then be

$$q' = 2\ \text{m} \times 11.12 = 22.24\ \text{kN/m}$$

And the maximum bending moment and shear force of simply supported secondary beam are estimated as

$$\begin{aligned} M'_{\max} &= q'L^2/8 = 22.24\ \text{kN/m} \times 12\ \text{m} \times 12\ \text{m}/8 = 400.32\ \text{kN m} \\ Q'_{\max} &= q'L/2 = 22.24\ \text{kN/m} \times 12\ \text{m}/2 = 133.44\ \text{kN} \end{aligned}$$

Surface load is imposed upon 12-m-spaced, rigidly supported primary beams and

$$\begin{aligned} q &= 12 \times 11.12 = 133.44\ \text{kN/m} \\ M_{\max} &= qL^2/12 = 133.44\ \text{kN/m} \times 12\ \text{m} \times 12\ \text{m}/12 = 1601.28\ \text{kN m} \\ Q_{\max} &= qL/2 = 133.44\ \text{kN/m} \times 12\ \text{m}/2 = 800.64\ \text{kN} \end{aligned}$$

The maximum axial compression of the central column is estimated as

$$N_{\max} = 11.12 \times 12 \times 12 + 158.4 = 1759.68\ \text{kN}$$

The maximum bending moment and axial force of the side column are estimated as

$$\begin{aligned} M_{\max} &= 0.5 \times 1601.28 = 800.64\ \text{kN m} \\ N_{\max} &= 0.5 \times 1759.68 = 879.84\ \text{kN} \end{aligned}$$

(3) The most unfavorable combination of vertical loads upon the 3rd floor
The structural system of the floor is a primary–secondary beam system. The secondary beams are simply supported, spanning over a distance of 12 m with an in-between spacing of 1.5 m. The primary beams are rigidly supported, spanning over a distance of 12 m with an in-between spacing of 12 m as well.
The most unfavorable combination of vertical loads is 1.2 × deadweight + 1.4 × overhanging load + 1.3 × live load on the roof.

$$w = 1.2 \times 3.5 + 1.4 \times 0.30 + 1.3 \times 7.5 = 14.37\ \text{kN/m}^2$$

The equivalent uniform linear load (since the spacing between secondary beams is 1.5 m) would then be

$$q' = 1.5 \text{ m} \times 14.37 = 21.56 \text{ kN/m}$$

The maximum bending moment and shear force of simply supported secondary beam are estimated as

$$M'_{\max} = q'L^2/8 = 21.56 \text{ kN/m} \times 12 \text{ m} \times 12 \text{ m}/8 = 388.08 \text{ kN m}$$
$$Q'_{\max} = q'L/2 = 21.56 \text{ kN/m} \times 12 \text{ m}/2 = 129.364 \text{ kN}$$

Surface load is imposed upon 12-m-spaced, rigidly supported primary beams and

$$q = 12 \times 14.37 = 172.44 \text{ kN/m}$$
$$M_{\max} = qL^2/12 = 172.44 \text{ kN/m} \times 12 \text{ m} \times 12 \text{ m}/12 = 2069.28 \text{ kN m}$$
$$Q_{\max} = qL/2 = 172.44 \text{ kN/m} \times 12 \text{ m}/2 = 1034.64 \text{ kN}$$

The maximum axial conpression of the central column is estimated as

$$N_{\max} = 14.37 \times 12 \times 12 + 1759.68 = 3824.64 \text{ kN}$$

The maximum bending moment and axial force of the side column are estimated as

$$M_{\max} = 0.5 \times 2069.28 = 1034.64 \text{ kN m}$$
$$N_{\max} = 0.5 \times 3824.64 = 1912.32 \text{ kN}$$

(4) The most unfavorable combination of vertical loads upon the 2nd floor
The structural system of the floor is a primary–secondary beam system. The secondary beams are simply supported, spanning over a distance of 12 m with an in-between spacing of 1.5 m. The primary beams are rigidly supported, spanning over a distance of 12 m with an in-between spacing of 12 m as well.
The most unfavorable combination of vertical loads is 1.2 × dead-weight + 1.4 × overhanging load + 1.3 × live load on the roof.

$$w = 1.2 \times 3.5 + 1.4 \times 0.30 + 1.3 \times 7.5 = 14.37 \text{ kN/m}^2$$

The equivalent uniform linear load (since the spacing between secondary beams is 1.5 m) would then be

$$q' = 1.5 \text{ m} \times 14.37 = 21.56 \text{ kN/m}$$

The maximum bending moment and shear force of simply supported secondary beam are estimated as

$$M'_{max} = q'L^2/8 = 21.56\,\text{kN/m} \times 12\,\text{m} \times 12\ \text{m}/8 = 388.08\,\text{kN m}$$
$$Q'_{max} = q'L/2 = 21.56\,\text{kN/m} \times 12\,\text{m}/2 = 129.364\,\text{kN}$$

Surface load is imposed upon 12-m-spaced, rigidly supported primary beams and

$$q = 12 \times 14.37 = 172.44\ \text{kN/m}$$
$$M_{max} = qL^2/12 = 172.44\ \text{kN/m} \times 12\ \text{m} \times 12\ \text{m}/12 = 2069.28\ \text{kN m}$$
$$Q_{max} = qL/2 = 172.44\ \text{kN/m} \times 12\ \text{m}/2 = 1034.64\ \text{kN}$$

The maximum axial compression of the central column is estimated as

$$N_{max} = 14.37 \times 12 \times 12 + 3824.64 = 5893.92\ \text{kN}$$

The maximum bending moment and axial force of the side column are estimated as

$$M_{max} = 0.5 \times 2069.28 = 1034.64\ \text{kN m}$$
$$N_{max} = 0.5 \times 5893.9 = 2946.96\ \text{kN}$$

Figure B.3 is a diagram of the distribution of bending moment in the main frame when it is under the action of the most unfavorable combination of vertical loads.

9. **The estimation of the internal forces of cross section under the action of the most unfavorable combination of horizontal loads**

(1) The most unfavorable combination of horizontal loads on the 4th floor
The calculation of wind load:

$$hw = 1.4 \times 0.55\ \text{kN/m}^2 \times 12\ \text{m} = 9.24\ \text{kN/m}$$

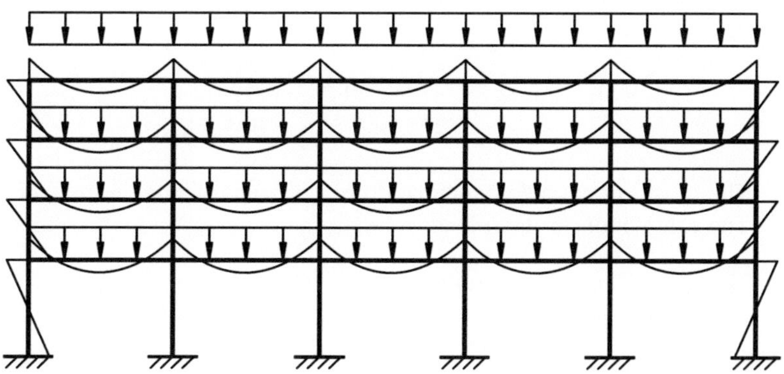

Fig. B.3 Diagram of vertical-load-induced distribution of the bending moment in the main frame

And the internal forces at the foot of the column are

$$Q_{\max} = 9.24 \times 5(1/6) = 7.7 \text{ kN}$$
$$M_{\max} = 7.7 \times 2.5 \text{ m} = 19.25 \text{ kN m}$$

The simplified calculation of seismic load: 7° seismic fortification intensity and total seismic-load-induced horizontal force on the 4th floor is

$$H4 = 0.1 \times 60 \times 60 \times 1.1 = 396 \text{ kN}$$

Since there are 36 columns, the shear force shared by each column would be

$$Q_{\max} = 396/36 = 11 \text{ kN}$$
$$M_{\max} = 11 \times 2.5 \text{ m} = 27.5 \text{ kN m}$$

(2) The most unfavorable combination of horizontal loads on the 3rd floor
The calculation of wind load:

$$hw = 1.4 \times 0.55 \text{ kN/m}^2 \times 12 \text{ m} = 9.24 \text{ kN/m}$$

And the internal forces at the foot of the column are

$$Q_{\max} = 9.24 \times 2 \times 5(1/6) = 15.4 \text{ kN}$$
$$M_{\max} = 15.4 \times 2.5 \text{ m} = 38.5 \text{ kN m}$$

The simplified calculation of seismic load: 7° seismic fortification intensity and total seismic-load-induced horizontal force on the 3rd floor is

$$H3 = 0.1 \times 60 \times 60 \times 1.1 + 0.1 \times 60 \times 60 \times 11.12 = 4399.2 \text{ kN}$$

Since there are 36 columns, the shear force shared by each column would be

$$Q_{\max} = 4399.2/36 = 122.2 \text{ kN}$$
$$M_{\max} = 122.2 \times 2.5 \text{ m} = 305.5 \text{ kN m}$$

(3) The most unfavorable combination of horizontal loads on the 2nd floor
The calculation of wind load:

$$hw = 1.4 \times 0.55 \text{ kN/m}^2 \times 12 \text{ m} = 9.24 \text{ kN/m}$$

And the internal forces at the foot of the column are

$$Q_{\max} = 9.24 \times 3 \times 5(1/6) = 23.1\ \text{kN}$$
$$M_{\max} = 23.1 \times 2.5\,\text{m} = 57.75\ \text{kN m}$$

The simplified calculation of seismic load: 7° seismic fortification intensity and total seismic-load-induced horizontal force on the 2nd floor is

$$\begin{aligned} H2 &= 0.1 \times 60 \times 60 \times 1.1 + 0.1 \times 60 \times 60 \times 11.12 + 0.1 \times 60 \times 60 \\ &\quad \times 14.37 \\ &= 9572.4\ \text{kN} \end{aligned}$$

Since there are 36 columns, the shear force shared by each column would be

$$Q_{\max} = 9572.4/36 = 265.9\ \text{kN}$$
$$M_{\max} = 265.9 \times 2.5\ \text{m} = 664.75\ \text{kN m}$$

(4) The most unfavorable combination of horizontal loads on the 1st floor
The calculation of wind load:

$$hw = 1.4 \times 0.55\ \text{kN/m}^2 \times 12\ \text{m} = 9.24\ \text{kN/m}$$

And the internal forces at the foot of the column are

$$Q_{\max} = 9.24(8+5+5+5)(1/6) = 35.42\ \text{kN}$$
$$M_{\max} = 35.42 \times 4\ \text{m} = 141.68\ \text{kN m}$$

The simplified calculation of seismic load: 7° seismic fortification intensity and total seismic-load-induced horizontal force on the 1st floor is

$$\begin{aligned} H1 &= 0.1 \times 60 \times 60 \times 1.1 + 0.1 \times 60 \times 60 \times 11.12 + 0.1 \times 60 \times 60 \\ &\quad \times 14.37 \times 2 \\ &= 14{,}745.6\,\text{kN} \end{aligned}$$

Since there are 36 columns, the shear force shared by each column would be

$$Q_{\max} = 14,745.6/36 = 409.6\ \text{kN}$$
$$M_{\max} = 409.6 \times 4\ \text{m} = 1638.4\ \text{kN m}$$

Figure B.4 is a diagram of the distribution of bending moment in the main frame under the action of the most unfavorable combination of horizontal loads (seismic load is the control load).

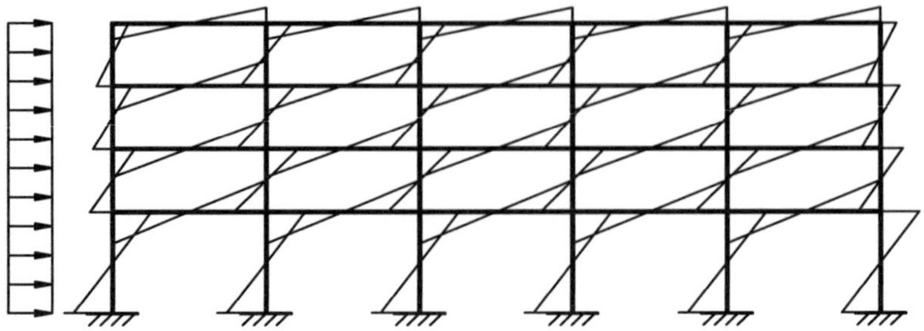

Fig. B.4 Distribution of horizontal-load-induced bending moment in the main frame

10. **Estimation and selection of sections for the beams and columns in the main frame**

From the calculation above, the combination of loads affecting first floor beams and columns should be the control forces.

As for the primary beams above the first floor,

$$M_{\max} = 2069.28\ \text{kN m}$$
$$Q_{\max} = 1034.64\ \text{kN}$$

As for the secondary beams above the first floor,

$$M'_{\max} = 388.08\ \text{kN m}$$
$$Q'_{\max} = 129.36\ \text{kN}$$

As for the columns on the first floor,

$$M_{\max} = 1638.4\ \text{kN m}$$
$$Q_{\max} = 409.6\ \text{kN}$$
$$N_{\max} = 5893.92\ \text{kN}$$

Symmetrical wide-flanged I-shaped section H800 × 400 × 12 × 20 is adopted for the primary beams above the ground floor;

Symmetrical wide-flanged I-shaped section H380 × 280 × 8 × 16 is adopted for the secondary beams;

As for the ground floor columns, box-shaped section 560 × 560 × 16 × 16 is adopted and the thickness of the box-shaped section is gradually decreased from the second floor to the top.

(1) The checking of the bending moment
 Assume steel Q345, strength of design being 310 MPa, is employed in the main frame.

Then, the sectional modulus of the selected section for the primary beams should meet the requirement of

$$W > M_{\max}/310 = 2{,}069{,}280{,}000/310 = 6{,}675{,}100\ \text{mm}^3$$

The moment of area of the primary beams shall be

$$\begin{aligned} W &= 2 \times 400 \times 20 \times 400 + 12 \times 800 \times 800/6 \\ &= 7{,}680{,}000\ \text{mm}^3 > 6{,}675{,}100\ \text{mm}^3 \\ &\text{OK!} \end{aligned}$$

The moment of area of the secondary beams shall be

$$W' = 2 \times 280 \times 16 \times 190 + 8 \times 380 \times 380/6 = 1{,}895{,}000\ \text{mm}^3$$
$$W' > 388{,}080{,}000/310 = 1{,}252{,}000\ \text{mm}^3 \quad \text{OK!}$$

And the moment of area of the columns shall be

$$W'' = 2 \times 560 \times 16 \times 280 + 2 \times 16 \times 560 \times 560/6 = 6{,}690{,}000\ \text{mm}^3$$
$$W'' > 1{,}638{,}400{,}000/310 = 5{,}290{,}000\ \text{mm}^3 \quad \text{OK!}$$

(2) The checking of the shear force
The maximum shear stress on the section of primary beams is

$$\tau_{\max} = 15 \times 1034.64\ \text{kN}/(12 \times 800) = 162\ \text{MPa} < [\tau] = 180\ \text{MPa} \qquad \text{OK!}$$

The maximum shear stress on the section of secondary beams is

$$\tau_{\max} = 1.5 \times 129.36\ \text{kN}/(8 \times 380) = 67.4\ \text{MPa} < [\tau] = 180\ \text{MPa} \qquad \text{OK!}$$

The maximum shear stress on the section of columns is

$$\tau_{\max} = 1.5 \times 409.6\ \text{kN}/(2 \times 16 \times 560) = 34.3\ \text{MPa} < [\tau] = 180\ \text{MPa} \qquad \text{OK!}$$

(3) The checking of axial compression
The maximum axial stress on the section of columns is

$$\sigma_{\max} = 5893.62\ \text{kN}/(4 \times 16 \times 560) = 164\ \text{MPa} < [\sigma] = 310\ \text{MPa} \quad \text{OK!}$$

(4) The checking of vertical deformation of primary beams
Sectional moment of inertia is first estimated.

$$Ix = 2 \times 400 \times 20 \times 400 \times 400 + 12 \times 800^3/12 = 3{,}072{,}000{,}000 \text{ mm}^4$$

The combination of loads selected for the checking of deformation (different from that for the checking of strength) is 1.0 deadweight + 0.5 overhanging load + 0.5 live load on the floor

$$w = 3.5 + 0.5 \times 0.3 + 0.5 \times 7.5 = 7.4 \text{ kN/m}^2$$
$$q = 12 \text{ m} \times 7.4 \text{ kN/m}^2 = 88.8 \text{ kN/m}$$

The maximum vertical deformation of primary beams is

$$\Delta b = (1/384)88.8 \text{ kN/m} \times 12^4 \text{ m}^4/\left(2.06 \times 10^5 \text{ N/mm}^2 \times 3{,}072{,}000{,}000 \text{ mm}^4\right)$$
$$= 0.008 \text{ m} < L/400 = 0.03 \text{ m} \quad \text{OK!}$$

The same method could be applied to the checking of the maximum vertical deformation of secondary beams, which by the way meets the requirement of design, and readers could conduct the checking to see for yourselves. With reference to Appendix A and Chap. 3 of this book, lateral displacement of columns could also be checked.

11. **The estimation of steel consumption in the main rigid frame**

Steel employed in each frame in factory building C

$$1.10 \times 0.8\left[(16 \times 560 \times 4)10^{-9} \times 7850 \times 23 \times 6 + (12 \times 800 + 2 \times 20 \times 400)10^{-9} \times 7850 \times 12 \times 20 + (8 \times 380 + 2 \times 16 \times 280)10^{-9} \times 7850 \times 12 \times 6 \times 20)\right] = 196 \text{ T}$$

In the equation above, coefficient 1.1 is for the steel needed by the connecting plates and the coefficient 0.8 is for the gradually reduced load with the elevation of story.

The average unit-employed steel is

$$196{,}000/(60 \times 12 \times 4) = 68.06 \text{ kg/m}^2$$

And the total steel employed in the main rigid frame in factory building C is

$$1.05 \times 68.06 \times 60 \times 60 \times 4 = 1029 \text{ T}$$

In the equation above, coefficient 1.05 has considered the extra steel employed in all diagonal braces in the main rigid frames.

12. **The simplified design and calculation of the supporting ground and foundation**

From the calculation conducted with the most unfavorable combination of loads we could have the most unfavorable internal forces of the central columns on the ground floor.

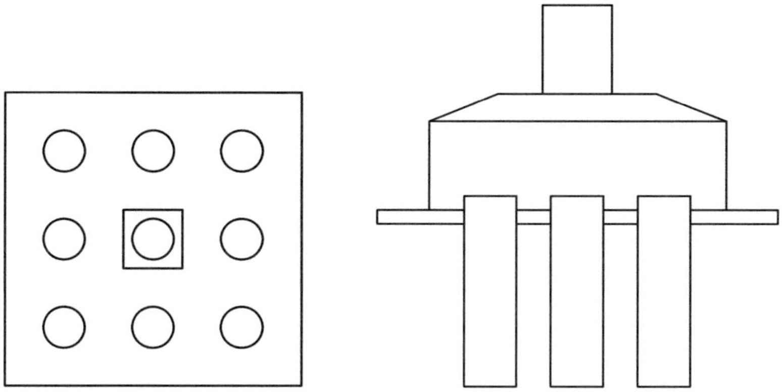

Fig. B.5 Diagram of the below-column pile foundation and the cap

$$M_{\max} = 1638 \text{ kN m}$$
$$Q_{\max} = 410 \text{ kN}$$
$$N_{\max} = 5894 \text{ kN}$$

According to the results of proprietor-provided preliminary survey of the supporting ground and taking into consideration the bearing capacity of the supporting ground and a rational form of force bearing, an independent below-column pile foundation is adopted, as shown in Fig. B.5. The independent pile foundation below the central column consists of a square pile cap and nine prestressed prefabricated supporting piles. The diameter is 1200 mm; the length of each side is no less than 3 m; and the thickness of the cap is no less than 1000 mm. According to the distribution of soil layers of the supporting ground, the length of the piles shall be 14 m (Ra = 810 kN).

The checking of the bending moment and compression of the cap below central column

$$Mx = My = 3N_{\max} 1.6 = 3 \times 810 \times 1.6 = 3888 \text{ kN m} > 1638 \text{ kN m} \quad \text{OK!}$$
$$N_{\max} = 9 \times 810 \text{ kN} = 7290 > 5894 \text{ kN} \quad \text{OK!}$$

13. **Simplified design and calculation of floor**

According to the soil conditions of the supporting ground and a strict restriction proposed by the proprietor upon the settlement of the floor, piles are needed in the supporting ground below the floor. Prestressed prefabricated piles with 300 mm diameter, 8 m length and 2.5 m × 2.5 m spacing could meet the requirement.

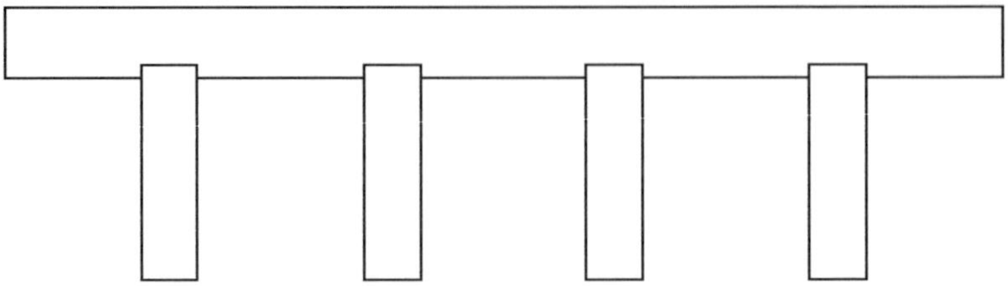

Fig. B.6 Diagram of floor

The checking of vertical loads

$$(30 \times 2.5 \times 2.5) = 188\ \text{kN} < Ra = 259\ \text{kN} \quad \text{OK!}$$

C30 reinforced concrete floor with a thickness of 300 mm is preselected, as shown in Fig. B.6.

The design yields two-way double-layered II reinforcements with 16 mm diameter and 150 mm spacing for the plate strip above the column and those with 12 mm diameter and 200 mm spacing for the mid-span plate strip. The average steel employed could be calculated as (42.13 + 17.8)/2 = 30 kg/m^2.

14. **Summary of design**

Conceptual concepts adopted in the design of this project are summarized as follows:

(1) The design of the main rigid frame is the key in this project. The estimation of loads, the simplification of calculation model, and the consideration of the most unfavorable combination of loads are to be conducted with careful attention. Afterward, practical construction experiences and simplified methods of calculation introduced in previous chapters could be referred to in the design and checking calculation. Figure B.7 is the diagram of selected sections for the main rigid frame. With the elevation of story and the decrease of load, the steel employed could be gradually reduced.

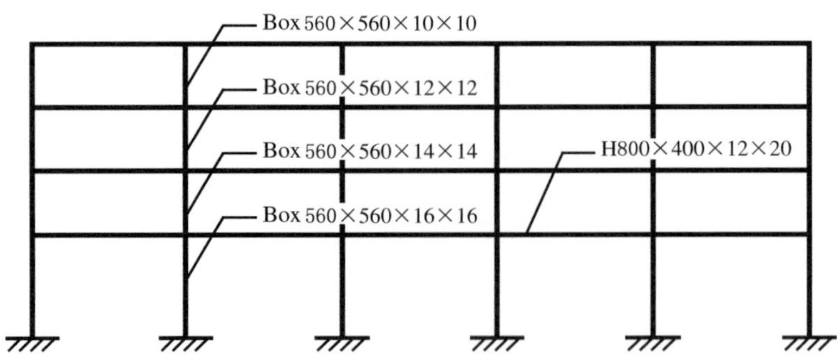

Fig. B.7 Diagram of selected sections for the main rigid frame

(2) Unit steel employed in the main frame of this factory building is 68.06 kg/m^2, on the basis of which the total steel employed in the main steel structure could be derived as 1029 t along with the probable cost of construction.
(3) The design of the supporting ground and foundation could determine the type of the pile foundation as well as the length and type of the pile. And this is of importance for the estimation of cost of construction concerning this part of the structure.
(4) A simple design of floor could determine whether or not to adopt pile, the technique entailed in the construction of floor and the estimation of reinforcements to be installed within. At the completion of this process, the cost of construction of this part could be estimated.

References

1. Lin SP (2002) Fuzzy-AI in design consideration. Lecture Presentation at Center for Design Informatics. Harvard University, 8 Nov 2002
2. Aamodt A, Plaza E (1994) Case-based reasoning: foundational issues, methodological variations, and system approaches. AI Commun 7(1):39–59
3. Zhang J, Xila L (1989) Expert system for preliminary design of frame shear-walled high-rise buildings. In: Proceedings of the 4th CCEA national conference on computer applications, Xiamen, China, pp 213–221, Oct 1989. (in Chinese)
4. http://www.cnbridge.cn/2011/0408/31094.html
5. Wilson EL (2005) Static and dynamic analysis of structures. Copyright 1996–2004. Computer and Science Inc. Chinese edition Copyright 2005 by Beijing Civil King Software Technology Co. Ltd.
6. Lin TY, Sidney D (1999) Stotesbury, structural concepts and systems for architects and engineers, 2nd edn. Copyright 1988 by Van Nostrand Reinhold. Chinese edition copyright by China Architecture & Building Press
7. Structural Special Award 2000-WS Atkins for the Arabian Tower, Dubai. The Institution of Structural Engineers (2000)
8. http://www.cnaec.com.cn/Info/Show.asp?ID=168703
9. Huang Z, Lin SP (2010) Conceptual design for modern structures. China Architecture & Building Press.
10. Chen L, Chen X, Huang Z (2005) The protect technology for concrete structures under hazards. Chemical Industry Press
11. Load code for the design of building structures (GB50009-2012)
12. Ou JP (2003) Vibration control of structures-active, semi-active and smart materials. Science Press, Beijing
13. Huang Z et al (2007) Proper placement and reasonable quantity of MR dampers for reduction of wind induced vibration control in spatial structures. Spat Struct 13(4):11–17
14. Wang L, Huang Z, Zhou D (2007) Semi-active control model with prospective control methods. Vibr Impacts 26(10):109–112
15. Yang G, Zhou D, Huang Z (2007) Optimization of damper amounts and positions for vibration reduction of structures. J Shanghai Jiao Tong Univ 41(6):955–959
16. Huang Z et al (2005) Elastoplastic analysis of knee braced frame. J Zhe Jiang Univ 6A (8):784–789
17. Li Q, Huang Z, Chen L (2005) Analysis on elastoplastic model of knee braced steel frame. J Comput Mech 22(3):360–365
18. Liu XL (1997) The status and future of structural engineering research. People's Traffic Press, Beijing
19. Liu XL (2007) The progress and prospects of structural engineering. China Architecture & Building Press, Beijing
20. Chinese National Design Code, Design Code of Concrete Structures, GB50010-2010, 2010

S. Lin and Z. Huang, *Comparative Design of Structures*,
DOI 10.1007/978-3-662-48044-1

21. Huang Z, Liu XL (2006) Modified skew bending model for segmental bridge with unbonded tendons. ASCE J Bridge Eng 11(1):59–63
22. Huang Z, Liu XL (2007) Unified approach for analysis of box-section members under combined actions. ASCE J Bridge Eng 12(4):494–499
23. Engel H (1997) Structure systems. Verlage Gerd Hatje
24. Lin TY, Burns NH (1983) Design of prestressed concrete structures, 3rd edn. China Railway Press. (Chinese Version)
25. Huang Z, Wu YJ, Zhang C (2012) Bamboo strengthened with carbon fiber reinforced polymer for the substitutes of steel substructures. Key Eng Mater 517:233–237
26. Wang QS, Wang MY, Huang Z (2013) Predictive control of vibration of spatial structure. ASME 2013 International Mechanical Engineering Congress and Exposition, San Diego
27. Wang XD, Zhang C, Huang Z (2013) Impact experimental research on hybrid bamboo fiber and steel fiber reinforced concrete. Appl Mech Mater (2):174–177
28. Zhang C, Huang Z, Chen GW (2013) Experimental research on bamboo fiber reinforced concrete. Appl Mech Mater (2):174–177
29. Wang P, Huang Z et al (2012) Performances of hybrid fiber reinforced concrete with steel fibers and polypropylene fibers. In: 2012 ASCE proceedings, pp 458–461
30. Huang GQ, Huang Z et al (2012) Experimental study on carbon fiber polymer reinforced bamboos. In: 2012 ASCE proceedings, pp 259–262
31. Huang GQ, Huang Z et al (2012) Performances of carbon fiber cloth reinforced bamboos. Appl Mech Mater 174–177:1459–1462
32. Chinese national specification: technical specification for concrete structures of tall Building. JGJ3-2010, 2010, Beijing
33. Chinese national specification: code for seismic design of buildings, GB50011-2010, 2010, Beijing
34. Lv XL (2005) Requirements of high-rise construction for seismic resistance. Tongji University Press
35. Shanghai World Financial Center Preliminary Report. Dec 2002
36. Clifton GC (2001) Collapse of the world financial centre towers. HERA Struct Eng:1–13
37. http://baike.baidu.com/view/1320056.htm?fromId=763212
38. Wang DS et al (2008) Design and thinking of CCTV tower structure. Build Struct 29(3):25–30
39. American Association of State Highway Transportation Official Code (AASHTO) 1999
40. Huang Z (2001) Large-span arch steel structures-design and construction [C]. In: PSSC 2001 sixth pacific steel structure proceeding, Beijing, Oct 2001, pp 451–456
41. Huang Z et al (2000) Design and construction of large span arch steel structures. Steel Struct 15(1):14–16
42. Lupu Bridge (2004) Shanghai construction and management committee. Shanghai Science and Technology Press
43. Lin TY, Cheng K, Mark K et al (2003) Design of the Ling-Tie Bridge. In: Proceedings of the 5th international conference on current and future trends of bridge design, Construction and Maintenance. Shanghai, China, July 2003
44. Feng YR, Wang JY (1998) Jiangyin Yangtze river bridge in construction. In: Conference proceedings of Institute of China Civil Engineering 8th annual conference, Qinghua University Press
45. Lin TY (1998) Selected paper of T.Y. Lin, 2nd edn. Science Press
46. Wang XL (2008) Conceptual design and analysis of large span steel structures. Mechanical Industry Press
47. Li ZJ (2006) Optimization methods of engineering. China Water Power Press
48. Liu XJ, Huang Z, Zhou D (2005) Construction of steel structure of Shanghai Expo Centre. Steel Struct 20(79):70–73
49. The 7 difficulties under construction of China Opera, CCTV.COM. 20 Sept 2007
50. Stretching dome design-Raise high the record roof, Engineering News Record, Mar 1992, pp 24–26

51. Song XB, Liu XL (2002) Kinetic membranous structure, Kunmin, China. IABSE Struct Eng Int 12(1):16–19
52. Lin SP, Dong B (1998) Modeling of fuzzy machine learning and fuzzy neural network in structural design. In: Ayyub BM (ed) Uncertainty modeling and analysis in civil engineering. CRC press, USA, pp 337–355
53. Garret JH Jr (1990) Knowledge-based expert systems: past, present and future. IABSE Periodica 3:21–40
54. Lin SP (1990) Fuzzy reasoning method—Its theoretical adaptability and code compatibility to design of earthquake-resistant structure. Research Report(1) under Contract NCEER-88-1006, School of Civil and Environmental Engineering, Cornell University, July 1990
55. Lin SP et al (1992) Fuzzy reasoning and machine learning of expert system for structural design, computational mechanics in structural engineering. Elsevier Applied Science, pp 450–463
56. Lin SP (1990) Fuzzy machine learning of design type expert system—an approach to Cu-strake for earthquake-resistant design of buildings. Research Report(4) under Contract NCEER-88-1006, School of Civil and Environmental Engineering, Cornell University, July 1990
57. Lin SP (1998) Fuzzy-AI model. In: Proceedings of 2nd international conference on artificial intelligence for engineering, Wuhan, China, May 1998, pp 56–72
58. Lin SP (2008) "Fuzzy-AI Model" for managerial science. In: Plenary session speech and proceedings of 4th PMI research conference, Warsaw, Poland, 13–16 July 2008
59. Hopfield JJ (1982) Neural networks and physical system with emergent collective computational abilities. Proc Nat Acad Sci 79:2554–2558
60. Dayhoff JE (1990) Neural network architectures, an introduction. Van Nostrand Reinhold, New York
61. Chen HC, Lin SP (1998) Fuzzy neural network model for structural design. In: Proceedings "Advances in Engineering Computational Technology", Civil-Comp Press, Edinburgh, UK, pp 189–197
62. Lin SP (2005) On paradox of fuzzy modeling: supervised learning for rectifying fuzzy membership function. Artif Intell Rev 23:395–405
63. Saaty TL (1980) The analytic hierarchy process, planning, priority setting, resource allocation. McGraw Hill, Inc.
64. Design Specification of Portal Frame Steel Structures, CECS102: 2002 Beijing, China Plan Press (2003)
65. Millais M (1997) Building structures. Taylor & Francis
66. Lin SP (1992) The reform in engineering education. Lecture in Sino-Polish symposium in computational mechanics, Cracow, Poland, Oct 1992
67. Arnold C, Reitherman R (1982) Building configuration and seismic design. Wiley, New York
68. Paulay T (1992) Seismic design of reinforced concrete and masonry buildings. Wiley, New York
69. Cowan HJ, Wilson F (1981) Structural system. Van Nostrand Reinhold, New York
70. Engle I (1980) Structural principles. Prentice-Hall, Englewood Cliffs, N.J
71. Hill LA, Jr (1975) Fundamentals of structural design: steel, concrete and timber. Intext Educational
72. Chen WF, Duan L (2000) Bridge engineering handbook. CRC Press
73. Neal BG (1963) The plastic methods of structural analysis, 2nd edn. Chapman & Hall Ltd., London
74. Johnston BG (1980) Guide to design criteria for metal structures. Column Research Council
75. Tall L (1974) Structural steel design, 2nd edn. Ronald Press, New York
76. Park KJ, Ang AHS, Wen YK (1985) Seismic damage analysis of reinforced concrete buildings. J Struct Eng ASCE 111(4):740–757
77. AISC manual of steel construction, 8th edn. American Institute of Steel Construction, New York (1980)

78. ACI Committee 318: building code requirements for reinforced concrete (ACI 318-89) and commentary. American Concrete Institute, Detroit
79. Load and resistance factors design specification for structural steel buildings. American Institute of Steel Construction, Inc. 27 Dec 1999
80. Chinese national code: design code for steel structures (GB50017-2002), Beijing (2002)
81. Chinese national code: design code for foundation of buildings (GB50007-2002), Beijing (2002)
82. Qingshen X et al (2003) Neural Network Methods in Mechanical Engineering, China Machine Press (in Chinese)
83. Kleindorfer PR et al (1993) Decision Science—An Integrative Perspective. Cambridge University Press
84. Kahneman D, Tversky A (2000) Choice, Values, and Frames. Cambridge University Press

MIX
Papier aus verantwortungsvollen Quellen
Paper from responsible sources
FSC® C105338

If you have any concerns about our products,
you can contact us on
ProductSafety@springernature.com

In case Publisher is established outside the EU,
the EU authorized representative is:
Springer Nature Customer Service Center GmbH
Europaplatz 3, 69115 Heidelberg, Germany

Printed by Libri Plureos GmbH
in Hamburg, Germany